Mathematical Works Printed in the Americas, 1554–1700

Johns Hopkins Studies in the History of Mathematics
Ronald Calinger, Series Editor

Mathematical Works Printed in the Americas 1554–1700

BRUCE STANLEY BURDICK

The Johns Hopkins University Press
BALTIMORE

© 2009 The Johns Hopkins University Press
All rights reserved. Published 2009
Printed in the United States of America on acid-free paper
2 4 6 8 9 7 5 3 1

The Johns Hopkins University Press
2715 North Charles Street
Baltimore, Maryland 21218-4363
www.press.jhu.edu

Library of Congress Cataloging-in-Publication Data

Burdick, Bruce Stanley.
  Mathematical works printed in the Americas, 1554–1700 / Bruce Stanley Burdick.
    p. cm.
  Includes bibliographical references and index.
  ISBN-13: 978-0-8018-8823-6 (hardcover : alk. paper)
  ISBN-10: 0-8018-8823-9 (hardcover : alk. paper)
  1. Mathematics—America—Early Works to 1800—Bibliography. 2. Mathematical literature—America—Early works to 1800—Bibliography. 3. Ethnomathematics—America—Early works to 1800—Bibliography. 4. America—Imprints—Early works to 1800—Bibliography. I. Title.
  Z6652.B87 2008
  [QA33]
  016.51—dc22   2008024009

A catalog record for this book is available from the British Library.

Special discounts are available for bulk purchases of this book. For more information, please contact Special Sales at 410-516-6936 or specialsales@press.jhu.edu.

The Johns Hopkins University Press uses environmentally friendly book materials, including recycled text paper that is composed of at least 30 percent post-consumer waste, whenever possible. All of our book papers are acid-free, and our jackets and covers are printed on paper with recycled content.

Contents

List of Illustrations   vii

Acknowledgments   ix

Introduction   1

PART ONE
Mathematical Works, Excluding Almanacs   19

PART TWO
Almanacs, Ephemerides, and Lunarios   183

Appendix A
A Guide to Astrological Symbols   321

Appendix B
Seventeenth-Century Comet Books Printed in the New World   322

Appendix C
Some Comments on the Place of Logic in Mathematics   325

Works Consulted   328

General Index   345

Index to Works in Part One   350

Other Indices   354

Illustrations

1. Title page of the *Recognitio Summularum*, 1554.   21
2. From the *Recognitio Summularum* of 1554. The square of opposition.   22
3. From the *Recognitio Summularum* of 1554. Modal statements that are equivalent are grouped together.   23
4. From the *Recognitio Summularum* of 1554. The square of opposition with modality.   24
5. Title page of the *Dialectica Resolutio*, 1554.   29
6. Marginal illustrations from the *Dialectica Resolutio*. Parallels and triangles.   30
7. Marginal figures from the *Dialectica Resolutio* illustrating Euclidean propositions.   32
8. Title page of the *Sumario Compendioso*, 1556.   35
9. Juan Diez Freyle's table of congruous and congruent numbers.   44
10. $c^2 \pm 2ab = (a \pm b)^2$. From either of these equations one could derive the Pythagorean Theorem.   58
11. Title page of the *Phisica Speculatio*, 1557.   67
12. Title page of the *Dialogos Militares*, 1583.   78
13. Title page of the *Instrucción Nautica*, 1587.   83
14. Garcia de Palacio's thumb diagrams for computing the epact from the golden number. Each diagram, after the first, works for the range of years in the heading.   84
15. Garcia de Palacio's thumb diagrams for computing the epact from the golden number. Each diagram works for the range of years in the heading.   85
16. Title page of the *Libro General*, 1597.   102

ILLUSTRATIONS

17. Joán de Belveder shows how dividing by a percentage can be converted into multiplication. Note the use of currency units to express fractions. 103
18. Title page of the *Reportorio de los Tiempos*, 1606. 114
19. The volvelle from the *Reportorio de los Tiempos*, 1606. The inner disk turns so that the date along its outer edge can be lined up with the time of day. Then the rising sign can be found on the offset circle. The circle of signs is offset so that the signs of fast ascension cross the arm faster than the signs of slow ascension. 119
20. The method that Martínez uses to calculate the length of the period of daylight makes use of right triangles on the celestial sphere. The time difference between sunrise and 6 A.M. is proportional to the length of arc EV. 125
21. Title page of the *Exposición Astronomica*, 1681. 147
22. Eusebio Kino's map of the heavens, from the *Exposition Astronómica*, showing the path of the comet of 1680–1681. The lower strip contains the ten figures to which Kino refers within the text. 148
23. Title page of the *Libra Astronómica*, 1690. 152
24. A "Compendium of Mensuration" is the second part of Jacob Taylor's *Tenebrae*, 1697. 175
25. Title page of John Foster's almanac of 1676, the first to appear from his own press. 231
26. Feliciana Ruiz, possibly the first woman to author a book printed in the New World, asks permission from the Inquisition to print the almanac that she has created. 233
27. Title page of Thomas Brattle's almanac for 1678. 237
28. Brattle's table of the length of the longest period of daylight of the year at various latitudes. 238
29. The "Man of Signs" from Foster's almanac of 1678. Aries is associated with the head and Pisces with the feet, with the other signs in order in between. The poem is offered as an aid to memory. Almanac users could check the moon tables to see what sign the moon was in and thus know which part of the body the moon was currently influencing. 243

Acknowledgments

This book has been shaped by many people. Back in 1996, my good friend Ed Sandifer gave me the idea of doing something with Spanish colonial mathematics books, and he has been involved all along the way. He was there the first time I requested rare books from a library, and he taught me what to look for; he read several early drafts of this book and provided valuable comments; he worked with me at both the Yale and the John Carter Brown libraries and helped me to understand some of the Latin texts; and he was generally full of encouragement and good ideas.

In the fall of 2000, Richard Ring made me realize the nature of my project. "You're doing a bibliography?" "Yes . . . I . . . am!" "Oh, that will be very useful."

My colleague Joel Silverberg always seemed able to tease out the answer to whatever I asked from the World Wide Web. Alex Scorpan was ever ready to help me with TeX. Thanks also to the rest of the Mathematics Department at Roger Williams University—Earl Gladue, Annela Kelly, Ruth Koelle, and Yajni Warnapala-Yehiya—for putting up with me.

One advantage of being on the faculty of a small- to medium-sized university is the chance to meet colleagues from diverse disciplines and exchange ideas. Peter Thompson read an early version of the manuscript and helped me make my Spanish (and English!) more correct. Many others had an influence on this book or were generally helpful and encouraging. A partial list includes Adam Braver, Avelina Espinosa, Christine Fagan, Ernie Greco, Tony Hollingsworth, Jeffrey Meriwether, Jim Tackach, and Mel Topf. There are many more that I'm unable to mention here.

Thanks go to some of my former students. Erin Smith assisted me with my research at the Biblioteca Nacional de México on a grant for under-

ACKNOWLEDGMENTS

graduate research from RWU. Rory Plante wrote a fine senior thesis on Carlos de Sigüenza y Góngora, which is cited in the references.

The history of mathematics community has been supportive, encouraging, and tremendously helpful. Ed Sandifer was already mentioned. Shirley Gray read an early version and made extensive and highly useful comments. Larry D'Antonio set me straight about congruent numbers. Pier Daniele Napolitani made me realize that I had overlooked Maurolico's *Computus Ecclesiasticus*, even though I had looked right at it. Some others are Rob Bradley, Sandro Caparrini, Ximena Catepillan, Tom Drucker, Barney Hughes, Clara Nucci, Danny Otero, Kim Plofker, Fred Rickey, Glen Van Brummelen, and Greisy Winicki-Landman.

Heather Vincent and Sin Guanci helped me with some Latin translations. Norton Starr and Norman Fiering read early drafts and made comments that resulted in significant improvements in this book. Thanks also go to Brian Raines and to Allison Sandman for their help.

My mother, Cathie Burdick, read a draft and helped me to pay more attention to my commas. Thanks, Mom!

Roger Williams University has supported this project in many ways. Two important ways are through the Professional Development Committee and the Foundation to Promote Scholarship and Teaching. Thanks to all who have served on these committees over the years. RWU also granted me a sabbatical for the academic year 2004–2005, and this greatly accelerated the progress of my research. In those fifteen months alone, I was able to visit more than fifty libraries and to rewrite the draft of this book several times.

Thanks especially to all of the wonderful librarians and curators at all of the many libraries and museums that I visited or contacted in the course of researching this book. It is impossible to mention everybody. Private collectors also have been universally helpful in supplying me with information about their holdings. There is an Index of Libraries and Collections beginning on page 357, listing everywhere that has an original copy of any item in my lists of sixteenth- and seventeenth-century American mathematical printed works. At the end of that, I have listed all of the other libraries that I visited during the course of this project. I hope that a general thank-you to all of the staff members at all of these institutions, and to all of the collectors, will assure everyone that their enthusiastic assistance is still greatly appreciated.

Mathematical Works Printed in the Americas, 1554–1700

Introduction

At the 1932 meeting of the American Historical Association in Toronto, Herbert Eugene Bolton gave a speech entitled "The Epic of Greater America"[1] in which he proposed a more pan-American approach to the writing and teaching of the history of the New World's nations. "There is a need," he said, "for a broader treatment of American history, to supplement the purely nationalistic presentation to which we are accustomed. European history cannot be learned from books dealing alone with England, or France, or Germany, or Italy, or Russia; nor can American history be adequately presented if confined to Brazil, or Chile, or Mexico, or Canada, or the United States." He proceeded to outline what history could look like if the events of the New World were laid side by side so that transnational movements could be more fully appreciated and not be obscured within a nationalistic context.

Bolton mentions the dates for the founding of institutions of higher learning in the Americas: universities in Mexico City[2] and in Lima[3] in 1551, the Jesuit College of Quebec in 1635, Harvard in 1636, William and Mary in 1695, and Yale in 1701. He tells us that until "near the end of the eighteenth century not Boston, not New York, not Charleston, not Quebec, but Mexico City was the metropolis of the entire Western Hemisphere."

Perhaps Louis Karpinski was influenced by Bolton when he chose to include in his *Bibliography of Mathematical Works Printed in America Through 1850* works from all quarters of North and South America, and

---

1. In Lewis Hanke, ed., *Do the Americas Have a Common History? A Critique of the Bolton Theory*, New York, 1964.
2. The present-day Universidad Nacional Autónoma de México.
3. The present-day Universidad Nacional Mayor de San Marcos.

even from the Philippines and the Hawaiian Islands, arranged in chronological order. Our efforts here are also in the spirit of Bolton's outline of the history of the hemisphere. We lay side by side the mathematical writings of three centers of activity: Mexico, Lima, and the English colonies of Massachusetts, Pennsylvania, and New York. The reader will see that some interests, for example, astronomy and especially comets, captured imaginations in Mexico, Peru, and New England, and that practical concerns, such as the need to do accurate transactions in silver and gold, caused information to flow between Mexico, Peru, and Spain.

The printing press has been in the New World for nearly five centuries. By looking at José Toribio Medina's series of bibliographic indices for cities and countries,[4] we see a first entry for Mexico in 1539; for Lima, 1584; for Puebla, 1640; for Guatemala, 1660; for Paraguay, 1705; for Havana, 1707; and for Oaxaca, 1720. In the case of the English colonies, the press first operated in Cambridge, Massachusetts, in 1639, according to Thomas[5] and others. (Not to slight our Canadian friends, we note that the first printing press in Canada began operating in Halifax in 1751.) Early entries for most sites in Latin America are religious works, sermons, and Christian doctrine, sometimes translated into native languages. Science and mathematics tended to be delayed until later, although the first work printed in Puerto Rico, in 1808, was an almanac. The press in Cambridge first printed the *Freeman's Oath*. The second work produced there (and the first book) was an almanac.

The New World was not long without printed science and mathematics, however, because many titles were imported from Europe. A single shipment inventory from 1600 discussed by Green and Leonard,[6] for example, has three copies of Euclid, one of Archimedes, and one of Apollonius. Of the 678 items listed in this shipment, Green and Leonard found 33 texts relating to mathematics, astronomy, and physics. Therefore the books included in this work—the ones printed in the New World—reflect perceptions, on the part of the denizens of the Americas, of needs that were not being met by the imported books.

Our intended audience includes diverse groups. We want this to be a useful reference work for both historians and bibliophiles. Some of the

---

4. José Toribio Medina, *La Imprenta en México*, *La Imprenta en Lima*, etc. (See "Works Consulted.")

5. Isaiah Thomas, *The History of Printing in America*, New York, 1970.

6. Otis H. Green and Irving A. Leonard, "On the Mexican Book Trade in 1600: A Chapter in Cultural History," *Hispanic Review* 9/1 (Jan. 1941).

INTRODUCTION

more interesting titles included here have moved us to write at length about certain mathematical points raised by the texts, and we raise some non-mathematical issues as well. It is hoped that these outpourings will be useful and enjoyable both to teachers and students of mathematics and to anyone who loves a good story.

For instance, a startling and perhaps controversial thing we say about the *Sumario Compendioso* of 1556 is that we believe that nearly everything that has been said in print in the last eighty-five years about its author, Juan Diez Freyle, is wrong. For this reason we have looked very closely at parts of the book that would otherwise be only marginally interesting for the history of mathematics in the hope that they would provide clues about the author. When information in the *Sumario Compendioso* relating to transactions in silver and gold is compared to that contained in similar works later on the list, there is some support for a theory of Joaquín García Icazbalceta concerning Diez Freyle's possible origins in Peru. However, the matter is certainly not laid to rest.

In contrast, in the case of Eusebio Francisco Kino and Carlos de Sigüenza y Góngora, authors of titles published in 1681 and 1690, so much has already been said, and so well, that we couldn't resist the urge to give a short biographical narrative sifted from the secondary material. We put the relevant chapter from the lives of these two extraordinary intellects into the context of the worldwide debate about the nature of comets, and discover how each man used mathematics in his arguments.

In the case of the almanacs included in Part 2, the mathematical content does not change much from one author or year to another. While some general observations are in order, the comments on the individual entries tend to focus on the non-mathematical quirks of each issue. One exception to this is the discussion of the table of durations of the longest day for different latitudes in Thomas Brattle's almanac of 1678, where we look at his numbers and try to guess how he got them.

Much of what is written here about history is heavily dependent on secondary sources, so readers should not regard this as a historical study in the usual sense. Nevertheless, it is hoped that the list of works provided here is something that historians of mathematics have not seen before in this form and that it will be useful for that reason.

With the hope of attracting as wide an audience as possible, we have endeavored in the mathematical notes not to assume too much. Familiarity with secondary school algebra, geometry, and trigonometry should suffice.

It is our wish that other researchers will be inspired to do further work on some of the texts we have discussed here. We have tried in many cases not to draw final conclusions but to leave open the possibility for further research. We note also that of the books that are in Spanish or Latin, only three have been translated into English to date. We would greatly enjoy seeing a few more of them in translation in the future.

What is offered here can be summarized by the observation that, although no new mathematics came out of the Americas during the period in question, mathematics was most definitely a part of the intellectual life of the Americas in the sixteenth and seventeenth centuries.

Overview: Criteria for Inclusion

What is a mathematical work? Rather than follow mathematical custom and state a definition, we choose to describe in some detail why some works were included and what was left out. It will become clear that in our pursuit of the category of "mathematical work" some idiosyncratic choices have been made. We are willing to defend these choices, but we recognize that someone else might have chosen differently.

Our study began with the list given by Louis Karpinski, whose *Bibliography* has eleven entries for the sixteenth and seventeenth centuries. We have rejected one of those, the *Pragmática sobre los diez días* of 1584, with originals at the John Carter Brown Library and at Harvard University. This is a decree from King Philip II on the topic of the change to the Gregorian calendar, a change that was already implemented in Spain in 1582 but was delayed in the colonies. The text deals with rents, terms of sentence, feast days, etc., in view of the fact that ten days must be skipped when changing from the old calendar to the new. Although it is an important document for the history of calendar reform, as well as the history of printing in Peru, it does not appear to have any mathematical content.

Karpinski's first entry, the *Sumario Compendioso* of 1556, is, on the other hand, a remarkable book. The sections on algebra and arithmetic, running for some 25 pages, represent a quantity of mathematical exposition unrivaled by other Spanish colonial works for decades after. Yet most of the book consists of accounting tables whose use would not demand much in the way of mathematical sophistication, although they should be of interest to a historian studying the colonial economy. These tables, though, in their own way, are also interesting for the mathematical historian, if only because of the mistakes and inconsistencies that are present.[7]

There are many other similar books of accounting tables from this period, and none of them has an amount of mathematical content comparable to that in the *Sumario Compendioso*. Nevertheless, all examples of such works that we could find were included, perhaps because of their "family resemblance" to the *Sumario Compendioso*. In addition to the *Sumario Compendioso* and the three other books of tables listed by Karpinski (which are our entries in Part 1 for 1597, 1607, and 1615), there are six more items on our list that are books of accounting tables. These were found by scanning the bibliographies, mentioned earlier, of Medina. All nine of these descendents of the *Sumario Compendioso* are briefly described and compared in the conclusion of the discussion of the *Sumario*, in Part 1 under 1556; they also have their own entries. Moreover, Part 1 also includes entries for six works on currency reform, also found by looking in Medina.

The *Sumario Compendioso* could also be called an arithmetic if one ignores the first ninety-one folios. In addition to the *Sumario Compendioso*, with its arithmetic and algebra sections in the back, Karpinski found three other arithmetics from these centuries (our entries in Part 1 for 1623, 1649, and 1675). This list has been harder to supplement. We can, however, point to two items that Karpinski missed: the *Declaración del Quadrante* of 1682, with its rules for dividing money among multiple parties, and the *Teórica, y Práctica de Esquadrones* of 1660, on the topic of arrangements of squadrons, which will take up much of Fernández de Belo's arithmetic fifteen years later. A somewhat more speculative item that is included is the *Declaración de los Puntos* of 1621. No copies are known, but we are guessing, based on the content of the *Arte para Aprender* by the same author two years later, that it is likely to contain some treatment of arithmetic.

Our inclusion of books on logic needs to be remarked upon. Logic has long been classified as a branch of philosophy, but, starting with the algebraic approach to logic in the nineteenth century, and continuing with the

---

7. D. E. Smith wrote in *The Sumario Compendioso of Brother Juan Diez, the Earliest Mathematical Work of the New World*, Boston, 1921: 11, that the tables were of no interest, but Karpinski countered in his review of Smith's book, "Books and Literature," *School and Society* XIV (Aug. 13, 1921): 84, that "A careful study of the work by one versed in sixteenth century Spanish and in the history of Spanish America will doubtless demonstrate that the tables are even yet of impo[r]tance..."

use of symbolic logic to shore up the foundations of mathematics in the early twentieth century, not to mention the overwhelming importance of binary logic for programming in the computer era of the last half century, it seems reasonable to say that formal logic is a field shared by philosophy and mathematics.[8] Therefore, we include several manuals on Aristotelian logic within our genre of mathematical works.[9] These are the *Recognitio, Summularum* of 1554, *Introductio in Dialecticam Aristotelis* of 1578, *Comentarii ac Quaestiones* of 1610, *The Logic Primer* of 1672, and *Summa Tripartita—in Logicam* of 1693.

The *Dialectica Resolutio* of 1554, a translation of and commentary on Aristotle, contains numerous examples drawn from geometry. Two books on Aristotelian physics are included because of their development of geometrical concepts for astronomy: the *Phisica, Speculatio* of 1557 and the *De Sphæra* of 1578. (We know that Karpinski knew of the *Phisica, Speculatio* because he mentioned it in two articles,[10] and in the first of these articles he noted that it included the *Tractatus de Sphera* by Companus of Novara, but for whatever reason he chose not to include the *Phisica, Speculatio* in his *Bibliography*.) Two works on astrology have been included because of their use of trigonometry: the *Reportorio de los Tiempos* of 1606 and the *Opúsculo de Astrología* of 1660. The *Computus Ecclesiasticus* of 1578 is a treatise on the Julian calendar.

In the seventeenth century there were many spectacular comets, and a flurry of books about them appeared in the New World. Two works considered here, Kino's *Exposición Astronomica* of 1681 and Sigüenza's *Libra Astronomica* of 1690, are noteworthy for their use of mathematics in arguments for and against the claim that comets were auguries of dire events.[11] Some other works that were part of the same debate were not judged to be sufficiently mathematical to be included in the main list. They are listed in Appendix B. Some are also described within the discussion of Sigüenza's work of 1690, and some are mentioned in Part 2 because their authors were also the authors of almanacs.

---

8. It has been pointed out to us that electrical engineering also makes heavy use of logic.
9. We will elaborate on our justification for this in our Appendix C.
10. Louis C. Karpinski, "Books and Literature," *School and Society* XIV (Aug. 13, 1921): 83–84; and "The Earliest Known American Arithmetic," *Science* 63 (Feb. 19, 1926): 193–195.
11. Cajori describes this as "the earliest instance of a clash of intellects in public print on a scientific question that occurred in America" in *The Early Mathematical Sciences in North and South America*, Boston, 1928.

INTRODUCTION

Jacob Taylor's *Tenebrae* of 1697 announces itself in its full title to be a book of tables of eclipses for the next twenty years. That would be enough for inclusion here, but it also has a section that covers the solving of triangles and other geometric figures. This book was forgotten for many years, but it is now available for study in a nearly complete copy.

In a review of Karpinski's *Bibliography*, Wroth noted the absence of any almanacs.[12] Since Karpinski's publication in 1940, we now have Drake's *Almanacs of the United States* of 1962,[13] as well as works by Quintana[14] and Schwab[15] on the almanacs of Mexico and Peru, respectively, so it is now easier to compile a list of almanacs in the Americas. Almanacs of the times under consideration here were mostly ephemerides, giving the positions of the sun, moon, and planets throughout the year. In colonial times a poor man with his almanac could tell the time more accurately than a rich man with his pocket watch.[16] As with the books of accounting tables discussed earlier, such works can give insight into the use of mathematics by their authors and their users, even if the algorithms they used are not made explicit.

Smith and Ginsberg, in their *A History of Mathematics in America Before 1900*, claim that among many almanacs of the seventeenth century, those of Samuel Danforth,[17] Urian Oakes, Thomas Brattle, and Henry Newman

---

12. Lawrence C. Wroth, "Notes for Bibliophiles," Review of *Bibliography of Mathematical Works Printed in America Through 1850* by Louis C. Karpinski, *New York Herald Tribune Books* (Aug. 25, 1940).

13. Robb Sagendorph reports that Milton Drake worked for twenty-five years on this guide to almanacs and that among his other credits is the song "Mares Eat Oats," in *America and her Almanacs*, Dublin, NH, 1970: 16.

14. José Miguel Quintana, *La Astrología en la Nueva España, en el Siglo XVII (De Enrico Martínez a Sigüenza y Góngora)*, México, 1969.

15. Federico Schwab, "Los Almanaques Peruanos," *Boletín Bibliográfico* XIX/1–2 (1948).

16. The following observation was made by Nathanael Low in his almanac of 1786: "Almanacks serve as clocks and watches for nine-tenths of mankind; and in fair weather are far more sure and regular than the best time-piece manufactur'd here or in London—twenty gentlemen in company will hardly be able, by the help of their thirty-guinea watches, to guess within two hours of the true time of night——one says it is nine o'clock, another half after eight—a third, half after ten; whilst the poor peasant, who never saw a watch, will tell the time to a fraction, by the rising and setting of the moon, and some particular stars, which he learns from his almanack."

17. There were two Samuel Danforths who wrote almanacs in the seventeenth century. We do not know which one was intended.

are more than just "prognostications," but are noteworthy for their mathematical content. In fact, nearly all almanacs from the period have about the same amount of information about the motions of the sun, moon, and planets, and so all of the almanacs given by Drake up to 1700 are included, with titles from the Spanish colonies merged in. (It appears that no Spanish colonial almanacs from the seventeenth century have survived in their printed form.) Wherever possible we have tried to give complete titles rather than short titles in the style of Drake. Because of the number of titles, we have put almanacs in a separate section, Part 2, after the other mathematical works in Part 1. (In the table of short titles in the following section these lists are merged in chronological order.)

A work that excited us when it again became available, but which ultimately turned out not to have any mathematics, was Juan Vásquez de Acuña's biography of Galileo, printed in Lima in 1650. It is just two leaves without a title page and was listed by Medina with its first few words, *Galileo Galilei, Filosofo, y Mathematico el mas celebre*, etc, in lieu of a title. This was out of sight for many years, but it is now at the John Carter Brown Library.

A final note is about another work that was not included. Augustín de Vetancurt's *Teatro Mexicano* of 1698, in Part 1, Treatise 1, Chapter 8, says that the world is 6,300 Spanish leagues in circumference (since there are 17.5 Spanish leagues to a degree at the equator) and is 2,004 leagues in diameter, thus implicitly giving the value 3.1437 for π. In Part 2, Treatise 2, Chapters 5 to 7, he discusses in detail the Mesoamerican calendar with its fifty-two-year cycle, known as the Calendar Round, and compares it to the Gregorian calendar. In spite of these interesting items, we could not justify calling this a mathematical work.

It seems inevitable that some titles have been missed. Our chief method for finding titles not in Karpinski was to page through all the listings in Medina's Mexico and Peru volumes looking for titles that seemed promising. We might not have spotted Martínez's *Reportorio de los Tiempos* by this method had we not already seen it discussed by Elías Trabulse.[18] How many other titles might hide some interesting uses of mathematics? We welcome recommendations from readers for consideration in future editions.

---

18. Elías Trabulse's *Historia de la Ciencia en México*, abridged version, México, 1994, was where we first learned of the *Reportorio*.

INTRODUCTION

Here is the list of works considered, with short titles (some of which have been altered to accord with modern Spanish or English spelling). The full titles (as originally printed) and bibliographic information will be given later. Almanacs and lunarios are listed together with the other works here; later, they will be placed in a separate list. Items are numbered according to the order in which they appear in each section, with the entries from Part 2 (the almanacs) having the prefix 'A'.

Multiple entries within each of the years 1554 and 1578 are typically bound together, so they are listed here in the order in which they are bound. Starting in 1610, multiple entries for the same year are alphabetized by author first and title second.

Short Title Index to Works in Parts One and Two

Alonso de la Vera Cruz, *Recognitio, Summularum*. México, 1554.	1.
Alonso de la Vera Cruz, *Dialectica Resolutio*. México, 1554.	2.
Juan Diez Freyle, *Sumario Compendioso*. México, 1556.	3.
Alonso de la Vera Cruz, *Phisica, Speculatio*. México, 1557.	4.
Francisco de Toledo, *Introductio in Dialecticam Aristotelis*. México, 1578.	5.
Francesco Maurolico, *De Sphœra*. México, 1578.	6.
Francesco Maurolico, *Computus Ecclesiasticus*. México, 1578.	7.
Diego García de Palacio, *Diálogos Militares*. México, 1583.	8.
Diego García de Palacio, *Instrucción Náutica*. México, 1587.	9.
Joán de Belveder, *Libro General*. Lima, 1597.	10.
Felipe de Echagoyan, *Tablas de Reducciones*. México, 1603.	11.
Enrico Martínez, *Lunario y Regimiento*. México, 1604.	A1.
Enrico Martínez, *Reportorio de los Tiempos*. México, 1606.	12.
Francisco Juan Garreguilla, *Libro de Plata*. Lima, 1607.	13.
Pedro de Aguilar Gordillo, *Alivio de Mercaderes*. México, 1610.	14.
Jerónimo Valera, *Commentarii ac Quæstiones*. Lima, 1610.	15.
Álvaro Fuentes y de la Cerda, *Libro de Cuentas*. México, 1615.	16.
Pedro de Paz, *Declaración de los Puntos*. México, 1621.	17.
Pedro de Paz, *Arte para Aprender*. México, 1623.	18.
William Pierce, *An Almanac*. Cambridge, 1639.	A2.
Anon., *An Almanac*. Cambridge, 1640.	A3.
Anon., *An Almanac*. Cambridge, 1641.	A4.
Anon., *An Almanac*. Cambridge, 1642.	A5.
Anon., *An Almanac*. Cambridge, 1643.	A6.

MATHEMATICAL WORKS PRINTED IN THE AMERICAS, 1554–1700

Anon., *An Almanac*. Cambridge, 1644.	A7.
Anon., *An Almanac*. Cambridge, 1645.	A8.
Samuel Danforth (1626–1674), *An Almanac*. Cambridge, 1646.	A9.
Samuel Danforth (1626–1674), *An Almanac*. Cambridge, 1647.	A10.
Samuel Danforth (1626–1674), *An Almanac*. Cambridge, 1648.	A11.
Felipe de Castro, *Lunario y Repertorio*. México, 1649.	A12.
Samuel Danforth (1626–1674), *An Almanac*. Cambridge, 1649.	A13.
Gabriel López de Bonilla, *El Diario y Discurso*. México, 1649.	A14.
Atanasio Reatón, *Arte Menor*. México, 1649.	19.
Pedro Diez de Atienza, *Sobre la Reducción de Monedas*. Lima, 1650.	20.
Urian Oakes, *An Almanac*. Cambridge, 1650.	A15.
Francisco de Villegas, *Repuesta*. Lima, 1650.	21.
Anon., *An Almanac*. Cambridge, 1651.	A16.
Francisco Ruiz Lozano, *Reportorio Anual*. México, 1651.	A17.
Anon., *An Almanac*. Cambridge, 1652.	A18.
Pedro Diez de Atienza, *Cerca de la Reformación de la Moneda*. Lima, 1652.	22.
Francisco Ruiz Lozano, *Reportorio Anual*. México, 1652.	A19.
Anon., *An Almanac*. Cambridge, 1653.	A20.
Juan Ruiz, *Reportorio*. México, 1653.	A21.
Anon., *An Almanac*. Cambridge, 1654.	A22.
Francisco Ruiz Lozano, *Reportorio Anual*. Lima, 1654.	A23.
Anon., *An Almanac*. Cambridge, 1655.	A24.
Martín de Córdoba, *Pronóstico*. México, 1655.	A25.
Francisco Ruiz Lozano, *Reportorio Anual*. Lima, 1655.	A26.
Gabriel López de Bonilla, *Lunario y Discurso Astronómico*. México, 1656.	A27.
Francisco Ruiz Lozano, *Reportorio Anual*. Lima, 1656.	A28.
Thomas Shepard, *An Almanac*. Cambridge, 1656.	A29.
Samuel Bradstreet, *An Almanac*. Cambridge, 1657.	A30.
Luis Enríquez de Guzmán, *Para que Pudiese Correr la Moneda*. Lima, 1657.	23.
Francisco Ruiz Lozano, *Reportorio Anual*. Lima, 1657.	A31.
Francisco de Villegas, *Reparos*. Lima, 1657.	24.
Anon., *An Almanac*. Cambridge, 1658.	A32.
Francisco Ruiz Lozano, *Reportorio Anual*. Lima, 1658.	A33.
Zecharia Brigden, *An Almanac*. Cambridge, 1659.	A34.
Juan Ruiz, *El Pronóstico*. México, 1659.	A35.

INTRODUCTION

Francisco Ruiz Lozano, *Reportorio Anual*. Lima, 1659.	A36.
Samuel Cheever, *An Almanac*. Cambridge, 1660.	A37.
Juan de Figueroa, *Opúsculo de Astrología*. Lima, 1660.	25.
Antonio de Heredia y Estupiñán, *Teórica, y Práctica de Esquadrones*. Lima, 1660.	26.
Juan Ruiz, *El Pronóstico*. México, 1660.	A38.
Samuel Cheever, *An Almanac*. Cambridge, 1661.	A39.
Juan Ruiz, *El Reportorio*. México, 1661.	A40.
Nathaniel Chauncy, *An Almanac*. Cambridge, 1662.	A41.
Martín de Córdoba, *El Pronóstico y Lunario*. México, 1662.	A42.
Gabriel López de Bonilla, *Diario y Discurso*. México, 1662.	A43.
Israel Chauncy, *An Almanac*. Cambridge, 1663.	A44.
Martín de Córdoba, *El Pronóstico de Temporales*. México, 1663.	A45.
Gabriel López de Bonilla, *Diario*. México, 1663.	A46.
Juan Ruiz, *El Lunario y Regimiento*. México, 1663.	A47.
Israel Chauncy, *An Almanac*. Cambridge, 1664.	A48.
Martín de Córdoba, *El Pronóstico*. México, 1665.	A49.
Gabriel López de Bonilla, *Diario y Discursos*. México, 1665.	A50.
Alexander Nowell, *An Almanac*. Cambridge, 1665.	A51.
Juan Ruiz, *El Lunario, Regimiento*. México, 1665.	A52.
Martín de Córdoba, *El Lunario, y Pronóstico*. México, 1666.	A53.
Josiah Flint, *An Almanac*. Cambridge, 1666.	A54.
Gabriel López de Bonilla, *El Diario y Discursos*. México, 1666.	A55.
Juan Ruiz, *Lunario, Regimiento*. México, 1666.	A56.
Samuel Brackenbury, *An Almanac*. Cambridge, 1667.	A57.
Gabriel López de Bonilla, *Diario y Discursos*. México, 1667.	A58.
Juan Ruiz, *El Pronóstico*. México, 1667.	A59.
Juan de Castañeda, *Reformación de las Tablas*. México, 1668.	27.
Joseph Dudley, *An Almanac*. Cambridge, 1668.	A60.
Gabriel López de Bonilla, *Diario y Discursos*. México, 1668.	A61.
Joseph Browne, *An Almanac*. Cambridge, 1669.	A62.
Juan Ruiz, *El Lunario, Regimiento*. México, 1669.	A63.
Nicolás de Matta, *Lunario y Pronóstico*. México, 1670.	A64.
John Richardson, *An Almanac*. Cambridge, 1670.	A65.
Juan Ruiz, *El Pronóstico*. México, 1670.	A66.
Daniel Russell, *An Almanac*. Cambridge, 1671.	A67.
Carlos de Sigüenza y Góngora, *El Lunario y Pronóstico*. México, 1671.	A68.

MATHEMATICAL WORKS PRINTED IN THE AMERICAS, 1554–1700

John Eliot, *The Logic Primer*. Cambridge, 1672.	28.
Jeremiah Shepard, *An Ephemeris*. Cambridge, 1672.	A69.
Carlos de Sigüenza y Góngora, *El Lunario y Pronóstico*. México, 1672.	A70.
Nehemiah Hobart, *An Almanac*. Cambridge, 1673.	A71.
Juan Ruiz, *Pronóstico*. México, 1673.	A72.
Juan de Saucedo, *El Juicio Astronómico*. México, 1673.	A73.
Carlos de Sigüenza y Góngora, *Lunario*. México, 1673.	A74.
Juan Ruiz, *Lunario*. México, 1674.	A75.
Juan de Saucedo, *El Pronóstico*. México, 1674.	A76.
John Sherman, *An Almanac*. Cambridge, 1674.	A77.
Carlos de Sigüenza y Góngora, *El Lunario*. México, 1674.	A78.
Benito Fernández de Belo, *Breve Aritmética*. México, 1675.	29.
John Foster, *An Almanac*. Cambridge, 1675.	A79.
John Sherman, *An Almanac*. Cambridge, 1675.	A80.
Carlos de Sigüenza y Góngora, *El Lunario y Pronóstico*. México, 1675.	A81.
John Foster, *An Almanac*. Boston, 1676.	A82.
Feliciana Ruiz, *El Lunario, Regimiento*. México, 1676.	A83.
John Sherman, *An Almanac*. Cambridge, 1676.	A84.
Carlos de Sigüenza y Góngora, *Lunario*. México, 1676.	A85.
John Foster, *An Almanac*. Cambridge, 1677.	A86.
Juan de Saucedo, *El Pronóstico*. México, 1677.	A87.
John Sherman, *An Almanac*. Cambridge, 1677.	A88.
Carlos de Sigüenza y Góngora, *El Lunario*. México, 1677.	A89.
Thomas Brattle, *An Almanac*. Cambridge, 1678.	A90.
John Danforth, *An Almanac*. Cambridge, 1678.	A91.
José de Escobar, Salmerón, y Castro, *El Lunario y Regimiento*. México, 1678.	A92.
John Foster, *An Almanac*. Boston, 1678.	A93.
Carlos de Sigüenza y Góngora, *El Pronóstico*. México, 1678.	A94.
John Danforth, *An Almanac*. Cambridge, 1679.	A95.
José de Escobar, Salmerón, y Castro, *Lunario y Regimiento*. México, 1679.	A96.
John Foster, *An Almanac*. Boston, 1679.	A97.
Carlos de Sigüenza y Góngora, *Lunario y Pronóstico*. México, 1679.	A98.
José de Escobar, Salmerón, y Castro, *El Lunario y Regimiento*. México, 1680.	A99.

INTRODUCTION

John Foster, *An Almanac*. Boston, 1680.	A100.
Juan Ramón Koenig, *El Conocimiento de los Tiempos*. Lima, 1680.	A101.
Carlos de Sigüenza y Góngora, *El Lunario y Pronóstico*. México, 1680.	A102.
John Foster, *An Almanac*. Boston, 1681.	A103.
Eusebio Francisco Kino, *Exposición Astronómica*. México, 1681.	30.
Juan Ramón Koenig, *El Conocimiento de los Tiempos*. Lima, 1681.	A104.
Carlos de Sigüenza y Góngora, *Lunario*. México, 1681.	A105.
Antonio Sebastián de Aguilar Cantú, *Pronóstico de los Temporales*. México, 1682.	A106.
William Brattle, *An Ephemeris*. Cambridge, 1682.	A107.
Martín de Echagaray, *Declaración del Quadrante*. México, 1682.	31.
José de Escobar, Salmerón, y Castro, *El Lunario y Pronóstico*. México, 1682.	A108.
Juan Ramón Koenig, *El Conocimiento de los Tiempos*. Lima, 1682.	A109.
Carlos de Sigüenza y Góngora, *El Lunario y Pronóstico*. México, 1682.	A110.
Anon., *The Cambridge Ephemeris*. Cambridge, 1683.	A111.
José de Escobar, Salmerón, y Castro, *El Lunario y Pronóstico*. México, 1683.	A112.
Juan Ramón Koenig, *El Conocimiento de los Tiempos*. Lima, 1683.	A113.
Cotton Mather, *The Boston Ephemeris*. Boston, 1683.	A114.
Carlos de Sigüenza y Góngora, *El Lunario y Pronóstico*. México, 1683.	A115.
José de Escobar, Salmerón, y Castro, *El Pronóstico o Lunario*. México, 1684.	A116.
Benjamin Gillam, *The Boston Ephemeris*. Boston, 1684.	A117.
Juan Ramón Koenig, *El Conocimiento de los Tiempos*. Lima, 1684.	A118.
Noadiah Russell, *The Cambridge Ephemeris*. Cambridge, 1684.	A119.
Carlos de Sigüenza y Góngora, *Repertorio*. México, 1684.	A120.
Juan Ramón Koenig, *El Conocimiento de los Tiempos*. Lima, 1685.	A121.
Nathanael Mather, *The Boston Ephemeris*. Boston, 1685.	A122.
Carlos de Sigüenza y Góngora, *Almanaque*. México, 1685.	A123.
William Williams, *The Cambridge Ephemeris*. Cambridge, 1685.	A124.
Samuel Atkins, *Kalendarium Pennsilvaniense*. Philadelphia, 1686.	A125.
Juan de Avilés Ramírez, *El Pronóstico de Temporales*. México, 1686.	A126.
Samuel Danforth (1666–1727), *The New England Almanac*. Cambridge, 1686.	A127.
Juan Ramón Koenig, *El Conocimiento de los Tiempos*. Lima, 1686.	A128.

Nathanael Mather, *The Boston Ephemeris*. Boston, 1686. **A129.**
Carlos de Sigüenza y Góngora, *El Pronóstico*. México, 1686. **A130.**
Antonio Sebastián de Aguilar Cantú, *El Pronóstico*. México, 1687. **A131.**
Juan de Avilés Ramírez, *Pronóstico de Temporales*. México, 1687. **A132.**
José de Campos, *El Pronóstico*. México, 1687. **A133.**
Juan Ramón Koenig, *El Conocimiento de los Tiempos*. Lima, 1687. **A134.**
Daniel Leeds, *An Almanac*. Pennsylvania, 1687. **A135.**
Carlos de Sigüenza y Góngora, *Almanaque*. México, 1687. **A136.**
John Tulley, *An Almanac*. Boston, 1687. **A137.**
William Williams, *The Cambridge Ephemeris*. Cambridge, 1687. **A138.**
Juan de Avilés Ramírez, *Pronóstico de Temporales*. México, 1688. **A139.**
José de Campos, *Pronóstico de Temporales*. México, 1688. **A140.**
Edward Eaton, *An Almanac*. Philadelphia, 1688. **A141.**
Juan Ramón Koenig, *El Conocimiento de los Tiempos*. Lima, 1688. **A142.**
Daniel Leeds, *An Almanac*. Philadelphia, 1688. **A143.**
Carlos de Sigüenza y Góngora, *El Pronóstico y Lunario*.
México, 1688. **A144.**
John Tulley, *An Almanac*. Boston, 1688. **A145.**
Antonio Sebastián de Aguilar Cantú, *El Pronóstico*. México, 1689. **A146.**
Juan de Avilés Ramírez, *Pronóstico de Temporales*. México, 1689. **A147.**
José de Campos, *El Pronóstico*. México, 1689. **A148.**
Juan Ramón Koenig, *El Conocimiento de los Tiempos*. Lima, 1689. **A149.**
Daniel Leeds, *An Almanac*. Philadelphia, 1689. **A150.**
Carlos de Sigüenza y Góngora, *El Lunario*. México, 1689. **A151.**
John Tulley, *An Almanac*. Boston, 1689. **A152.**
Antonio Sebastián de Aguilar Cantú, *El Pronóstico de los
Temporales*. México, 1690. **A153.**
Juan de Avilés Ramírez, *El Pronóstico de Temporales*. México, 1690. **A154.**
Juan Ramón Koenig, *El Conocimiento de los Tiempos*. Lima, 1690. **A155.**
Daniel Leeds, *An Almanac*. Philadelphia, 1690. **A156.**
Henry Newman, *Harvard's Ephemeris*. Cambridge, 1690. **A157.**
Carlos de Sigüenza y Góngora, *Almanaque*. México, 1690. **A158.**
Carlos de Sigüenza y Góngora, *Libra Astronómica*. México, 1690. **32.**
John Tulley, *An Almanac*. Boston, 1690. **A159.**
Antonio Sebastián de Aguilar Cantú, *El Lunario y Pronóstico*.
México, 1691. **A160.**
Juan de Avilés Ramírez, *El Pronóstico de Temporales*.
México, 1691. **A161.**

INTRODUCTION

Juan Ramón Koenig, *El Conocimiento de los Tiempos*. Lima, 1691.	A162.
Daniel Leeds, *An Almanac*. Philadelphia, 1691.	A163.
Henry Newman, *News from the Stars*. Boston, 1691.	A164.
Diego Pérez de Lazcano, *Proposición, y Manifiesto*. Lima, 1691.	33.
Carlos de Sigüenza y Góngora, *El Almanaque*. México, 1691.	A165.
John Tulley, *An Almanac*. Boston, 1691.	A166.
Antonio Sebastián de Aguilar Cantú, *Pronóstico de los Temporales*. México, 1692.	A167.
Juan de Avilés Ramírez, *Pronóstico de Temporales*. México, 1692.	A168.
Benjamin Harris, *The Boston Almanac*. Boston, 1692.	A169.
Benjamin Harris, *Monthly Observations*. Boston, 1692.	A170.
Juan Ramón Koenig, *El Conocimiento de los Tiempos*. Lima, 1692.	A171.
Daniel Leeds, *An Almanac*. Philadelphia, 1692.	A172.
Carlos de Sigüenza y Góngora, *Almanaque*. México, 1692.	A173.
John Tulley, *An Almanac*. Boston, 1692.	A174.
Antonio Sebastián de Aguilar Cantú, *El Pronóstico de los Temporales*. México, 1693.	A175.
Juan de Avilés Ramírez, *El Pronóstico de Temporales*. México, 1693.	A176.
Juan Ramón Koenig, *El Conocimiento de los Tiempos*. Lima, 1693.	A177.
Daniel Leeds, *An Almanac*. Philadelphia, 1693.	A178.
Nicolás de Olea, *Summa Tripartita—in Logicam*. Lima, 1693.	34.
Carlos de Sigüenza y Góngora, *Almanaque*. México, 1693.	A179.
John Tulley, *An Almanac*. Boston, 1693.	A180.
Juan Ramón Koenig, *El Conocimiento de los Tiempos*. Lima, 1694.	A181.
Daniel Leeds, *An Almanac*. New York, 1694.	A182.
Christian Lodowick, *An Almanac*. Boston, 1694.	A183.
Carlos de Sigüenza y Góngora, *Almanaque y Lunario*. México, 1694.	A184.
John Tulley, *An Almanac*. Boston, 1694.	A185.
Antonio Sebastián de Aguilar Cantú, *El Pronóstico de los Temporales*. México, 1695.	A186.
Juan Ramón Koenig, *El Conocimiento de los Tiempos*. Lima, 1695.	A187.
Daniel Leeds, *An Almanac*. New York, 1695.	A188.
Christian Lodowick, *The New England Almanac*. Boston, 1695.	A189.
Carlos de Sigüenza y Góngora, *El Almanaque y Lunario*. México, 1695.	A190.
John Tulley, *An Almanac*. Boston, 1695.	A191.

Antonio Sebastián de Aguilar Cantú, *El Pronóstico de los*
Temporales. México, 1696. **A192.**
Juan de Avilés Ramírez, *El Pronóstico de Temporales.*
México, 1696. **A193.**
Juan Ramón Koenig, *Cubus*. Lima, 1696. **35.**
Juan Ramón Koenig, *El Conocimiento de los Tiempos*. Lima, 1696. **A194.**
Daniel Leeds, *An Almanac*. New York, 1696. **A195.**
José Martí, *Tabla General*. Lima, 1696. **36.**
Carlos de Sigüenza y Góngora, *Almanaque y Lunario.*
México, 1696. **A196.**
John Tulley, *An Almanac*. Boston, 1696. **A197.**
Antonio Sebastián de Aguilar Cantú, *El Pronóstico de los*
Temporales. México, 1697. **A198.**
John Clapp, *New York Almanac*. New York, 1697. **A199.**
Juan Ramón Koenig, *El Conocimiento de los Tiempos*. Lima, 1697. **A200.**
Daniel Leeds, *An Almanac*. New York, 1697. **A201.**
Carlos de Sigüenza y Góngora, *Almanaque*. México, 1697. **A202.**
Jacob Taylor, *Tenebræ*. New York, 1697. **37.**
John Tulley, *An Almanac*. Boston, 1697. **A203.**
Manuel de Zuaza y Aranguren, *Reducciones de Plata*. México, 1697. **38.**
Antonio Sebastián de Aguilar Cantú, *El Pronóstico de los*
Temporales. México, 1698. **A204.**
Marco Antonio de Gamboa y Riaño, *Lunario y Pronóstico.*
México, 1698. **A205.**
Juan Ramón Koenig, *El Conocimiento de los Tiempos*. Lima, 1698. **A206.**
Daniel Leeds, *An Almanac*. New York, 1698. **A207.**
Carlos de Sigüenza y Góngora, *Almanaque*. México, 1698. **A208.**
John Tulley, *An Almanac*. Boston, 1698. **A209.**
Juan Ramón Koenig, *El Conocimiento de los Tiempos*. Lima, 1699. **A210.**
Daniel Leeds, *An Almanac*. New York, 1699. **A211.**
Carlos de Sigüenza y Góngora, *El Almanaque y Lunario.*
México, 1699. **A212.**
Jacob Taylor, *An Almanac*. New York, 1699. **A213.**
John Tulley, *An Almanac*. Boston, 1699. **A214.**
Antonio Sebastián de Aguilar Cantú, *El Pronóstico de los*
Temporales. México, 1700. **A215.**
Samuel Clough, *The New England Almanac*. Boston, 1700. **A216.**
Francisco de Fagoaga, *Reducción de Oro*. México, 1700. **39.**

INTRODUCTION

Juan Ramón Koenig, *El Conocimiento de los Tiempos*. Lima, 1700. A217.
Daniel Leeds, *An Almanac*. New York, 1700. A218.
Carlos de Sigüenza y Góngora, *El Almanaque*. México, 1700. A219.
John Tulley, *An Almanac*. Boston, 1700. A220.

Breakdown of Works Considered

Number of Mathematical Works (Excluding Almanacs)

México	23
Perú	14
Massachusetts	1
New York	1
Total	39

Number of Almanacs

México	95
Perú	27
Massachusetts	80
Pennsylvania	9
New York	9
Total	220

N.B. The most likely undercount is with the Mexican almanacs.

Some day, every book in every library of the world will be cataloged on the World Wide Web. We may have world peace first, however. While many of the original copies listed in the next section were found using the Web, principally through the use of the catalog search engines WorldCat and RLIN (the Research Libraries Information Network), many others were found by going to the libraries in question and using either a local electronic catalog or a card catalog. We have tried to examine personally as many of the originals as possible.[19] If any readers know of the locations of original copies that we have not listed here, we would be pleased to be notified and would hope to include such information in future editions.

---

19. At the Centro de Estudios de Historia de México Condumex, access to originals was not permitted in cases where they had a microfilm or facsimile copy, which was true for every book we were interested in.

At the Biblioteca Nacional de Chile, roughly half of the books we were interested in seeing were not made available because of the poor condition of the originals, and microfilm copies were provided instead.

Also listed are locations of copies for some early titles by Alonso de la Vera Cruz, Francisco de Toledo, Francesco Maurolico, and Francisco de Fagoaga that are not main entries. For the first three authors, these are European printings of titles printed once in Mexico, and the Mexican printing is the main entry. For the last author, there are also Mexican reprints that cross over our cutoff date of 1700. We have tried less hard to verify either the completeness or the accuracy of these lists of copies, oftentimes simply relying on WorldCat without visiting the libraries. Nevertheless, we would appreciate any information that would make these lists more accurate.

In the following two sections, each entry includes (when known) the author; a transciption of the title page; the printer and city of publication; bibliographic citations; the location of originals; the existence of microfilm, microfiche, or electronic copies; and the paging. Following this are discussed, in no particular order, the existence of either partial or complete reprints or translations, the existence of images of partial or complete text on Web sites, the condition of the originals, and, in some cases, anything unusual about what the originals are bound with or how they are catalogued. Also included, sporadically, and with no consistency as to length, are notes about the author, the printer, the content of the text, the history of the mathematical ideas present, and anything else that seems appropriate.

PART ONE

Mathematical Works, Excluding Almanacs

All the history of every people is symbolic.
—Octavio Paz, *Posdata*, 1970

Digo que el número sea una cosa (lit., "I say that the number is a thing")
—Juan Diez Freyle, 1556

1554

Alonso de la Vera Cruz.   1
RECOGNITIO,SVM / mularumReuerendi / PATRIS ILLDEPHONSIA VERA / CRVCE AVGVSTINIANI ARTIVM / ac sacræ Theologiæ Doctoris apud indorum in- / clytam Mexicum primarij in Academia / Theologiæ moderatoris. / Sagitaueras tu dñe[1] / cor meũ charitate tua.[2] / MEXICI. / Excudebat Ioannes Paulus Brissensis. / 1554.

Juan Pablos of Brescia, México.

Icazbalceta 20. Sabin 98918. Medina, *México*, 22. Wagner 28.

Originals located at: Biblioteca Cervantina, Biblioteca Nacional de Chile, Biblioteca National de México (two copies), British Library, Centro de Estudios de Historia de México Condumex, Hispanic Society of America, Huntington Library, Indiana University, John Carter Brown Library, New York Public Library, University of Texas at Austin.

Microfilm copies are available.

Paging: Ninety-six folios. The first eighty-eight folios are numbered except for 1 and 2, 15 is numbered xv, 33 is misnumbered 35, and they are

1. Line printed vertically.
2. Line printed vertically.

followed by an eight-folio summary of the work with numbering on the bottom of folios 2 to 5 only.

The title page is printed in red and black.

Henry Wagner reported that thirteen copies existed in 1940.[3] Earlier he wrote that the *Recognitio* represents the first use of Roman type in Mexico.[4]

The three Vera Cruz titles on this list are bound together at the John Carter Brown Library. In many instances it is the first two that are bound together. This is the case at the British Library where, according to the catalog, two flyleaves at the beginning of the volume and one flyleaf at the end are of paper made from the agave plant. Copy 1 at the Biblioteca Nacional de México has the first leaf mutilated, and in Copy 2 it is missing. In the Hispanic Society of America copy, the title page is restored and bound with a copy of the British Library title page. The Chile copy has a facsimile title page and much insect damage.

According to José Toribio Medina this work was reprinted in Salamanca in 1561, 1569, 1572, and 1593. Ennis gives 1562, 1569, 1573, 1595—posthumous.[5] The source of the discrepancy may be that the second edition, for instance, has 1562 on the title page but 1561 in the printer's colophon at the end. The following table summarizes the editions we have seen and some of the libraries known to hold them.

Year	Edition	Held by
1562	Second	Biblioteca Nacional de España, Universidad de Salamanca, Universidad de Sevilla
1569	Third	Biblioteca Nacional de Chile (two copies), Biblioteca Nacional de México, Centro de Estudios de Historia de México Condumex, Stanford University, Universidad de Barcelona, Universidad de Sevilla, University of Cambridge, University of Michigan
1573	Fourth	Biblioteca Universitaria Alessandrina, Biblioteca Nacional de México, California State Library, Escorial, Indiana University, Universidad de Barcelona, Universidad de Sevilla, University of Oxford, University of Pennsylvania

3. Henry R. Wagner, *Nueva Bibliografía Mexicana*, México, 1940.

4. Henry R. Wagner, *Mexican Imprints, 1544–1600, in the Huntington Library*, San Marino, 1939.

5. Arthur Ennis, *Fray Alonso*, offprint from *Augustiniana*, Louvain, 1957: 200.

Figure 1. Title page of the *Recognitio Summularum*, 1554. The John Carter Brown Library at Brown University, Providence, RI.

DE OPPOSITIONE PROPOSITIONVM. Fo. 30.

dinem, vt homo est animal, & homo non est animal, aliæ ordine conuerso, vt homo est animal, animal est homo. Et sic similiter de alijs.
¶ Propositionū participātium vtroqȝ termino, & ordine eodem: aliæ sunt cō tradictoriæ, aliæ contrariæ, aliæ sub cō trariæ, aliæ sub alternæ.
¶ Cōtradictoriȩ sunt, vniuersalis affirmatiua, & particularis negatiua: vel e contra, particularis affirmatiua, & vni uersalis negatiua: eiusdem subiecti, & eiusdem prȩdicati. Vt hȩc, omnis homo est animal, & aliquis homo nō est animal. Et nullus homo est animal, aliquis homo est animal.
¶ Contrariȩ sunt, vniuersalis affirmatiua, & vniuersalis negatiua: eiusdē sub iecti, & eiusdem prȩdicati. Vt omnis homo est animal, nullus homo est aial.
¶ Sub contrariȩ sunt, vtraqȝ particularis, vna negatiua: & alia affirmatiua, eiusdem subiecti, & eiusdem prȩdicari. Vel in difinita vtraqȝ, homo est animal homo non est animal.
¶ Sub alternȩ sunt, vniuersalis affirma tiua, & particularis assūmatiua: eiusdē subiecti, & eiusdem prȩdicati. Vel vniuersalis negatiua, & particularis negatiua, omnis homo est animal, aliquis ho mo est animal. Nullȝ homo est animal aliquis homo non est animal. Et hæc ad oculum patent, in figura sequenti.

FIGVRA OPPOSITIONVM.

Figure 2. From the *Recognitio Summularum* of 1554. The square of opposition. The John Carter Brown Library at Brown University, Providence, RI.

CAPITVLVM.X.

vel dissimiliter se habere: quantum ad affirmationem, vel negationem & sic de verbo dicendum.

¶ In prædictis regulis non fit mentio de contingenti: quia conuertitur cum possibili.

¶ Propositionum modalium, aliæ sunt contrariæ, aliæ sub contrariæ, aliæ contradictoriæ, aliæ subalternæ. Vnde omnes propositiones, quæ sunt in quarta linea de Purpurea, contrariantur omnibus illis: quæ sunt in tertia linea de Illiace, & omnes propositiones quæ sunt in prima linea de Amabimus; sub contrariantur illis quæ sunt in secunda linea de Edentuli. Item primus ordo, & tertius: contradicunt. Similiter secundus, & quartus. Item primus sub alternatur, quarto: & similiter secundus tertio.

Figure 3. From the Recognitio Summularum of 1554. Modal statements that are equivalent are grouped together. The John Carter Brown Library at Brown University, Providence, RI.

DE MODALIBVS.　　Fo.　36.

ducam,prius refoluam refolutorie fic, hoc fuit nigrum. &c.
¶ Circa modales diuifas,oportet notare: q̄- litatem poſſe attendi, vel penes verbū prin cipale, vel penes modum. Vt Petrus poſsi- biliter nō diſputat,dicitur negatiua. Et affir matiua de modo. Et hęc.Petrus non poſsibi biliter non diſputat. dicitur affirmatiua de copula & negatiua de modo.
¶ Quantitas etiam fumi poteſt, vel ex par- te ſubiecti, ſicut in alijs cathegoricis : vel ex parte modi : vt propoſitio de ly neceſſario, & impoſsibiliter,affirmatis & de ly poſsibi

le,& contingens, negatis : ſint vniuerſales de modo. Et econtrario, ſi iſti modi ſuman tur: ſit particularis modalis.(ſicut diximus) in modali compoſita. Et ſic hæc propoſitio omnis homo poſsibiliter diſputat, eſt vni- uerſalis cathegorica, & particularis de mo- do. Et hęc, Petrus impoſsibiliter diſputat, ſingularis Cathegorica : & vniuerſalis de modo. Et ſic poteſt poni omne genus oppo ſitionis, ſicut in ſimplicibus propoſitioni- bus,ex parte ſubiecti, Et ex parte modi, Vt patet in figura.

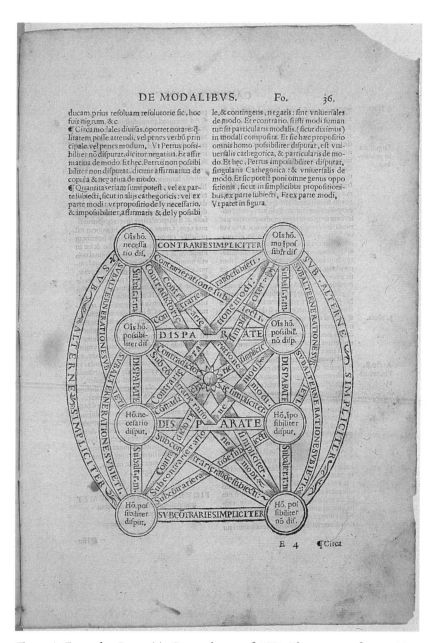

E 4　　　¶ Circa

Figure 4. From the *Recognitio Summularum* of 1554. The square of opposition with modality. The John Carter Brown Library at Brown University, Providence, RI.

One source suspects that the fifth edition doesn't exist.[6]

Five excerpts from this book were translated into Spanish and appeared in *Antología de Fray Alonso de la Vera Cruz* in 1988. Two of these five were reprinted the following year as separate books, the *Tratado de los Tópicos Dialécticos* and the *Libro de los Elencos Sofísticos*, with Latin and Spanish on facing pages. These two books encompass the last two sections of the *Recognitio*, the *Tractatus de Locis Dialecticis* and the *Elenchorum Liber*, respectively.

A copy of the signature of Alonso de la Vera Cruz is reproduced in the biography by Ramírez López.[7]

The following biographical data are drawn from the article by Arthur Ennis.[8]

Alonso Gutiérrez (1507[9]–1584) was born in Caspueñas, in Spain. He did university studies first at Alcalá de Henares and later at Salamanca. He graduated from the University of Salamanca and held a teaching position there when he was convinced, about 1535, to bring his teaching skills to the New World. Soon after his arrival in Veracruz, Mexico, on June 22, 1536, he joined the Augustinian order and changed his name. (Some library catalogs use Gutiérrez as the preferred name for this author, so researchers also should check under that name. Many other libraries alphabetize using his first name, Alonso, or even Alfonso.)

Ennis writes that Alonso founded four libraries, in Tiripetío, in Tacámbaro, and at San Agustín and San Pablo, both in Mexico City.[10] The library in Tiripetío, founded in 1544 and whose collection is now housed at

---

6. "Nadie ha visto la edición de 1593 citada por Nicolás Antonio, *Bibliotheca hispano nova*, I, p. 53, Eguiara, *Bibliotheca mexicana*: p. 103, y Berestáin, *Biblioteca hispanoamericana septentrional*, III, p. 267. Probablamente no existe." Amancio Bolaño e Isla, *Contribución al Estudio Biobibliográfico de Fray Alonso de la Vera Cruz*, México, 1947: 51.

7. Ignacio Ramírez López, *Tres Biografías*, México, 1948.

8. Arthur Ennis, *Fray Alonso de la Vera Cruz*, offprint from *Augustiana*, Louvain, 1957.

9. Many sources say Alonso was born in 1504, but Ennis argues that "he wrote in 1565 that he was then 58; therefore, he was born in 1507" ("Fray Alonso de la Vera Cruz," Introduction and chaps. I–III, *Augustiniana* V Fasc. 1–2 [Apr. 1955]: 64). Also, "Grijalva says he was eighty [at the time of his death], but his exact age was seventy-six. Vera Cruz himself had written on Dec. 23, 1565, that he was then approaching his fifty-eighth birthday" ("Fray Alonso de la Vera Cruz," chaps. VI–VII, *Augustiniana* VII Fasc. 1–2 [Apr. 1957]: 194).

10. Arthur Ennis, "Fray Alonso de la Vera Cruz," Introduction and chaps. I–III, *Augustiniana* V Fasc. 1–2 (Apr. 1955): 69, 108.

the Museo Michoacano de Morelia, has been called the first library in the New World, but there is a rival claim for the library of the College of Santa Cruz de Tlatelolco, founded in 1535.

When the Universidad Real y Pontifical de México (now Universidad Nacional Autónoma de México) began operating in 1553,[11] Vera Cruz was made the simultaneous holder of two chairs on the faculty: Professor of Sacred Scripture, and the more prestigious Chair of St. Thomas, created especially for him. The Chair of St. Thomas was declared to be equal to the First Chair of Theology, already occupied by someone else.[12] Alonso refers to himself as First Chair of Theology, in various ways, on the title pages of the three works listed here.

"It appears that, at least in the beginning, he taught only one course to fulfill the duties of the two chairs."[13] He taught there only a few years, but it can be inferred that his position was held for him for some time. In 1562, when Vera Cruz returned to Spain for a period of eleven years, he renounced his Chair of Theology.[14]

According to Ennis, the three Vera Cruz works we consider here were the basis for the courses taught by Vera Cruz at the new university in Mexico. He may have begun composing them in the 1540s when he was teaching his first philosophy course in Tiripetío. The aim of the courses and the texts, says Ennis, was the reform of scholastic philosophy.[15] Among the works penned by Vera Cruz, four titles were published under his name during his lifetime. (See Ennis's Appendix I: 200–203 in the offprint for a list of Alonso's publications and manuscripts.) As well as the three that appear on the list here, the fourth printed text by Vera Cruz is the *Speculum Coniugiorum*, in which he attempts to give guidelines for settling the

---

11. We have already used the year 1551 for the founding of the university (from the speech by Bolton) and this turns out to be consistent. Ennis writes, "The University of Mexico was thus legally founded, but the actual beginning did not take place until two years later . . . The formal foundation took place on the Feast of the Conversion of St. Paul, Jan. 25, 1553 . . ." ("Fray Alonso de la Vera Cruz," Introduction and chaps. I–III, *Augustiniana* V Fasc. 1–2 [Apr. 1955]: 99.)

12. Ibid.: 101.

13. Ibid.: 50.

14. Ibid.: 103–104.

15. Ibid.: 114–118. Mauricio Beuchot offers the same opinion with an explication of the title at hand: "En ese trabajo fray Alonso pretende hacer una revisión (*recognitio*) de los manuales de lógica formal o de 'súmulas' (es decir, compendios lógicos)." Alonso de la Vera Cruz, *Libro de los Elencos Sofísticos*, notes by Beuchot, México, 1989: ix.

problems arising from the former polygamy of the converted natives. If an indigenous man had several wives before converting to Christianity, which one was to be regarded as his true wife after he converted? This is the sort of question that Alonso wrestled with in this work.

Missionaries in the New World usually learned one or more native languages. Ennis writes, "Vera Cruz apparently spoke only Tarascan, but he knew it well enough to be able to preach."[16]

The *encomienda* system in the Spanish colonies could give a person the right to tax a specific group of native people or to sell their labor to pay the tax. Members of religious orders typically worked directly with the natives and witnessed the cruelty of this system. Some were moved to struggle with secular authority on behalf of these people.

To show what kind of thinker Alonso de la Vera Cruz was, we quote from Ernest J. Burrus's summary of the argument made by Vera Cruz in one of his manuscripts. Burrus writes (paraphrasing Alonso):

Dominion resides primarily and principally in the community itself; and, in order to be exercised by one (a king, emperor or other monarch) or more (aristocracy, oligarchy or other groups), it must be conferred on the ruler by the community. This transference of a right, Vera Cruz terms the *title* by which the ruler may govern.

The transference is just—and consequently also the title—if it is for the good of the community, and is made with the consent of the community; otherwise the transference and title are unjust and illegitimate. He who does not govern with the consent of the people governed and for the common good is not a ruler but a tyrant.

What is said of dominion in the sense of jurisdiction is also true of dominion in the sense of ownership. Hence, if the Spaniards have not acquired the natives justly, they have no just title to them and may not justly exact tribute from them, but must make restitution of the tribute already collected from them; and, in order not to persist in their unlawful possession, they must free the natives. The same principles apply to the property taken or acquired from the Indians.

. . . Supposing for the sake of argument that the emperor has just dominion in the Americas, he may justly transfer this dominion to the

---

16. Arthur Ennis, "Fray Alonso de la Vera Cruz," Introduction and chaps. I–III, *Augustiniana* V Fasc. 1–2 (Apr. 1955): 107.

conquerors with the consent of the governed and for their common good. But should the Spaniards rule tyrannically over the natives, they would lose all dominion over them, and hence the latter would be justified in revolting against their usurping masters and in deposing them.[17]

This is two centuries before Rousseau and Jefferson![18]
The present work treats logic, including syllogisms and modal logic. Appendix C includes a discussion of why we feel such books should be included here.

In the text at hand, the diagrams on folio 30r, folio 34v, and folio 36r reflect complex classifications of the various types of opposition among simple statements. The terms "contrary," "subcontrary," and "contradictory," illustrated in the first diagram, are modified by the point of the third diagram to accommodate statements that contain the modality of "necessity." The second diagram groups together modal statements that are equivalent to each other. Diagrams of this type are a standard part of the exposition of Aristotelian logic, but they also represent a sort of structured abstraction, which makes them akin to mathematics.

Alonso de la Vera Cruz. 2
Other Authors: Aristotle, Porphyry.
DIALECTICA / resolutio cum textu / ARISTOTELIS EDITA PER / REVERENDVM PATREM / ALPHONSVM AVERA CRVCE / Augustinianum , Artium at[que] sacrę Theo / logię magistrum in achademia Me / xicana in noua Hispa- / nia cathedræ pri / mæ in Theo / logia / modera torem. / MEXICI / Excudebat Ioannes paulus Brissensis. / Anno.1554.
Juan Pablos of Brescia, México.
Icazbalceta 21. Sabin 98912. Medina, *México*, 23. Wagner 29.
Originals located at: Biblioteca Cervantina, Biblioteca Nacional de Chile, Biblioteca National de México, British Library, Centro de Estudios de Historia de México Condumex, Hispanic Society of America, Huntington

---

17. Vera Cruz, *The Writings of Alonso de la Veracruz: II . . . Defense of the Indians: Their Rights*, ed. and trans. Ernest J. Burrus, Rome, 1968: 25–26.
18. Ennis argues that Alonso was influenced in these matters by his teacher, Francisco de Vitoria, a Dominican, in "Fray Alonso de la Vera Cruz," Introduction and chaps. I–III, *Augustiniana* V Fasc. 1–2 (Apr. 1955): 65, 110–113.

Figure 5. Title page of the *Dialectica Resolutio*, 1554. The John Carter Brown Library at Brown University, Providence, RI.

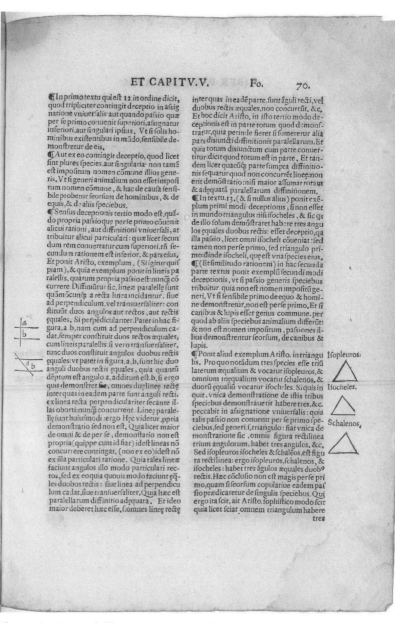

Figure 6. Marginal illustrations from the *Dialectica Resolutio*. Parallels and triangles. The John Carter Brown Library at Brown University, Providence, RI.

Library, Indiana University, Institución Colombina, John Carter Brown Library, New York Public Library, Southern Methodist University, University of Texas at Austin, Williams College.

Microfilm copies are available. Some catalogs use Aristotle as the primary author.

Paging: Ninety-eight folios, with 1 and 2 unnumbered, 79 misnumbered as 76, and 82–87 misnumbered 81–86; after 88 the numbering starts over with a ten-folio summary of the work numbered on the bottom of folios 2 through 6 only.

The British Library copy has two identical copies of each of folios 75 and 78.

The copy at the John Carter Brown has a note in Latin on the title page: "From that of the author, now in second edition and in some places added to."

Henry R. Wagner reported that thirteen copies existed in 1940.[19]

This entry was reprinted in Salamanca in 1562, 1569, and 1573 according to José Toribio Medina. Ennis gives 1562, 1569, and 1572,[20] but both the title page and the colophon of the fourth edition say 1573.

Year	Edition	Held by
1562	Second	Biblioteca Nacional de España, Universidad de Salamanca, Universidad de Sevilla
1569	Third	Biblioteca Nacional de Chile, Biblioteca Nacional de México, Centro de Estudios de Historia de México Condumex, Stanford University, Universidad de Barcelona, (two copies), Universidad de Sevilla, University of Cambridge, University of Michigan
1573	Fourth	Biblioteca Universitaria Alessandrina, Biblioteca Nacional de México, California State Library, Escorial, Indiana University, Universidad de Barcelona (two copies), Universidad de Sevilla, University of Oxford, University of Pennsylvania

Strangely, the Mexico edition is listed in F. Edward Cranz's *A Bibliography of Aristotle Editions, 1501–1600*, but none of Salamanca editions is.

19. Henry R. Wagner, *Nueva Bibliografía Mexicana*, México, 1940.
20. Arthur Ennis, *Fray Alonso*, offprint from *Augustiniana*, Louvain, 1957: 201.

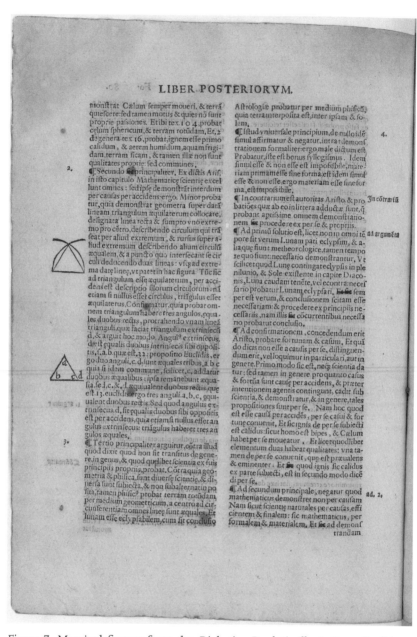

Figure 7. Marginal figures from the *Dialectica Resolutio* illustrating Euclidean propositions. The John Carter Brown Library at Brown University, Providence, RI.

A facsimile edition with the summary missing was published in 1945. An excerpt translated into Spanish is included in the *Antología de Fray Alonso de la Veracruz* published in 1988.

Medina quotes a letter from Angel Nuñez Ortega to Joaquín García Icazbalceta to the effect that the title-page woodcut is a reuse of the same design on the title page of a prayer book of Edward VI of England printed by Edward Whitechurch in 1549. This would explain the initials "E.W." at the bottom. The catalog at the British Library says that there are two changes from the original woodcut: the coat of arms at the top and the heart with arrows at the bottom. Wagner opines in 1939 that the woodcut was redone in Seville.[21]

This work contains a Latin translation of Aristotle's *Categories* as well as various commentaries. *Liber Prædicabilium* is a translation of *Isagoge*, an introduction to Aristotles's *Categories* by Porphyry (leaves 3–25). *Seguiter Resolutio Libri . . .* is the *Categories* with commentary by Alonso (leaves 26–86, i.e., 87).[22] The Aristotle text is marked *textus* and the commentary is marked *glosa*. Regarded as a reprint of Aristotle, this has been called the first printing, in the New World, of one of the classics.

Also here is a discussion of the incommensurability of the diagonal of the square on folio 63r and some figures relating to triangles and parallels on folio 70r. On 80v there is a construction of an equilateral triangle with propositions from Euclid mentioned.

Folio 60r contains the story of Socrates and the Slave from Plato's *Meno*, together with a diagram illustrating the geometrical points that Socrates makes. For an English translation of this story, see Clifton Fadiman's *Fantasia Mathematica* (translation by Benjamin Jowett).

1556

Juan Diez Freyle. 3

Sumario cõpēdioso delas quētas / de plata y oro q̃ en los reynos del Piru son necessarias a / los mercaderes:y todo genero de tratantes. Cõ algunas / reglas tocantes al Arithmetica. / Fecho por Juan Diez freyle.

---

21. Henry R. Wagner, *Mexican Imprints, 1544–1600, in the Huntington Library*, San Marino, 1939.

22. Notes from the John Carter Brown Library catalog card.

MATHEMATICAL WORKS PRINTED IN THE AMERICAS, 1554–1700

Juan Pablos of Brescia, México.
Icazbalceta 25. Medina, *México*, 27. Wagner 34. Karpinski [1].
Originals located at: British Library, Duke University, Huntington Library, Universidad de Salamanca.

Paging: 105 folios, numbered (with lowercase Roman numerals) except for i, lxxxi, and cv (the second folio has ij at bottom.) Folio vij is misnumbered vi, and folio lxxi is misnumbered lxi. Folio ciiij recto is a table of contents and the verso is blank. Folio cv is an errata.

The shield on the title page is identical to that of Alonso de la Vera Cruz's *Speculum Coniugiorum*, printed in the same year by same printer.

Most of the book consists of tables for calculating percentages and for converting between various monetary units and various amounts and purity levels of silver and gold. The back of the book (folios 91v–103v) is a treatise on arithmetic and algebra, with numerous examples. In 1921, D. E. Smith produced a facsimile of this latter part of the book, with each facsimile page facing an English translation with explanatory notes. Parts of this book coincide with Smith's paper of 1921. Smith's facsimile and translation was reprinted in 1996.

In 1985, in Madrid, the Instituto de Cooperación Iberoamericana produced a facsimile edition of the entirety of the Salamanca copy; this included even the handwritten notes at the beginning and end. The first handwritten page is upside down in the facsimile compared to the way it is bound into the original in Salamanca. This edition was printed to commemorate the visit of the president of Mexico, according to the colophon on the last page.

Among the works we consider here, this item has generated the greatest interest over the years among historians of mathematics. We discuss the condition and provenance of the known copies and the likelihood of there being a fifth copy extant. Then we address what is known about the author, in view of what we believe to be a misattribution that has been pervasive for the past eighty-five years. To make sense of the many tables that we find here and in later books on our list, we review the coins and currencies in use in the Spanish American colonies in the sixteenth and seventeenth centuries. The sections on arithmetic and algebra are surveyed. Finally, we look at whether the clues in the text and tables can tell us anything more about the author.

Figure 8. Title page of the *Sumario Compendioso*, 1556. Duke University, Durham, NC.

Extant Copies

The copy in Salamanca is complete. In the Duke copy, folio lxxiij is three quarters missing, and folios lxxxviiij and xc are gone. The copy at the British Library is missing folio cv, and the copy at the Huntington Library is missing folios ij, xli, xlviij, ciiij, and cv.

In the Salamanca copy we see the author's signature on folio 105v. (We will switch to Hindu-Arabic numerals for the folio numbers for the rest of this discussion so we will be able to attach "r" for recto and "v" for verso without causing confusion.) Also, the initials "JDF" are among the handwritten notes that appear on the leaf preceding the title page.

José Toribio Medina wrote in 1912 that there were copies in the British Museum, the Biblioteca del Ministerio de Fomento (in Madrid) and the Biblioteca de Don Jacobo Parga (in Madrid). Henry Wagner, in 1924, reported that by his time the Biblioteca del Ministerio de Fomento had been incorporated into the Biblioteca Nacional de España.[23] D. E. Smith reports seeing a copy sometime before 1921 at the Biblioteca Nacional that went up to folio ciij but was missing three folios along the way. In fact he may have examined two different copies in a short period of time, since his article in 1921 shows a title page with the stamp of the Ministerio de Fomento on it and his facsimile edition of the same year shows a title page without that stamp. There is a reasonable probability that the copy used by Smith for his book of 1921 is the one now at the Huntington. A comparison of Smith's book and the copy at the Huntington yields several points of similarity, and the latter is missing exactly the same number of pages that Smith reports. Maggs Brothers, from whom the Huntington got its copy (1922, £120), speculate in their catalog that their copy came from either the Ministerio de Fomento or the Jacobo Parga, probably based on a glance at Medina. We conclude that it must have been the Parga copy because it does not have the Fomento stamp.[24]

Smith also mentions a copy at the Convento de la Merced; this may be the copy that Duke has now. There is such a convent in Lima, and the Duke copy was acquired from Lima in 1929. It has a partially legible handwritten note on the title page about a Convento Grande in Lima.

---

23. Henry R. Wagner, "Sixteenth Century Mexican Imprints," in *Bibliographic Essays*, Cambridge, 1924.

24. Bartolomé José Gallardo's bibliography of 1866 indicates that the Parga copy was at that time missing only folios ij, ciiij, and cv.

According to Wagner, and also Augustín Millares Carlo and Julián Calvo, the copy that was in the Ramírez auction (1880, £24) is the one now in the British Library.[25] Indeed, the binding of the British Library copy has the initials J.F.R.[26] along the base of the spine.

The copy in Salamanca came from the Librería del Colegio Mayor de Cuenca, according to a handwritten note on the title page. This may refer to the city of Cuenca, but probably refers to a division of the Universidad de Salamanca by that name.

The Fomento copy may still exist somewhere. Louis Karpinski mentions in his *The History of Arithmetic* in 1925 that there is a copy at the Escorial in Spain, and he repeats this in his *Bibliography* of 1940. There is no such copy there now. Wagner writes in his *Nueva Bibliografía* in 1940 that the Fomento copy has disappeared. Millares Carlo and Calvo, in their *Juan Pablos* in 1953, say that Ugarte had a copy, and Millares Carlo repeats this in his new edition of Joaquín García Icazbalceta's *Bibliografía* in 1954.

The Author

There has been a great deal of confusion about the author. We will try to sort out the facts here, but there is still much that is unknown.

D. E. Smith wrote: "The author of the *Sumario* was Juan Diez, a native of the Spanish province of Galicia, a companion of Cortés in the conquest of New Spain, and the editor of the works of Juan de Avila, known as 'the apostle of Andalusia,' and of the *Itinerario* of the Spanish fleet to Yucatan in 1518. He is sometimes confused with Juan Diaz, a contemporary theologian and author."[27] This was in the introduction to his facsimile edition of the *Sumario* in 1921. Yet in his textbook on the history of mathematics, from 1923, he says, "It should be stated, however, that there were several writers at this time by the name of Juan Diez (Diaz), two of them apparently being in Mexico, and there is much uncertainty among the best Spanish biographers as to which was the author of the *Sumario*."[28]

---

25. Joaquín García Icazbalceta wrote in 1886 that the copy in the Ramírez auction was the one that he knew of in the Convento de la Merced. Since he thought that there was only one copy in existence it might be natural that he would assume that.

26. José Fernando Ramírez?

27. D. E. Smith, *The Sumario of Brother Juan Diez*, Boston, 1921: 6.

28. D. E. Smith, *History of Mathematics, Volume I, General Survey of the History of Elementary Mathematics*, Boston, 1923: 353–354.

It is true that he prefaced the latter quote with a repetition of the barest details of the claims from 1921. In 1923 he threw those claims into serious doubt, if not entirely refuting them. What is the cause of this change?

Smith's one-time coauthor, Karpinski, blasted Smith for a number of errors when he reviewed Smith's facsimile edition of the *Sumario*. "Professor Smith makes the author of the Sumario to be Juan Diaz, the first parish priest of the new world, who wrote the 'Itinerario' of the Spanish fleet to Yucatan in 1518, published in 1521. This Juan Diaz was not only killed soon after 1520 but eaten, according to Bancroft ('History of Mexico,' Vol. 2, pp. 158–159, San Francisco, 1883) . . ."[29] It is likely that Smith was aware of this criticism by the time he changed his position in his *History*.

It could also be that in the interval between these projects his attention was drawn, possibly by Karpinski, to the following passage in Medina's *Imprenta en México*: "Neither in the books printed in Lima in the 16th century, nor in the Archivo de Indias have we been able to find the least mention of Diez Freyle." Medina adds to this a quote from García Icazbalceta. "I do not have any biographical knowledge of Diez Freyle," writes García Icazbalceta, who adds that "[his knowledge] seems superior to that of a simple merchant." These comments would suggest that neither of these scholars identified Diez Freyle with any well-known Mexican priest or author. García Icazbalceta should have been familiar with the Juan Díaz who wrote the *Itinerario* since, as the original Spanish text of the *Itinerario* was lost, he was the translator who took it from Italian back to Spanish.

Smith's earlier statements about the author of the *Sumario* could have been the result of using the catalog at the British Library (then the British Museum). There, all entries for the *Itinerario* are assigned to Juan Díaz Freile, and the entry for Juan Díaz, Clerigo, uses the phrase "apostle of Andalucia." The British Library copy of the Sumario has "Diaz (Juan) Clérigo" penciled inside.

In addition to Karpinski's correction of Smith and Smith's guarded correction of himself, the only other time we know of that Smith's mistake was noted in print was by Millares Carlo in 1954.[30] We now examine why Smith's earlier position is extremely unlikely.

---

29. Louis C. Karpinski, "Books and Literature," *School and Society* XIV (Aug. 13, 1921): 83–84.

30. Joaquín García Icazbelceta, *Bibliografía Mexicana del Siglo XVI*, New Edition by Augustín Millares Carlo, Mexico, 1954.

The word *frey* is a form of address used by members of military religious orders, as opposed to other religious orders, which use *fray*.³¹ A *freile* is then a member of such a military religious order. Commentators have taken the second last name *Freyle* as descriptive—most conspicuously, this gives Smith the "Brother Juan Diez" in the title of his facsimile in 1921. However, Karpinski suggests in his review that we quoted earlier, "That the author was a priest is by no means certain as the 'Freyle' may well be a portion of his name." In the introduction to a later work on our list, that of Francisco Juan Garreguilla in 1607, a Contador Luís de Morales Figueroa makes reference to "Fray Joan Diaz Freyle, de la orden de Sancto Domingo," and Medina quotes this without comment in his Peru series. (This is in spite of the quote we used here from Medina's Mexico volume; Medina's mention of Diez Freyle in his Peru series was published eight years earlier.) We see, therefore, that the nearest commentator to Juan Diez's own time makes him a *fraile*, not a *freile*, and confirms that Freyle is part of his name.

It is a matter of record that there was a priest named Juan Díaz with Cortés during the conquest of Mexico. This Juan Díaz was born in 1480 in Seville, came to the New World in 1514, and was chaplain to the Yucatan expedition under Juan de Grijalva in 1518 and wrote his *Itinerario* about this outing. He was later one of the two chaplains to accompany Cortés.³² (The other was a Mercedarian, Bartolomé de Olmedo.³³) Díaz was a secular priest, so he would not have been addressed as either *frey* or a *fray*. Juan de Zumárraga, first bishop of Mexico, writes that he was parish priest of

---

31. Real Academia Española, *Diccionario de la Lengua Española*, 21st ed, Madrid, 1992.

32. Bernal Díaz del Castillo, in his *Historia Verdadera*, relates that Juan Díaz was among the conspirators who plotted against Cortés in favor of the governor of Cuba, Diego Velázquez. (Prescott, in his *History of the Conquest of Mexico*, elevates Díaz to leader of the plot.) When the plot was discovered, two of the conspirators were hanged, and others were put under the lash, but Díaz was spared. "Padre Juan Díaz would also have been punished, but he was needed to say Mass. He was given a good fright, anyway." (Translation by Albert Idell.) This was alleged to be the incident that led Cortés to his famous burning of his ships, but according to Prescott, in his *History of the Conquest of Mexico*, and also Restall, in his *Seven Myths of the Spanish Conquest*, the ships were merely scuttled. Cohen, in his translation of Díaz del Castillo's account, indicates that perhaps none of the sentences was carried out, since one of the condemned men was alive in the following year.

33. Charles S. Braden calls Olmedo a Franciscan on page 87 of *Religious Aspects of the Conquest of Mexico* but refers to him as a Mercedarian on page 140.

the cathedral in Mexico City in 1533, using the phrase "clérigo anciano y honrado" to describe him. Hubert Bancroft, writing in his *History of Mexico* in 1890, says that Díaz "was killed during a tumult between the Popolucas at Quecholac" sometime soon after June 1529, and he surveys what earlier writers had said about this.[34] The anonymous introduction[35] to the 1972 edition of the *Itinerario* (the edition that contains García Icazbalceta's Spanish translation) also surveys the various accounts of the death of Díaz and makes no speculation about the year of his demise. Millares Carlo, in his new edition of García Icazbalceta's bibliography, writes that Díaz had already fallen 25 years before 1556, i.e., 1531. Germán Vásquez, on the other hand, in his 1988 edition of the *Itinerario*, states that Díaz died in 1549 and was buried (according to tradition) in a chapel of the cathedral at Puebla.[36]

Smith, as stated, identified Díaz and Diez Freyle in his earlier remarks of 1921 while at the same time stating that Juan Diez is sometimes confused with a contemporary named Juan Díaz. If Smith had been right in this identification, and if the death date of 1549 can be accepted, then the 1556 publication of the *Sumario compendioso* would have been posthumous. The introductory matter in the text at hand does not tell us much about the author, but it does contain the phrase "por quãto Juã diez freyle estãte al presente en esta ciudad de Mexico . . ." (inasmuch as Juan Diez Freyle being at present in this city of Mexico . . .) on folio 1v, written by the secretary to Viceroy Luys de Velasco and dated April 15, 1556. So either the most commonly used death date for Juan Díaz is wrong, or he is not the same person as Juan Diez Freyle.[37] We lean strongly toward the second of these possibilities.

Smith also claims in 1921 that Juan Diez is the editor of the works of Juan de Ávila—this may be yet a third person. In the *Obras del Padre Maestro Ivan de Avila*, published in Madrid in 1588, there is a reference to a Juan Díaz but

---

34. We did not see, in the 1890 edition, the reference to Díaz's being eaten that Karpinski cites from the 1883 edition.

35. Probably by Jorge Gurría Lacroix and/or Alfredo Hlito.

36. The online catalogs of numerous libraries give the years 1480 to 1549 for Juan Díaz, based presumably on whatever Vásquez's source was.

37. Karpinski also noted in his review in 1921 that the *Sumario Compendioso* could not be posthumous for the very reason that we have given. The catalogs of the Biblioteca Nacional de México, Duke University, Harvard University, Ohio State University, and the University of Texas at Austin, just to name a few, all give the years 1480 to 1549 for Juan Díez, in spite of the fact that they are refering to a 1556 publication of this author. In view of our bringing to light, once again, the passage from the forward to the *Sumario*, we hope that these libraries and others will amend their records.

it is not clear what his role is. He is not mentioned on the title page, but he signed the dedication. Another part of the introductory material says that the work was printed at the petition of Juan Díaz, clérigo, who was then in charge of the printing. Thus the catalog entry at the British Library calls him the editor. In any event, no one except Smith and the British Library connect this individual to Juan Diez Freyle.

Smith's note about another Juan Díaz, theologian and author, could refer to the Protestant theologian Juan Díaz (1510–1546), who was born in Cuenca, Spain.

It is not even certain whether Juan Diez Freyle wrote his name with an accent on the i in Diez. This is because the printer has chosen to render almost every i in the text as an í. While the name Díaz is almost always accented, the name Diez occurs in both forms in, for example, the *Enciclopedia Universal Ilustrada*. We follow Medina's example and leave Diez unaccented, but this is not universal, even among sources written in Spanish. Díez is the apellative form of Diego, ultimately from Santiago according to the *Enciclopedia Universal Ilustrada*.

One sign that the notion of identifying Juan Diez Freyle with Juan Díaz arose with an English speaker is that, outside of the many sources in English, there appear to be only three other authors who make such an identification, and all of them clearly have Smith as their source.[38] As

38. We have found one example in French. Paul ver Eecke, in his notes to his French translation of Fibonacci's *Book of Squares*, says that one finds an interesting solution to one of Fibonacci's problems "dans le plus ancien ouvrage mathématique publié dans le Nouveau-Monde, à Mexico, en 1556, par Juan Diez, compagnon de Cortès dans la conquête de la Nouvelle-Espagne, sous le titre: *Sumario Compendioso*. Voir la réimpression de ce rarissime ouvrage, texte espagnol en fac-similé et traduction anglaise par David Eugene Smith."

Another example is the entry for Juan Díez in the *Gran Enciclopedia Gallega*, the encyclopedia of the autonomous region of Galicia in northwestern Spain. Alberto Vilanova Rodríguez, author of this entry, claims that Juan Díez was Galician, citing Smith, and also Juan de Torquemada's *Monarchia Indiana*, in spite of the claim by other sources in Spanish that Juan Díaz was born in 1480 in Seville. (We could not find any such claim one way or the other in Torquemada, but the *Enciclopedia Universal Ilustrada* agrees that Juan Díaz came from Galicia without calling him Juan Díez.) He identifies this Díez as the author of both the *Sumario* and the *Itinerario* and as the chaplain to Cortés, even mentioning his complicity in the plot against Cortés. He notes that the *Sumario* must have been printed posthumously, a claim we have already rejected (see earlier discussion).

Finally there is "¡Cosas de Gallegos!," a regular feature of *Cultura Gallega*, a magazine published in Havana in the 1930s. The June–July 1937 article is devoted to Juan Díez and uses phrases that mark it as inspired by Smith, although it has no citations. It attempts to reproduce for the reader the quadratic formula, but this is badly botched by the typesetter.

noted, two bibliographers, García Icazbalceta and Medina, take a completely different position from the one espoused in Smith's earlier (1921) statements and in various articles by people whom Smith has influenced. García Icazbalceta even speculates that Diez Freyle was a young man at the time of writing, which would make it impossible for him to have accompanied Cortés in the 1520s.[39] Karpinski also mentions the passage in the *Sumario*, where Diez Freyle apologizes for his youthfulness, that led García Icazbalceta to say this.[40]

García Icazbalceta also surmises (perhaps from the mention of Peru in the title, among other things) that Diez Freyle may have been from Peru and had his book published in Mexico because the press had not yet been brought to Peru (the first printed work in Peru appeared in 1584).[41] This piece of speculation will be seen to be consistent with our later observations on the text (see following). We note that when Mexico got a mint in 1535, silver coins began to slowly replace assayed gold as a medium of exchange. Peru did not get a mint until 1568, so this process of replacing one medium by another would have been farther along in Mexico than in Peru when Diez Freyle was writing. On folio 85r, Diez Freyle says that he has exposed his reader to those things needed to do accounting in Peru, and now will direct the reader to a few items relevant to Mexico (*Nueva España*). Only then does he mention the *peso* based on the silver coin minted in Mexico City. (Before that point he uses only *pesos de minas*, an imaginary currency.) This ordering of topics suggests that he thought his main readership to be in Peru and his secondary audience to be in Mexico. The bold may at this point conjecture (with García Icazbalceta) that Juan

---

39. "Sigue inmediamente el prólogo *Al Letor:* de él se deduce que el autor era Joven aún" (Joaquín García Icazbalceta, *Bibliografía Mexicana*, México, 1886).

40. ". . . the Juan Diez Freyle who wrote this work expressly apologizes for his youth in the perface [sic] . . ." (Louis C. Karpinski, "Books and Literature," *School and Society* XIV [Aug. 13, 1921]: 84).

41. "No tengo ningun noticia biográfica de Diez Freile. De la licencia del virrey se deduce, que al tiempo de imprimirse la obra estaba en México el autor. Tal vez sería algún comerciante del Perú, y como aun no se introducía la imprenta en aquel país, vino á imprimir su libro en México; si bien los conocimientos que manifiesta parecen superiores á los de un simple mercader. Es probable que la mayor parte de la edición se llevarse al Perú, y así se explica la falta de ejemplares en México: á lo menos, yo no conozco más que el descrito." (Joaquín García Icazbalceta, *Bibliografía Mexicana*, México, 1886.) The one copy he knew about at the time was the one at the Convento de la Merced, probably in Lima. We think that this is the very copy that is now at Duke.

Diez Freyle came from Peru to Mexico, had his book printed there, and then took the bulk of the copies back to Peru for his Peruvian readership. Without direct historical evidence, however, this must remain merely a plausible speculation.

It was mentioned earlier that the historical record contains the claim that Juan Diez Freyle was a Dominican. There exists a counterclaim, without any reference to the historical record, that he was a Franciscan. Indeed, a recent Google search for "Juan Diez Freyle" gave one Web site out of the top ten hits that made this claim. If this was based only on who was in Mexico at the time, then there are more possibilities than just these two. For instance, we have already seen that the first author on our list was an Augustinian.

Here is a table of when, in the sixteenth century, priests and religious of various affiliations first entered Mexico. It is based on information in Charles S. Braden's *Religious Aspects of the Conquest of Mexico*.

Secular priest, i.e., no order	1518	(Juan Díaz, who performed the first baptism on the American mainland, was part of the first expedition to the Yucatán.)
Mercedarian	1519	(Bartolomé de Olmedo, together with Juan Díaz, accompanied Cortés to Tenochtitlán.)
Franciscan	1524	
Dominican	1526	
Augustinian	1533	
Jesuit	1572	
Carmelite	1586	

So we see that Juan Diez Freyle, if he was indeed a member of a religious order in Mexico in 1556, could have been from any one of four orders. Probably the hypothesis that he came from Peru would not change that number by much.

There is always the possibility that the name Juan Diez Freyle was a pseudonym. At a recent exhibition at the Biblioteca Nacional de España, it was suggested that books on accounting and arithmetic from the sixteenth century were sometimes attributed to a priest to avoid the taint that the church's ban on usury might give to the book.

Quiſtiones por los numeros.

¶ Cōgruos.   **¶ Cōgruentes.**   **¶ Exemplo.**

25. Re. y da.	24.
100. Re. y da.	96.
169. Re. y da.	120.
225. Re. y da.	216.
289. Re. y da.	240.
400. Re. y da.	384.
625. Re. y da.	336 \ y. 600.
676. Re. y da.	480.
841. Re. y da.	840.
900. Re. y da.	864.
1156. Re. y da.	960.
1225. Re. y da.	1176.
1212. Re. y da.	1080.
1681. Re. y da.	720.
2025. Re. y da.	1944.
2500. Re. y da.	1344 \ 2400.
2602. Re. y da.	2160.
2704. Re. y da.	1920.
2809. Re. y da.	2520.
3025. Re. y da.	2905.
3364. Re. y da.	3360.
3600. Re. y da.	3456.
3221. Re. y da.	1320.
4225. Re. y da.	2026.

¶ Da me vn numero q̄drado tal que ajuſtādole, 6, haga numero q̄drado y reſtando del, 6, quede numero quadrado, para lo qual has de buſcar vn tal numero congruē te que partiendole por, 6, vēga nu mero quadrado el qual como a fue ra veys el pmero es, 24. pues par te, 24, por, 6, viene, 4, que es qua drado (y ſu rayz es dos) y luego to ma el numero congruo quadrado correſpōdiente deſte numero con gruente que es, 25, partale por los 4, que es el aduenimiento de el pmero vienen, 6, $\frac{1}{4}$ y aqueſte es el numero demādado que ſi le aju ſtas ſeys haze, 12, $\frac{1}{4}$ que es numero quadrado y ſu rayz es, 3, $\frac{1}{2}$ y ſi reſtas del, 6, queda $\frac{1}{4}$ y ſu rayz es $\frac{1}{2}$ porque media vez media es $\frac{1}{4}$ y el meſmo es quadrado que ſu rayz es dos y medio.

Jta in alijs.

Figure 9. Juan Diez Freyle's table of congruous and congruent numbers. Duke University, Durham, NC.

To bring this note on the author to a close, we have concluded that Juan Diez Freyle is probably not the same person as either of the individuals named Juan Díaz that have been made to fit him by some. Consequently, we are left in the unhappy situation of having nothing to say about the life of Juan Diez Freyle that can be supported by the historical record, other than that he was a Dominican and was present in Mexico at least in April of 1556.

Spanish Colonial Coins, Taxation, and Accounting Practice

To fully understand the tables, it is important to survey the monetary system and tax structure in place in Spain and its colonies in the sixteenth century. For this Clarence Haring's 1915 article in the *Quarterly Journal of Economics* is instructive. This article is based largely on accounting ledgers held at the Archivo de Indias in Seville and does not directly use any of the printed texts considered here. It clarifies much that would otherwise be confusing in these texts, but sometimes it directly contradicts them. John McCusker's *Money and Exchange in Europe and America, 1660–1775* is also useful. This handbook is especially relevant for the seventeenth century, and so it will be essential when looking at later works in the tradition of the *Sumario*. Another interesting book is Octavio Gil Farrés's *Historia de la Moneda Española*, which emphasizes coins.

The principal gold coin of Spain for a long time was the *castellano*, which contained $\frac{1}{50}$ of a *marco* of gold at $23\frac{3}{4}$ carats. (Haring gives the remarkably precise figure of 230.0675 grams to the Spanish *marco*,[42] while Farrés says that a *castellano* massed 4.6 grams, which implies that a *marco* was 230 grams.[43]) After 1497, the ducat (*ducado*) was introduced, in imitation of another European coin, at $65\frac{1}{3}$ to the *marco* of $23\frac{3}{4}$ carat gold. This coin was set to be equal to 375 *maravedis*,[44] thus defining for us a gold equivalent for

---

42. Clarence H. Haring, "American Gold and Silver Production in the First Half of the Sixteenth Century," *The Quarterly Journal of Economics* 29 (May 1915): 435.

43. Octavio Gil Farrés, *Historia de la Moneda Española*, Madrid, 1976: 405. The *Diccionario de la Lengua Española* gives the weight of approximately 596 milligrams for a *tomín*, with eight *tomines* to the *castellano* and 50 *castellanos* to the *marco*. This implies a weight of 238.4 grams for the *marco*.

44. Smith, in his facsimile edition of the *Sumario*, says that this word comes from the Moorish dynasty Murābitīn, from whose time the coin is known. Wikipedia refers to the dynasty as *al-Murabitun*, which has been Latinized as Almoravides. The span of the dynasty's rule over parts of Africa and Iberia was 1061–1147. Fans of Gilbert and Sullivan have been quick to point out to us that the word *maravedi* appears in "*Iolanthe.*"

the latter coin. A ducat had the same value as 3.485 grams of pure gold, so a *maravedi* carried the value of 0.009293 grams of gold. This is 2.988×10^{-4} troy ounces, so at the rate used at www.xe.com/ucc/ on April 26, 2008, a *maravedi* in gold would be worth 26.4 cents (U.S.) today.

A silver coin, the *real*, was set to be $\frac{1}{67}$ of a *marco* of silver at a fineness of $\frac{67}{72}$. Haring says that the *real* was fixed at 34 *maravedis* after 1497.[45] Therefore a *maravedi* had the value of 0.09398 grams of pure silver. A *maravedi* of pure silver is 0.003022 troy ounces, with a value of 5.08 cents (U.S.) today (lower than in the last paragraph since the ratio of values of gold to silver today is 56.625 to 1).

Haring's numbers imply a ratio of 10.1135 to 1 for the values of gold to silver in the sixteenth century. Haring himself calculates a ratio of 10.18 to 1, but he seems to be comparing gold and silver at different concentrations.[46] Farrés calculates a value of 10.7 to 1 by dividing 49.5 grams of silver by 4.6 grams of gold (this actually gives 10.76), but it is not clear that he, either, has taken into account the differing concentrations of silver and gold in the coins.[47] Haring adds to his calculation that the legal ratio was 10.11. Later entries will state that Joán de Belveder in 1597 and Felipe Echagoyan in 1603 used exactly 10 to 1, and that Diez Freyle did, too, as long as it is assumed that his *pesos* of weight are 50 to the *marco*.

There was no mint in the New World until 1535. In the meantime, silver coins were actually shipped from Spain to the Indies. The silver *real* was circulated at 44 *maravedis* in Hispaniola, while it was still worth only 34 in Spain, and this caused problems when the crown tried to restore its true rate after the mint was established.[48]

Soon after the conquest, a gold *peso* was circulated in Mexico that was debased from its legal purity. According to Farrés, it came to be known as a *peso de tipuzque* after a native word, *teputzli*, for copper. In 1536, the viceroy, Antonio de Mendoza, made the *peso de tipuzque* officially equal to 8 *reales*, or 272 *maravedis*.[49] Haring writes that in 1537 a silver coin, the *peso*

---

45. John McCusker says after 1536. (*Money and Exchange in Europe and America, 1600–1775*, Chapel Hill, 1978: 99).

46. Clarence H. Haring, "American Gold and Silver Production in the First Half of the Sixteenth Century," *The Quarterly Journal of Economics* 29 (May 1915): 477.

47. Octavio Gil Farrés, *Historia de la Moneda Española*, Madrid, 1976: 406.

48. Clarence C. Haring, "American Gold and Silver Production in the First Half of the Sixteenth Century," *The Quarterly Journal of Economics* 29 (May 1915): 475–476.

49. Octavio Gil Farrés, *Historia de la Moneda Española*, Madrid, 1976: 405.

fuerte, was introduced with the same value. As silver gradually replaced gold as a medium of exchange in Mexico, the name *peso de tipuzque* came to be applied to the silver *peso* of 8 *reales*. This name is used by several of the authors of accounting tables. Two of the authors discussed here refer to *pesos de tipuzque* as *patacones*, and the next three books of tables on the list also make use of a *peso de plata corriente* valued at 9 *reales*.

The king was, in different places and times, due a *quinto* (one fifth), or a *diezmo* (one tenth), of all gold and silver assayed. We see calculations of the *quinto* and *diezmo* in many of the texts on our list. Sometimes even one eighth or one ninth was the king's share, according to Haring. Haring indicates that there also was a charge of 1% for "smelting, assaying, and stamping" that was applied to both the crown's share and the net. "This 1 per cent was first deducted from the bullion, and then the quinto."[50] Haring says he is taking this rule from a decree of Philip II from 1579. The total tax to the customer, in the case of the *quinto*, is $20\frac{4}{5}$% by this reckoning, 1% for the assayer and $19\frac{4}{5}$% for the king. (The amount left after tax can be calculated multiplicatively, i.e., $0.99 \times 0.8 = 0.792 = 79.2\%$.) Haring also records that in 1552 Charles V had already raised the assayer's fee to $1\frac{1}{2}$%, but that this order was not put into practice in Mexico, leading to a repetition of the order in 1578. The $1\frac{1}{2}$% was not used in Potosí in Peru until 1585. See the entry for 1597 for an explanation (in Chap. 23 of that work) of how the quinto plus $1\frac{1}{2}$% was calculated some 40 years later than the work at hand.

McCusker tells us that European powers and their colonies used both real money and money of account. He says, for instance, that "the Royal Mint of Great Britain produced no coin equivalent to the 'pound' until the nineteenth century."[51] The pound was an imaginary currency, money of account, equal to 20 shillings. Shillings and guineas (equal to 21 shillings) were actual coins, real money.

So also with Spain and its colonies. We will see in the *Sumario* and later texts the *peso de minas*, which was used for bookkeeping purposes. This *peso* was set at 450 *maravedis*. Haring indicates that this was supposed to represent, like the *castellano*, $\frac{1}{50}$ of a *marco* of gold, but at about 21.81

---

50. Clarence C. Haring, "American Gold and Silver Production in the First Half of the Sixteenth Century," *The Quarterly Journal of Economics* 29 (May 1915): 445.

51. John J. McCusker, *Money and Exchange in Europe and America, 1600–1775*, Chapel Hill, 1978: 6.

carats.⁵² (Farrés does not confirm this, using instead 22.5 carats, which matches the value of gold given in the *Sumario* and most of the other books of tables on our list.⁵³) Haring mentions, and the texts surveyed here confirm, that the two kinds of *pesos*—*pesos de minas* and *pesos de tipuzque*—coexisted, even though one was imaginary, into the seventeenth century. But McCusker asserts that in the seventeenth century the *peso de tipuzque* also became a money of account. This would explain the disappearance of the *peso de minas* from the later books on this list. The imaginary *peso de tipuzque* stayed at 8 *reales*, i.e., 272 *maravedis*, as the *real* and *maravedi* were devalued.⁵⁴ The real coin, the Spanish *peso*, *piastra*, or "piece of eight," was a universal currency in Europe and the Americas, becoming equal to 10 *reales* after 1686 and 11 *reales* after 1772.⁵⁵

The Tables

Several writers have discussed the sections on arithmetic and algebra, but we know of no one who has attempted an analysis, however thorough, of the tables. We will give here the beginnings of such a study, in the hope that others will do more.

After two folios of introductory material, Diez Freyle begins his first table on folio 3r, with an explanation of this table on the lower half of 2v. Folios 3r to 48r give the values of various weights of silver from purity levels of 1,500 to 2,400 *de ley* (2,400 represents pure silver according to Echagoyan in 1603). The weights are in *marcos* and ounces, with 8 ounces to the *marco*, and the values are in *pesos*, *tomines*,⁵⁶ and *maravedis*, to the nearest quarter of a *maravedi*, with $56\frac{1}{4}$ *maravedis* to the *tomín* and 8 *tomines* to the *peso* (and therefore these are *pesos* of 450 *maravedis*). The value of 1 *marco* of silver at concentration x *de ley* is x *maravedis*. For example, at 2,200 *de ley* we see 1 *marco* is worth 4 *pesos*, 7 *tomines*, $6\frac{1}{4}$ *maravedis*, or a total of 2,200 *maravedis*. Thus a *marco* of pure silver is worth 2,400 *maravedis* in Diez Freyle's reckoning. (This figure stays consistent throughout the texts

---

52. Clarence H. Haring, "American Gold and Silver Production in the First Half of the Sixteenth Century," *The Quarterly Journal of Economics* XXIX (May 1915): 477.

53. Octavio Gil Farrés, *Historia de la Moneda Española*, Madrid, 1976: 405.

54. McCusker says that the *ducado* also became a money of account, remaining at 375 *maravedis*. *Money and Exchange in Europe and America, 1600–1775*, Chapel Hill, 1978: 99.

55. Ibid.: 99.

56. According to Smith, from Arabic *tomn*, an eight part.

surveyed here, in spite of the fact that the numbers taken from Haring imply that a *marco* of pure silver is 72 *reales*, or 2,448 *maravedis*.) Later in the book, on 94v, Diez Freyle tells readers to "reckon the *marco* at 4 *pesos*."[57] This corresponds to 1,800 *de ley*, which may have been a standard purity level, although this is not supported anywhere else.

Folios 49r to 57r are a table of percentages, with the explanation of the table on 48v. One finds, as Smith has noted, an entry that tells us that 3% of 100 *pesos* is 3 *pesos*. Percentages run from 3 to 30. Answers are given in *pesos*, *tomines*, and *granos*, to the nearest $\frac{1}{2}$ *grano*, with $12\frac{1}{2}$ *granos* to the *tomín* and 8 *tomines* to the *peso*. The $12\frac{1}{2}$ figure can be confirmed by checking the following entries:

$$12\% \text{ of } 1 \text{ } peso = 12 \text{ } granos$$
$$12\tfrac{1}{2}\% \text{ of } 1 \text{ } peso = 1 \text{ } tomín$$
$$13\% \text{ of } 1 \text{ } peso = 1 \text{ } tomín \tfrac{1}{2} \text{ } grano.$$

Thus 100 *granos* make a *peso*, and so a *grano* is equivalent to a *centavo*. The odd thing about this is that in the majority of the tables in this book there are 12 *granos* to the *tomín*, and therefore 96 *granos* to the *peso*. Diez Freyle makes it clear that he knows that there is an inconsistency, but does not give a satisfactory explanation of why these tables are inconsistent.[58]

At the end of that table, in the right-hand column of 57r, is a table for converting *pesos* to *maravedis*. The rate is the same as mentioned earlier, at 450 *maravedis* to the *peso*. This third table of the book is not listed among the contents on folio 104.

Folios 58r to 81r give values of gold. The explanation of the table is on 57v. Purity levels go from 1 to 24 *quilates*, or carats, by quarter carats, which are called *granos*. Weights are given in *pesos* and *tomines*, with 8 *tomines* to the *peso*. Values are given in *pesos*, *tomines*, and *granos*, to the nearest half *grano*. Here there are 12 *granos* to the *tomín* and 8 *tomines* to the *peso*. At $22\frac{1}{2}$

---

57. Smith's translation.

58. The author writes in his introduction to this table, ". . . interesse por pesos tomines y granos aun que no son perfe[c]tos: porque no vale cada uno masq̃ quatro maravedis y medio: y assi doze y medio valē un tomin: pueden se nombrar perfectamente centavos: porq̃ ciento valen un peso . . ." We read this as, ". . . interest by *pesos, tomines*, and *granos* although they [the *granos*?] are not perfect, because each one is not valued at more than $4\frac{1}{2}$ *maravedis*, and thus $12\frac{1}{2}$ make a *tomín*; they can be called *centavos* exactly, because 100 make a *peso* . . ." Henry R. Wagner has written, "Diez suggested a division of the peso into one hundred centavos, with the object of simplifying calculations" (*Mexican Imprints, 1544–1600, in the Huntington Library*, San Marino, 1939).

carats, 1 *peso* (weight) of gold is worth 1 *peso* (value).[59] Consequently, assuming that this value is directly proportional to both concentration and weight, as was true for silver in the first table, the equation

$$\frac{1}{22.5} \times \text{carats} \times pesos \text{ (weight)} = pesos \text{ (value)}$$

should hold. At 4 *granos* (purity) to the carat and 96 *granos* (value) to the peso this means that

$$\frac{16}{15} \times granos \text{ (purity)} \times pesos \text{ (weight)} = granos \text{ (value)}$$

To check this we may look to see if at 15 *pesos* weight at 1 *grano* of purity adds 16 *granos* of value. This is true in almost every instance and over the long term.

Postulating that Diez Freyle uses 50 *pesos* of weight to the *marco*, a ratio supported by other texts, that implies exactly 10 to 1 as the ratio of values of gold to silver. For what it is worth, this creates a certain harmony to the numbers: 1 *marco* of silver at 2,250 *de ley*, i.e., 93.75% pure, is worth 5 *pesos*, while 1 *marco* of gold at the same purity level ($22\frac{1}{2}$ carats) would be worth 50 *pesos*. This would mean that the *peso de minas* is now valued at one fifth of a *marco* of $22\frac{1}{2}$ carat gold, up from the 21.81 that can be inferred from Haring's figures for the earliest years of the colonial period, and that a *marco* of $23\frac{3}{4}$ carat gold is now worth 23,750 *maravedis*, down from the value of $65\frac{1}{3}$ ducats, or 24,500 *maravedis*, given by Haring.

Folios 81v to 84v are another table of percentages. The heading is *de corriente ensayado*, and we take this to mean "from [silver] *currency* to *assayed* [silver]." The explanation of the table is on the lower half of 81r (but there are further hints later in the text—see subsequent discussion). Percentages run from 8 to 20. Given a value A and a rate r, the table allows us to look up

$$\frac{A}{1+r},$$

in other words, the value that, when increased by the factor $1 + r$, becomes the value A. Input values are in *pesos* and *tomines*, and outputs are in *pesos*,

---

59. This double use for the word *peso* can be put in context by observing that in later texts the value of gold is often given by translating a weight and a concentration into a weight of some standard concentration, like $22\frac{1}{2}$ carats.

tomines, and *granos*, to the nearest half *grano*, with 12 *granos* to the *tomín* and 8 *tomines* to the *peso*. One sees for instance in the 20% table that

$$1 \text{ } tomín \div 1.20 = 10 \text{ } granos,$$
$$2 \text{ } tomines \div 1.20 = 1 \text{ } tomín \text{ } 8 \text{ } granos,$$
$$4 \text{ } tomines \div 1.20 = 3 \text{ } tomines \text{ } 4 \text{ } granos,$$
$$1 \text{ } peso \div 1.20 = 6 \text{ } tomines \text{ } 8 \text{ } granos.$$

Diez Freyle refers to this table as giving the value of "silver currency bought as assayed"[60] on folio 91v at the end of the tables, in his summary of what he has done up to that point. The author expands upon this explanation on folio 92r. There he says that when the commission exceeds 20%, one must make a calculation. In his illustration, with a commission of 24%, he says that one brings an amount of silver currency and buys with it an amount of assayed gold or silver, i.e., one brings [A] and gets $[\frac{A}{1+r}]$. His explanation there of how to deal with 24% may show us how he calculated the values in this table. Similar tables appear in the later works of Belveder (1597) and Garreguilla (1607). Although there the labeling suggests a conversion between different kinds of *pesos*, we believe there is also a tax involved. By a tax, several things might be meant:

1. It could be a real tax, payable to a government.
2. It could be a tax plus an assayer's fee. We see examples of this in the following.
3. It could be both of these, plus a "rounding down" at the expense of the buyer. Our entry for 1603 gives examples of this for the reverse transaction, i.e., one brings silver to receive currency.
4. McCusker indicates that debasing of coins by "clipping, shaving, or leaching" out the silver was so common that many merchants accepted coins by weight rather than by face value.[61] This could be a reason for discounting currency.

These considerations may explain why there is such a range of percentages. From 85v to 91r there is a series of short tables, with a brief and incomplete explanation on 85r. Some of these tables are for converting between

---

60. Smith's translation.
61. John J. McCusker, *Money and Exchange in Europe and America, 1600–1775*, Chapel Hill, 1978: 8.

various currencies. Note that the only currencies mentioned in the *Sumario* up to this point have been the *peso* (and its divisions into *tomines* and *granos*) and the *maravedi*. The numbers in the first table (values of silver, 3r–48r) and the third table (conversion between *pesos* and *maravedis*, 57r) show that this *peso* was the *peso de minas*. Now conversion tables are given for the following currencies and coins, in various pairs: *pesos de minas* (valued at 450 *maravedis*), *pesos de tipuzque* (valued at 272 *maravedis*), *ducados* (valued at 375 *maravedis*), *coronas* (valued at 350 *maravedis*), and *maravedis*.

Diez Freyle says in his explanation on 85r, "Ya que como aueys visto tengo puesto lo necessario para en los reynos del Peru: de aqui adelante pondre todo lo mas necessario de las cosas tocantes a esta nueva españa . . ." or, "Since, as you have seen, I have set down [that which is] necessary for the kingdoms of Peru, from here on I will set down [that which is] all the more necessary among the things concerning New Spain." We are inclined to conclude from these remarks of the author, and from the organization of the tables, that *pesos de tipuzque*, *ducados*, and *coronas* were not yet very common in Peru in 1556, but they were common enough to cause confusion. We guess that a Peruvian accountant dealing in silver and gold would not need to transact with them on a day-to-day basis. (This comment is biased toward the difficult-to-prove hypotheses of a Peruvian origin for Juan Diez Freyle and an intended Peruvian audience for the *Sumario Compendioso*.) Further down he writes that "many in these parts, and principally in Peru, send money to Spain," and that these people do not seem to know the difference between *pesos*, ducats, and crowns.

Here is an index to this section of short tables.

[Table 6]	85v	*Maravedis* to *pesos de minas*.
[Table 7]	85v	*Maravedis* to *pesos de tepuzque*.
[Table 8]	86r	*Pesos de minas* to *pesos de tepuzque*.
[Table 9]	86r	*Pesos de tepuzque* to *pesos de minas*.
[Table 10]	86v	*Plata quintada* to *pesos de minas*.
[Table 11]	86v	*Plata quintada* to *pesos de tepuzque*.
[Table 12]	87r	*Plata del diezmo* to *pesos de minas*.
[Table 13]	87r	*Plata del diezmo* to *pesos de tepuzque*.
[Table 14]	87v	*Plata del quinto* to *pesos de minas*.
[Table 15]	87v	*Plata del quinto* to *pesos de tepuzque*.
[Table 16]	88r	*Plata del rescate* (6) to *pesos de tepuzque*.
[Table 17]	88r	*Plata del rescate* ($6\frac{1}{4}$) to *pesos de tepuzque*.

[Table 18]	88v	*Diezmo* and 1%.
[Table 19]	89r	*Quinto* and 1%.
[Table 20]	89v	*Cacaos*.
[Table 21]	89v	*Flete*.
[Table 22]	90r	*Peso de minas* to *ducados*.
[Table 23]	90r	*Ducados* to *pesos de minas*.
[Table 24]	90v	*Pesos de minas* to *coronas*.
[Table 25]	90v	*Coronas* to *pesos de minas*.
[Table 26]	91r	*Ducados* to *coronas*.
[Table 27]	91r	*Coronas* to *ducados*.

In addition to the currency conversions, values for silver are given under the labels *plata quintada*, *plata del diezmo*, *plata del quinto*, and *plata del rescate*. We believe that these four pages of tables refer to values of silver in different states of having been taxed and about to be taxed.

Recall that the king was commonly due one fifth of all silver and gold mined. The verb *quintar* appears to be etymologically linked to the act of paying the king his fifth. Diez Freyle uses the term freely for the paying of any kind of tax, although on folio 85r he uses the parallel verbs *quintar* and *dezmar*. Let's assume that *plata quintada* means silver already taxed at the time of weighing and thus given values in accordance with the first table of the book. Under that assumption, the values given here correspond to a fineness of 2,210 *de ley*[62] in that 90 *marcos* comes to exactly 442 *pesos de minas*. One *marco* should be 4 *pesos* 7 *tomines* $3\frac{7}{15}$ *granos* (at 12 *granos* to the *tomín*), and the last number is rounded to $3\frac{1}{2}$ by Diez Freyle.

The next table on the same page gives *plata quintada* in *pesos de tipuzque*. Here, 2,210 *maravedis* is exactly 65 *reales*[63] and so 1 *marco* has the value $8\frac{1}{8}$ *pesos de tipuzque*, or 8 *pesos* 1 *tomín* exactly. We believe that these two tables are here on this page to provide a standard to which the tables on the next three pages can be compared.

On the next page, *plata del diezmo* is given values that are about 10.9% lower than for *plata quintada*. Even for a weight of 500 *marcos*, answers are given to a precision of half a *grano*, so one would think that the tax rate could

---

62. The standard of 2,210 *de ley* was used in the later Spanish work, the 1620 *Libro Intitulado Reduciones* of Vázquez de Serna, in his tables for finding the *señorage*. A standard of 2,220 *de ley* would be a fineness of 0.925, a standard used today.

63. Maybe 2,210 was chosen as a standard because it is 65 × 34.

be computed to six significant figures. However, the table for *minas* leads to a rate of 10.9162% and the table for *tipuzque* gives a rate of 10.8889%. Thus, precision does not imply accuracy. To three significant figures, though, these two tables agree with the 10.9% rate that Haring implies would be charged if the king were taking his tenth after the assayer got his 1%.

Values for *plata del quinto*, compared to those for *plata quintada*, are 20.8% lower, the rate we discussed earlier. The agreement between the two tables for *minas* and *tipuzque* is much better than that between the *diezmo* tables on the previous page, with the *minas* table implying a rate of 20.8001% for a weight of 900 *marcos* and the *tipuzque* table being in agreement for weights of 50 or 500 *marcos*.

There are two kinds of *plata del rescate*[64] on the next page, and values for both are given in *pesos de tipuzque* only. Both of these tables give nice round numbers for the value of 1 *marco*, and later lines of the tables show no sign that rounding was done in the 1 *marco* value. For the first table, 1 *marco* is 6 *pesos* and for the second it is 6 *pesos*, 2 *tomines*. The numbers in the first case are exactly $26\frac{2}{13}$% lower than *plata quintada*, and in the second case they are exactly $23\frac{1}{13}$% lower. There is an example of a 26% tax later in the *Sumario*. A 23% tax could be the result of the assayer taking 1% followed by the king taking $\frac{2}{9}$, but looking ahead to Echagoyan's work of 1603 we see the same values as in this table, with the explanation that they have been rounded down from the values that appear here in the table for *plata del quinto*.

There are separate tables for the king's tax plus the 1% for the assayer. In these the amount to be taxed is given in *marcos* and the payments are in *marcos*, ounces, R[eales?], and *quartillos*. (8 R is 1 *onça* and 4 *quartillos* is 1 R.) One table gives for each weight the *diezmo* (9.9%) and then the 1% for the assayer. The next table replaces the middle column with the *quinto* (19.8%).

Where they occur in this section, *granos* are 12 to the *tomín*, except in one table in the second column of 89v. This table has to do with weights and cargo. At the bottom of the column is the note that $12\frac{1}{2}$ *granos* make one *tomín*.

The Arithmetic

The section on arithmetic and algebra, as noted, starts on 91v. One observes that the tables use Roman numerals throughout but that in the

---

64. *Rescate* is "rescued, redeemed, or ransomed." In that this refers to silver remaining after a tax, it appears that this silver has been "rescued" from the tax man.

sections on arithmetic and algebra the author switches to Hindu-Arabic numerals.

Previously mentioned was the example on folio 94r of dividing a number by 124%. On folio 93r this is described as an application of the "rule of three," which Smith calls "the most popular of all the medieval commercial rules." A discussion of the history of the rule of three can be found in Frank Swetz's *Capitalism and Arithmetic*. In D. E. Smith's translation of the *Treviso Arithmetic*, included in the book by Swetz, one can read, "Furthermore, that the rule of the three things, which is of utmost importance in this art, may be at your command, . . . The rule of the three things is this: that you should multiply the thing which you wish to know, by that which is not like it, and divide by the other. And the quotient which arises will be of the nature of the thing which has no term like it." Such a passage is difficult reading for the modern practitioner of algebra, who would explain the same thing symbolically. What it says is that, to solve for A in

$$\frac{A}{B} = \frac{C}{D},$$

one should perform the calculation

$$\frac{BC}{D}.$$

Diez Freyle's example on folio 93r illustrates this. To paraphrase somewhat, the question is how many large jugs of wine are equivalent to 1,000 small jugs if 125 small jugs contain the same as 100 large jugs. Since we want to know large jugs, we take the 1,000 small jugs (equivalent to what we want to know), multiply by the 100 large jugs (which is not like the 1,000 small jugs), and divide by the 125 small jugs (the other). The answer is 800 large jugs. Since the rule of three was so well known, many problems involving fractions or percentages were restated in terms of the rule of three. See our entry for 1597 for another example.

On folio 94v, Diez Freyle gives a rule for converting *maravedis* to *pesos* at 450 *maravedis* to the *peso*. He starts his example with 120,000 *maravedis*. We are first to divide the number of *maravedis* by 1,000 and double the result, so now the number is 240. So far we have divided by 500, and so the answer must now be increased by the factor $\frac{10}{9}$. The author has a curious means of doing this. He writes 240 and below it 24 and below that 2. The numbers shift right with each line and the last digit is truncated each time.

240
24
2

He totals these numbers and gets 266, to which he says we must add 50 *maravedis* for each unit in the sum.[65] This is 6 × 50, or 300, and so the value of 120,000 *maravedis* is 266 pesos and 300 *maravedis*.

This method, if misunderstood, fails in some other examples. Start with, say, 41,000 *maravedis*. After dividing by 500 we have 82. Write 8 below the 82.

82
8

The total is 90, and since the units digit is 0 we do not add the correction, so 90 *pesos* is the result. But the correct conversion is 91 *pesos*, 50 *maravedis*. What went wrong?

The answer is that we should take not the units digit in the sum, but the sum of the units digits in the units column, as the number to be multiplied by 50 *maravedis*. The units digit in the sum will only agree with the sum of the digits in the units column if the latter is less than 10. Diez Freyle in his example says only "and of the 6, 300 *maravedis*"[66] which does not clarify whether he realizes where the 6 has to come from. Smith, in his English translation, silently corrects the method by saying, "In the units column there is 6," where Diez Freyle says nothing about a units column.

The reason this method works is that

$$\frac{10}{9} = 1.1111111 \cdots$$

with the 1s repeating forever. So multiplying 240 by $\frac{10}{9}$ is equivalent to summing the numbers

240.
24.0
2.40
0.240
0.0240
. . .

65. . . . *en el suma por cada unidad toma 50 maravedis* . . .
66. . . . *y de los seys. 300. maravedis* . . .

with the right shift continuing forever. This is equivalent to summing the three numbers

$$240.240240\cdots$$
$$24.024024\cdots$$
$$2.402402\cdots$$

and this sum in turn is equal to the sum of

$$240.000000\cdots$$
$$24.444444\cdots$$
$$2.222222\cdots$$

Thus the fractional part will be represented if we add one ninth of a peso (50 *maravedis*) for each unit in the units column.

In the very next problem, Diez Freyle explains that the tax on silver brought to be assayed is 26%: one quarter for the king and 1% for the *marcador* or assayer. He briefly notes that the assayer is due one quarter more than this one percent but he likewise pays this back to the king as tax. These numbers, and the method of adding the two taxes, are at variance with several features of what Haring says was standard practice in the sixteenth century (see earlier discussion). What makes this more disturbing is that Diez Freyle uses the verb *quintar* (to pay the *quinto*, or one fifth) for the act of paying this tax of 26%.

Folio 95v states that ". . . there are and have been many to whom his majesty owes money."[67] Apparently, in Diez Freyle's world, when the king owes one money, one brings gold or silver to be taxed in such an amount that the tax and the debt cancel each other out. He shows how to find the right amount to bring by dividing by 0.26 (multiplying by 100 and dividing by 26). In other words, to cancel out a debt of 1,144 *pesos*, one brings $1,144 \div 0.26 = 4,400$ *pesos*. (The method of division is the galley method, which we discuss below in the section on the 1587 *Instrucción Náutica*.) One then walks away with the 4400 *pesos* free of obligation. We assume, although it is not mentioned in the text, that the king now owes the assayer his 1%, or 44 *pesos*.

On folio 96r we can find an example of the $20\frac{4}{5}$% tax reported by Haring and which appears in some of the tables we have discussed. In both the "First Problem" and the "Second Problem" on this page, an amount of

---

67. Smith's translation.

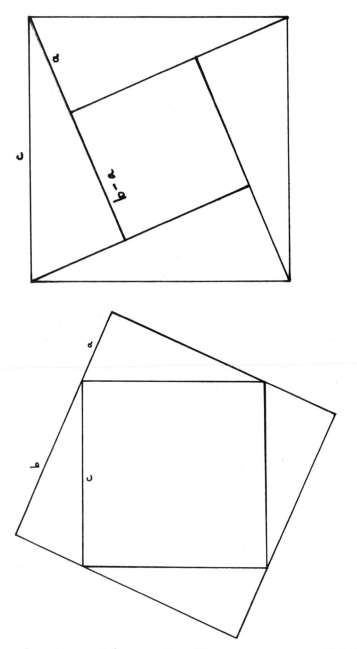

Figure 10. $c^2 \pm 2ab = (a \pm b)^2$. From either of these equations one could derive the Pythagorean Theorem.

2,000 *pesos* of gold is taken to be taxed. The tax is 416 *pesos*, exactly $20\frac{4}{5}$% of the 2,000, leaving 1,584. (In the "First Problem" one calculates the 416 and the 2,000 given the 1,584, and in the "Second Problem" one calculates the 1,584 from the 416.)

In contrast to these problems, there are a few questions from number theory that are not directly applied to commerce. One of the more advanced of these is the problem of finding three whole numbers in arithmetic progression that are perfect squares. Smith says that Leonardo of Pisa (later known as Fibonacci) had solved this problem using the identities

$$(x^2 + y^2)^2 - 4xy(x^2 - y^2) = (y^2 + 2xy - x^2)^2$$

and

$$(x^2 + y^2)^2 + 4xy(x^2 - y^2) = (x^2 + 2xy - y^2)^2.$$

In fact one finds this in Fibonacci's *Liber Quadratorum*, or *Book of Squares*, of 1225. L. E. Sigler, in his English translation of the *Liber Quadratorum*, points out that these formulas are related to an observation of Diophantus that if a, b, and c are the sides of a right triangle, with c the hypotenuse, then

$$c^2 \pm 2ab = (a \pm b)^2.$$

The figures that illustrate these identities are reminiscent of well-known dissection proofs of the Pythagorean Theorem.

It might be natural to suppose that Diez Freyle had direct access to the *Book of Squares*. Our colleague Lawrence D'Antonio has offered the following rebuttal of this notion in a message to us, for which we are deeply grateful:

> Fibonacci's *Liber Quadratorum* is certainly not the source. There are very few surviving copies of the work, it was never widely disseminated, and it is meant as a research work, not a textbook. More importantly, Fibonacci uses different terminology.
>
> In present-day terminology a congruent number is a number k such that there exists an x for which $x^2 + k$ and $x^2 - k$ are both

squares (we usually take k, x to be integers). The number x^2 is called the congruous square.

This is the terminology used by Juan Diez, but Fibonacci calls k a congruous number.

The most likely source for congruent numbers in the *Sumario* is the *Summa Arithmetica* of Luca Pacioli. The first edition of this was published in 1494, the second in 1523. This work was a popular textbook of its time (there are many surviving copies and it was widely referred to). Furthermore, it is Pacioli who defines 'congruent' and 'congruous' as we use them today. Pacioli also gives a table of congruent numbers.

I can't guarantee that Juan Diez actually read Pacioli. There are several popular texts of the early 16th century that are modeled on Pacioli. Works by Ghaligai, Feliciano, Forcadel, Tartaglia, and Gosselin all discuss congruent numbers and use Pacioli.

Here is a table of the solutions given by Diez Freyle on 98v with the numbers $(x^2 + y^2)^2$ called *congruos* and the numbers $4xy(x^2 - y^2)$ called *congruentes*, as D'Antonio points out in the just-quoted material. The values of x and y are given in the next two columns, followed by the three squares associated with each row.

Congruos	Congruentes	x	y	Squares		
25	24	2	1	1^2	5^2	7^2
100	96	3	1	2^2	10^2	14^2
169	120	3	2	7^2	13^2	17^2
225	216	–	–	3^2	15^2	21^2
289	240	4	1	7^2	17^2	23^2
400	384	4	2	4^2	20^2	28^2
625	336	4	3	17^2	25^2	31^2
625	600	–	–	5^2	25^2	35^2
676	480	5	1	14^2	26^2	34^2
841	840	5	2	1^2	29^2	41^2
900	864	–	–	6^2	30^2	42^2
1,156	960	5	3	14^2	34^2	46^2
1,225	1,176	–	–	7^2	35^2	49^2
1,521	1,080	–	–	21^2	39^2	51^2

(continued)

Congruos	Congruentes	x	y	Squares		
1,681	720	5	4	31^2	41^2	49^2
2,025	1,944	6	3	9^2	45^2	63^2
2,500	1,344	7	1	34^2	50^2	62^2
2,500	2,400	–	–	10^2	50^2	70^2
2,601	2,160	–	–	21^2	51^2	69^2
2,704	1,920	6	4	28^2	52^2	68^2
2,809	2,520	7	2	17^2	53^2	73^2
3,025	2,904	–	–	11^2	55^2	77^2
3,364	3,360	7	3	2^2	58^2	82^2
3,600	3,456	–	–	12^2	60^2	84^2
3,721	1,320	6	5	49^2	61^2	71^2
4,225	2,016	8	1	47^2	65^2	79^2

Five corrections made by Smith in his translation have been incorporated into this table. In the fifth row from the end, Diez Freyle gives 2,905 as the *congruente*, and Smith says it should be 2,925, but the correct number is 2,904. In lines where x and y are omitted, the *congruo* and *congruente* may be found by multiplying a perfect square through the numbers in some line above. The last *congruo*, 4,225, is actually a *congruo* in four different ways, as it can be paired with 2,016, 3,000, 3,696, or 4,056.

The reader familiar with some results in elementary number theory might wonder, at this point, whether all examples of three squares in arithmetic progression arise as in the table above. This turns out to be true. In fact, Leonardo of Pisa gave a proof of this in his *Liber Quadratorum*, Proposition 14. Leonardo's proof, based on the fact that any square number is the sum of an arithmetic series of odd numbers, can by found in L. E. Sigler's translation. Here is another proof.

Theorem. *Any sequence of three square numbers in arithmetic progression is given by the formulae $w^2(y^2 + 2xy - x^2)^2$, $w^2(x^2 + y^2)^2$, $w^2(x^2 + 2xy - y^2)^2$ where the positive integers x and y are relatively prime and one of them is even, one of them is odd.*

Proof. Given $a^2 < b^2 < c^2$ in arithmetic progression, without loss of generality we may assume that they are relatively prime, since any common factor the three numbers had would have to be a square, and so it could be factored out leaving three squares. Then it suffices to show that a^2, b^2, and c^2 are given by the formulae $(y^2 + 2xy - x^2)^2$, $(x^2 + y^2)^2$, and $(x^2 + 2xy - y^2)^2$. $c^2 - a^2$ is an even number so a and c have the same parity (they are both even or both odd). Then 4 divides $(c - a)(c + a) = c^2 - a^2$ so 2 divides $\frac{1}{2}(c^2 - a^2)$, which is the common difference $b^2 - a^2$ and $c^2 - b^2$. So all three of a^2, b^2, and c^2 have the same parity. Therefore they are all odd.

We are free to choose the signs of a, b, and c. $\frac{c+a}{2}$ and $\frac{c-a}{2}$ differ by the odd number a so exactly one of them is even. We make b and c positive and we choose the sign of a so that 4 divides $c + a$. Then $c - a$ is 2 times an odd number so 4 divides both $2b + (c-a)$ and $2b - (c-a)$. Using the fact that $2b^2 = a^2 + c^2$ we have

$$\frac{2b + (c - a)}{4} \cdot \frac{2b - (c - a)}{4} = \frac{4b^2 - (a^2 - 2ac + c^2)}{16}$$
$$= \frac{a^2 + 2ac + c^2}{16}$$
$$= \left(\frac{a + c}{4}\right)^2.$$

Any prime that divided both $\frac{2b + (c - a)}{4}$ and $\frac{2b - (c - a)}{4}$ would have to divide their sum, b, and their difference, $\frac{c-a}{2}$, and thus would be a divisor of a^2, b^2, and c^2. So $\frac{2b + (c - a)}{4}$ and $\frac{2b - (c - a)}{4}$ are relatively prime. Since their product is a square and their sum is positive then they are both squares.

Let

$$x = \sqrt{\frac{2b + (c - a)}{4}}, \quad y = \sqrt{\frac{2b - (c - a)}{4}}.$$

The difference between $\frac{2b + (c - a)}{4}$ and $\frac{2b - (c - a)}{4}$ is $\frac{c - a}{2}$, which is odd, as noted. So $\frac{2b + (c - a)}{4}$ and $\frac{2b - (c - a)}{4}$ have opposite parity and therefore so do x and y. Also $2xy = \frac{c + a}{2}$, $x^2 - y^2 = \frac{c - a}{2}$, and so we have the equations $y^2 + 2xy - x^2 = a$, $x^2 + y^2 = b$, and $x^2 + 2xy - y^2 = c$. □

A variation on this question yields a problem that is still at the forefront of research in number theory. What if the three squares, a^2, b^2, c^2, in the arithmetic progression are allowed to be the squares of rational numbers, while their common difference, $b^2 - a^2 = c^2 - b^2 = n$, is a positive integer? For what values of n will there be a solution? Such values are now called congruent numbers, and the problem of characterizing them is called the congruent number problem.

The smallest congruent number is 5. Fibonacci demonstrates that 5 is congruent by giving the squares, $(2\frac{7}{12})^2$, $(3\frac{5}{12})^2$, $(4\frac{1}{12})^2$, in Proposition 17 of the *Liber Quadratorum*. Diez Freyle demonstrates that 6 is a congruent number with the squares $(\frac{1}{2})^2$, $(2\frac{1}{2})^2$, $(3\frac{1}{2})^2$, which he finds by dividing the *congruo* and *congruente* in the first row of his table by 4.

Two alternate descriptions of congruent numbers are easily derivable. On the one hand, a positive integer n is a congruent number if and only if it is the area of a right triangle with rational sides. On the other, n is congruent if and only if the curve $y^2 = x^3 - n^2 x$ admits a point (x,y) with x and y rational and $y \neq 0$. The details of the arguments that both of these alternative definitions for congruent are equivalent to the one given earlier are given in Sigler's translation of the *Liber Quadratorum*, although one can quickly see one direction for the latter equivalence by noting that if x, $x - n$, and $x + n$ are all squares of rationals, then $x(x - n)(x + n) = x^3 - n^2 x = y^2$ is the square of a rational.[68]

The Web site http://modular.math.washington.edu/simuw06/table-1000.txt gives a list of the first one thousand congruent numbers. The first ten are 5, 6, 7, 13, 14, 15, 20, 21, 22, 23. Sigler states a conjecture about congruent numbers that is contradicted by this table.[69] The intended form of the conjecture is probably the one stated in a recent article, on congruent numbers, by Chahal: "If a is a square-free natural number such that $a \equiv 5$, 6, or 7 (mod 8), then a is the area of a right triangle with all sides rational," i.e., then a is congruent.[70] (A square-free number is one that has no perfect square other than 1 as a divisor. See the discussion of the *Instrucción Náutica* of 1587 for more about the $\equiv$ relation.)

Neal Koblitz points out in his *Introduction to Elliptic Curves and Modular Forms* that whereas 101 and 157 are both congruent, the simplest rational numbers that demonstrate this directly have numerators and denominators in excess of 20 digits. Thus an alternate characterization of congruent numbers, one that is easily checked, would be desirable. At the end of

---

68. Curves of the form $y^2 = ax^3 + bx^2 + cx + d$ are called elliptic curves. The existence of rational points on these curves has long been a central topic in number theory. Such considerations were an important component in the celebrated proof by Andrew Wiles of Fermat's Last Theorem just over a decade ago.

69. Leonardo Pisano [Fibonacci], *The Book of Squares, An Annotated Translation into Modern English by L. E. Sigler*, Boston, 1987: 81.

70. Jasbir S. Chahal, "Congruent Numbers and Elliptic Curves," *The American Mathematical Monthly* 113/4 (Apr. 2006): 308.

Koblitz's book, one can find a characterization of congruent numbers that was conjectured in 1983 by Jerrold B. Tunnell. This characterization provides, in Koblitz's words, "an effective and rapid algorithm for determining whether n is a congruent number." The truth of this conjecture depends on another famous conjecture, the Birch and Swinnerton-Dyer Conjecture, which is one of the seven Millennium Prize Problems whose solutions are each worth $1,000,000. (See www.claymath.org/millennium/ for more information about these problems.) Tunnell proved that if n is a congruent number, then the algorithm in question gives a positive answer, and that if the Birch and Swinnerton-Dyer Conjecture is true, the converse holds, i.e., the algorithm never gives a false positive.

The Algebra

As Smith shows us in his notes to the English translation, Diez Freyle uses the word *cosa* (thing) for the unknown x, and *zenso* (Mod. Sp. *censo*, tax) for x^2. This makes it possible to discuss problems involving quadratics without the use of algebraic equations as we know them. This mirrors the practice seen in early Arabic texts on algebra, where the word *shay* (thing) is used for x and the word *mal* (assets) is used for x^2.

The very first problem in the algebra section, on folio 100v, is "Give me a squared number that, subtracting from it $15\frac{3}{4}$, leaves its own root." We then see a *regla*, or rule, for getting the answer. "*Digo que el numero sea una cosa*" is how Diez Freyle begins his solution, literally, "I say that the number is a thing." We should paraphrase this as, "Let the number be x," for, not only is *sea* in the subjunctive, suggesting "let it be," but recall that *cosa*, thing, is a word to be manipulated as we would today manipulate the x in an algebra problem.

The method laid out in the *regla* is an example of what we now call completing the square. The *regla* is followed by a *prueva*, or proof. Diez Freyle uses *regla* for what we would now call a "worked example" and *prueva* for what we would now call "checking the answer." This pattern continues through the series of algebra problems. No general rules are stated, in the sense of giving algebraic formulae. The reader is expected, we suppose, to do a new problem by looking at the worked examples given and copying the method.

The explanations in the *reglas* are wordy without saying much, and this puts a burden on the reader who is trying to learn algebra from this text. For instance, the problem just considered amounts to solving the equation $x^2 - x = 15\frac{3}{4}$. The very next problem considered by Diez Freyle amounts to solving the equation $x^2 + x = 1{,}260$. Using almost the same words in both cases, he tells the reader to add $\frac{1}{4}$ to the number, $15\frac{3}{4}$ or 1,260, respectively, and take the square root of the result. Then, in the first case, $\frac{1}{2}$ is to be added, and in the second case $\frac{1}{2}$ is to be subtracted, with no explanation as to why. The careful reader must perceive that the slight difference between the two problems leads to a different operation, subtracting versus adding, at this step.

Conclusions

All in all, this is a unique book for its era, and it deserves the attention it has been given over the years.

The author clearly is not the same person as the Juan Díaz of the conquest. This notion almost certainly was first published by D. E. Smith, and it may have come to him from the British Library card catalog.

We wish we could come to a definite conclusion about the hypothesis that Diez Freyle was from Peru. Whenever we discuss other books of accounting tables in the following entries, we will give enough details so that a picture can be drawn of how practices differed in Mexico and Peru. Readers then can wrestle with the question of whether the *Sumario Compendioso* resembles more the accounting books from Peru or those from Mexico.

Here is a short summary of how the other accounting books on this list are similar to or differ from the *Sumario Compendioso*.

- Lima, 1597—*Libro General*—Similar in structure to the *Sumario Compendioso*, but with many more tables involving other currencies besides the *pesos de minas*. Algebra rules and example problems are found at the back.
- México, 1603—*Tablas de Reducciones*—Very different from the two previous accounting books. Values of silver are given in *pesos de tipuzque*. Two mutually incompatible tables give values of gold in *castellanos*.
- Lima, 1607—*Libro de Plata*—No gold, only silver. Roughly equal attention given to *pesos de minas*, *pesos* of 8 *reales*, and *pesos* of 9 *reales*.
- México, 1610—*Alivio de Mercaderes*—Gives values of silver in *pesos de tipuzque*. Three separate tables give values with tax taken out (*plata del diezmo* and *plata del rescate*) and already taxed (*plata quintada*).
- México, 1615—*Libro de Cuentas*—No copies known.
- México, 1668—*Reformación de las Tablas*—Begins the same as *Alivio de Mercaderes*, with three tables for silver taxed in various ways (with the same three names).

- Lima, 1696—*Tabla General*—A new way to calculate the value of gold requires the user to multiply, where earlier tables only counted on the user to be able to add. The price of gold changes for the first time.
- México, 1697—*Reducciones de Plata*—The price of silver is the same as in earlier books. *Quinto, diezmo,* and *plata quintada* have the same meanings as before. The main new twist is the use of a fineness of 2,376 (99%) as a standard.
- México, 1700— *Reducción de Oro*—Values of gold are given in *castellanos,* as in 1603. Floating point notation is used in the silver tables, and decimal fractions are employed in the explanation of those tables.

It is tempting to use the somewhat greater similarity of the early Lima books to the *Sumario Compendioso* to support the hypothesis of a Peruvian origin for Juan Diez Freyle. Unfortunately, the difference between the Mexico and Lima books is not sharp enough. With just the evidence that we have so far, we are not ready to conclude once and for all that the *Sumario Compendioso* was written in Peru, although it seems likely that it was written for Peru.

1557

Alonso de la Vera Cruz. 4
  Other Author: Campanus of Novara.
  PHISICA,SPECV= / latio,ÆditaperR. / P. F. ALPHONSVM A VERA CRVCE, AV- / gustinianæ familiæ Prouintialẽ, artiũ,& sacrę Theologiæ Doctorem, at[que] */ cathedræ primæ in Academia Mexicana in noua Hispania moderatorẽ / Accessit cõpendium spheræ Cãpani ad complementũ tractatus de cælo / Excudebat Mexici Ioã.Pau.Brissẽ.Anno Dñicę incarnationis.1557*
  Juan Pablos of Brescia, México.
  Icazbalceta 30. Sabin 98914. Medina, *México,* 33. Wagner 39.
  Originals located at: Biblioteca Cervantina, Biblioteca Nacional de Chile, Biblioteca National de México, British Library, Centro de Estudios de Historia de México Condumex, Hispanic Society of America, Indiana University, Institución Colombina, John Carter Brown Library, New York Public Library, University of Texas at Austin, Wellcome Library.
  Microfilm copies are available.

Figure 11. Title page of the *Phisica Speculatio*, 1557. The John Carter Brown Library at Brown University, Providence, RI.

Paging: a total of 205 leaves as follows: Four unnumbered leaves, 380 numbered pages, and eleven numbered leaves. The first eight pages are the title page, two letters, a prologue, and a four-page index. In the next 380 pages there are the following irregularities: page 1 is unnumbered, page 95 is misnumbered 96, page 132 is misnumbered 232, page 160 is misnumbered 150, page 181 is misnumbered 18, pages 370 and 371 are misnumbered 360 and 361, and pages 372 to 379 are misnumbered 366 to 373. The last eleven leaves are numbered 1 to 12, skipping 8. The copies held by the Biblioteca Nacional de Chile, Biblioteca Nacional de México, Indiana University, and the New York Public Library have a blank leaf between leaves 6 and 7 of this last section.

The three Vera Cruz titles on this list are bound together at the John Carter Brown Library. This copy has folio 12 at the end of the *Phisica Speculatio* bound after the title page of *Recognitio, Summularum*. The copy at the Biblioteca Nacional de México is missing the first four leaves, [2] through [4] of which are bound in at the end of their copy of the *Dialectica Resolutio* (1554). The copy in Chile has insect damage. The title page of the Hispanic Society copy is a copy of the British Library title page.

This work was reprinted in Salamanca in 1562, 1569, and 1573. These editions do not include the *Campanus de Sphera*, which contains pretty much all of the mathematical content of the first edition (see subsequent discussion). The 1569 edition does have some work on sequences of numbers at the end that does not appear in the Mexico edition.

Year	Edition	Held by
1562	Second	Biblioteca Nacional de España, Universidad de Salamanca (two copies), Universidad de Sevilla
1569	Third	Biblioteca Nacional de Chile (two copies), British Library, Centro de Estudios de Historia de México Condumex, Indiana University, John Carter Brown Library, Stanford University, Universidad de Barcelona, Universidad de Sevilla, University of Cambridge, University of Michigan
1573	Fourth	Biblioteca Casanatense, Biblioteca Universitaria Alessandrina, Biblioteca Nacional de México, California State Library, Escorial, Indiana University, Universidad de Barcelona (two copies), Universidad de Sevilla, University of Oxford, University of Pennsylvania

Parts of this work, Books I and II on the nature of the soul, were reprinted as *Investigación Filosófico-natural: Los Libros del Alma*, translated from Latin to Spanish by Oswaldo Robles, in 1942. Three such books appeared in Spanish in Robledo's *El Magisterio Filosófico* in 1984. The Robles translation was included in *Antología de Fray Alonso* in 1988, and part of Book II was in *Antología sobre el Hombre y la Libertad* in 2002.

This is widely regarded as the first work on physical science printed in the New World. It is firmly in the Aristotelian tradition and therefore does not much resemble modern physical science. There is nothing mathematical until the short tract at the end, written by another author.

The last eleven leaves are a separate treatise with the caption title *Campanus de Sphera* and referred to on the title page of the main work by the phrase below the plate of St. Augustine. The original manuscript was the *Tractatus de sphera* by Campanus of Novara. According to Francis Benjamin and G. J. Toomer, in their edition of *Theorica Planetarum*, another work by Campanus, the *Tractatus* cites two previous works of Campanus, the more recent of which can be dated to 1268, and so the *Tractatus* should be dated somewhat later. They also give the following printing history: Venice, L. A. de Giunta, 1518; Venice, Octavius Scotus, 1518; Venice, L. A. de Giunta, 1531; and then México, 1557, the item at hand.

The *Tractatus de sphera* is divided into fifty-four short chapters, and this chapter numbering has been preserved here. The three dimensions are treated in Chapter 1. Dimensions are described in terms of possible motions, and we see the statement, "a fourth dimension cannot be adjoined." Circles are treated, and the terms "diameter" and "circumference" are introduced. This is followed by the definition of the sphere in Chapter 2 as a rotated semicircle.

Epicycles are used to explain the retrograde motion of the planets, and eclipses of the sun and moon are discussed.

Lynn Thorndike[71] says that this work of Campanus was an attempt to rewrite the *Tractatus de sphera* of Iohannes de Sacrobosco, one of several such attempts by various medieval manuscript writers. Thorndike claims that little new has been added, except for the introduction of the terms "azimuth" and "almucantarath," two terms that come up in locating a position in the sky in the horizon system. An almucantar, in modern terminology, is a circle on the celestial sphere that is parallel to the horizon,

---

71. Lynn Thorndike, *The SPHERE of Sacrobosco and Its Commentators*, Chicago, 1949.

and the azimuth is the coordinate, measured in degrees, which progresses around such a circle.

1578

Francisco de Toledo. 5
INTRODVCTIO / *IN DIALECTICAM* / *ARISTOTELIS,* / *PER MAG-ISTRVM FRAN-* / *ciscum Toletū Sacerdotem societatis Iesu,* / *ac Philosophiæ in Romano Societatis* / *Collegio professore.* / [Four lines wrapped around the Jesuit emblem:] *DVLCE TVVM NOSTRO* / *FIGAS IMPECTORE NO MEN* / *NAMQ3 TVO CONSTAT* / *NOMINE NOSTRA SALVS.* / *MEXICI.* / *In Collegio Sanctorum Petri & Pauli,* / *Apud Antonium Ricardum.* / *M. D.LXXVIII.*
Antonio Ricardo, México.
Icazbalceta 80. Sabin 96108. Medina, *México,* 86. Wagner 114.

Originals located at: Biblioteca Cervantina, Biblioteca Nacional de Chile, Centro de Estudios de Historia de México Condumex, Hispanic Society of America, Huntington Library, Indiana University, John Carter Brown Library, New York Public Library, Universidad de Salamanca, Yale University.

Microfilm copies are available.

Paging: 171 folios, numbered except for 1 to 6; 14 is misnumbered 16, 16 is misnumbered 14, 50 is misnumbered 05, 59 is misnumbered 56, 100 is unnumbered, 112 is unreadable, and 120 is misnumbered 130.

Folio 129 is missing in the copy at the John Carter Brown Library, and the copy at Yale is missing the title page. The copy in Indiana is missing leaves [1], 8, and 104. All copies except the ones at the John Carter Brown and Yale libraries are bound with the next two entries.

The printing history of this work is fairly rich. This is not limited to just this book; a search of WorldCat shows that all of Toledo's books went through many printings in the sixteenth and early seventeenth centuries. We have found the following examples for the *Introductio.* Editions marked by an asterisk (*) are those that are listed in the *Bibliothèque de la Compagnie de Jésus* of 1898. The Web site sbnonline.sbn.it listed so many hits for one of the editions here, and one for the second entry for this year, that this table notes only, in those cases, the number of copies that are supposed to be in Italy. (We have not yet visited any of the German, Italian, Austrian, or Swiss libraries.)

MATHEMATICAL WORKS, EXCLUDING ALMANACS 1578

Year	City	Held by
*1561	Rome	Biblioteca Comunale Giosué Carducci, Biblioteca Nazionale Centrale di Roma (three copies), Biblioteca Universitaria di Bologna
*1562	Vienna	Österreichische Nationalebibliotek
*1565	Rome	Biblioteca Comunale Giosué Carducci, Biblioteca Comunale Planettiana, Biblioteca Nazionale Centrale di Roma
1568	Lyon	Bibliothèque Nationale de France
*1569	Rome	Biblioteca Apostolica Vaticana, Biblioteca Comunale (Sarnano), Biblioteca Nazionale Centrale di Roma
1574	Köln	Biblioteca Nacional de España
*1574	Venice	Biblioteca Nazionale Centrale di Roma, Escorial, University of San Francisco
1575	Köln	Bibliothèque Nationale de France
1577	Sevilla	Universidad de Sevilla
*1578	Venice	
*1578	Mexico	See above
1579	Venice	Biblioteca Nazionale Braidense, Biblioteca Nazionale Centrale di Firenze (two copies), Biblioteca Oliveriana, Biblioteca Universitaria di Bologna, Bibliothèque Nationale de France, Universitäts und Stadtbibliothek Köln
*1581	Alcalá	
1581	Lyon	Biblioteca Universitaria Alessandrina, Universidad de Barcelona
1582	Venice	Biblioteca Comunale Planettiana, Biblioteca di Area Umanistica dell'Università degli Studi di Urbino, Biblioteca Nazionale Centrale di Roma (two copies), Indiana University, University of Chicago
1584	Venice	Biblioteca Comunale (Terni), Biblioteca Comunale & Planettiana, Biblioteca Comunale Roberto Ardigó, Biblioteca Lancisiana, Biblioteca Nazionale Centrale di Roma, Indiana University
*1586	Lyon	Biblioteca Nacional de España, Real Biblioteca, Universidad de Barcelona, Universidad de Salamanca, Universidad de Sevilla

Year	City	Held by
1587	Lyon	Universidad de Barcelona
1587	Lyon	Biblioteca Nacional de México, Biblioteca Nazionale Centrale di Roma, Bibliothèque Nationale de France
1587	Venice	Escorial, Universidad de Barcelona, University of Montréal
*1588	Venice	Biblioteca Nazionale Centrale di Roma, Indiana University, University of Chicago, Yale University
*1592	Lyon	Biblioteca Nazionale Centrale di Roma (two copies), Biblioteca Nacional de España, Biblioteca Nacional de México (two copies), California State Library
1594	Rome	Biblioteca Civica (Nove Ligure), Duke University
1595	Venice	Biblioteca Diocesana Piervissani
1596	Venice	Biblioteca del Centro Teologico (Torino), Biblioteca dell'Instituto di Filosofia San Tommasso d'Aquino, Bibliothèque Nationale de France
*1601	Rome	University of Cambridge, University of Oxford; fourteen other copies in Italy
*1602	Venice	
1607	Milan	Biblioteca Nacional de España
*1607	Venice	
*1608	Lyon	
1613	Venice	Biblioteca Apostolica Vaticana
1615/6	Köln	
1620	Madrid	Biblioteca Nacional de Chile, Universidad Complutense de Madrid
*1620	Paris	
1621	Milan	Bibliothèque Nationale de France
1629	Torino	Biblioteca del Centro Teologico (Torino)

The Lyon editions of 1581, 1586, 1587 (Barcelona copy only), 1592, and 1608 were actually included with other material under the titles *Omnia Quæ Hucusque Extant Opera* for 1581, 1586, and 1587, and *Omnia Quæ Hucusque Edita Sunt Opera Philosophica* for 1592 and 1608. The Barcelona copy of the 1581 edition is identical with the *Omnia Quæ* anthology except that it has the *Introductio* title page.

The Alcalá edition of 1581 listed in the catalog at the Biblioteca Nacional de España turns out, when it is requested, to be the Köln edition of 1574. Henry Wagner made reference to a printing in 1577 in Alcalá.[72]

The 1615–1616 edition was part of a project in those two years to reprint the entire works of Toledo on Aristotle. This collection was reprinted as the *Opera Omnia Philosphica* in 1985.

Francisco de Toledo was born in Córdoba in 1532 and died in Rome in 1596. He was named the first Jesuit cardinal by Pope Clement VIII on September 17, 1593.[73] The *Introductio in Dialecticam Aristotelis* in 1561 was his first publication. Researchers should beware of another Francisco de Toledo whose years are 1515–1582. He was also an author and was the Viceroy of Peru.

José Toribio Medina reports that the printer Antonio Ricciardi of Turin printed in Mexico from 1577 to 1579 and then went to Peru in 1580. Ricciardi produced Lima's first printed work in 1584 and continued printing there until 1605, printing our entry for 1597 along the way. In the preliminary pages of this work, in the only two sections that are in Spanish, the Viceroy and the Archbishop grant Antonio Ricardo Piamontes license to print freely whatever is deemed necessary by the Provincial of the Company of Jesus.

The *Introductio* is the second work on formal logic on our list and the fourth on Aristotelian thought. It is divided into Books I through V. Most of the figures are in Book III. As with the logic book discussed earlier, the square of opposition and modality are treated.

Francesco Maurolico. 6
REVERENDI DO. / *FRANCISCI MAVRO-* / *LYCI, ABBATIS* / Messanensis, atque mathe- / matici celeberrimi. / De Sphæra. Liber vnus. / [Four lines wrapped around the Jesuit emblem:] *DVLCE TVVM NOSTRO* / *FIGAS IMPECTORE NO MEN* / *NAMQ3 TVO CONSTAT* / *NOMINE NOSTRA SALVS.* / *Mexici apud Antonium Ricardum* / *in Collegio diui Petri & Pauli.*
Antonio Ricardo, México.
Icazbalceta 80. Sabin 96108. Medina, *México*, 86. Wagner 114.

Originals located at: Biblioteca Cervantina, Biblioteca Nacional de Chile, Centro de Estudios de Historia de México Condumex, Hispanic Society of

---

72. Henry R. Wagner, *Mexican Imprints, 1544–1600, in the Huntington Library*, San Marino, 1939.
73. According to Augustin and Aloys De Backer, *Bibliothèque de la Compagnie de Jésus*, new edition by Carlos Sommervogel, Bruxelles, 1898.

America, Huntington Library, Indiana University, New York Public Library, Universidad de Salamanca.

Microfilm copies are available.

Paging is continuous with the next entry. Altogether there are fifty-three folios. Folios 1, 2, 4, 5, and 36 are unnumbered, the number 38 is skipped, 45 and 46 are unnumbered, 48 is misnumbered 46, 50 is misnumbered 48, and 51 through 54 are unnumbered. This item is folios 1 through 27 of the whole.

All known copies are bound with the other two entries for this year. Frequently this item and the next are not catalogued separately, so they can be found at some libraries only by searching for the entry by Toledo.

This work was first printed in Venice in 1575 as part of the *Opuscula Mathematica*, a collection of works by Maurolico on various topics: arithmetic, geometry, astronomy, the calendar, etc. It contains an incipit for the collected titles. *De Sphæra* is found there on pages 1 through 25. A few of the libraries where we have seen copies of this 1575 printing are Biblioteca Nacional de España (two copies), British Library, Brown University, Harvard University (two copies), Huntington Library, Indiana University, Universidad de Salamanca, University of California Berkeley, University of Cambridge (two copies), University of Glasgow (two copies), University of Michigan, University of Oxford (three copies), and Yale University; a great many more copies are listed by WorldCat and by the national catalogs of Germany and Italy.

The Web site www.maurolico.unipi.it is devoted to the works of Maurolico. There is some discussion of this text there under the heading "Astronomia."

Francesco Maurolico (1494–1575) is the subject of a thorough treatment by Arnaldo Masotti in the *Dictionary of Scientific Biography*. Maurolico was from a Greek family that had escaped the Turks by fleeing to Sicily. He became a priest in 1521 and became a professor at the University of Messina in 1569. He wrote extensively on optics, astronomy, and geometry and participated in the contemporary tradition of trying to reconstruct lost texts, including books V and VI of the *Conics* of Apollonius. He even made a weather forecast for the knights leaving Sicily for the Battle of Lepanto in 1571.

Maurolico is best known in mathematical circles for his *Arithmeticorum Libri Duo* (another item included in the *Opuscula Mathematica* of 1575), which is a book about certain classes of numbers. The set of perfect numbers is one example. It had been asserted to contain the first instance of a proof by the method of induction.[74] More recently, Victor Katz's *A History of Mathematics* states that the first European to use proof by induction was Levi ben Gerson (1288–1344).

In the text at hand, folios 4r to 5r give a diagram of knowledge. "Geometria Principia" runs from 5v to 9r and is a review of terms and concepts from geometry. Masotti asserts that Maurolico criticizes Copernicus in this text. This is found on folio 27v. (It seems to go with the *De Sphæra* even though it appears right before the title of the *Computus*.) This page also mentions John Sacrobosco, Robert Grosseteste, and Campanus of Novara, three authors of other *De Sphæras*. (See the notes in the entry discussing the *Phisica, Speculatio* of 1557.)

Francesco Maurolico.   7

COMPVTVS ECCLESI- / ASTICVS. IN SVMMAM COL- / lectus. Et primū d̄ tēporis diuisione.

Antonio Ricardo, México.

Icazbalceta 80. Sabin 96108. Medina, *México*, 86. Wagner 114.

Originals located at: Biblioteca Cervantina, Biblioteca Nacional de Chile, Centro de Estudios de Historia de México Condumex, Hispanic Society of America, Huntington Library, Indiana University, New York Public Library, Universidad de Salamanca.

Microfilm copies are available.

All known copies are bound with the previous two items.

Paging: Folios 28r to 50r immediately following the previous entry (see previous entry for paging details). The title given is a caption title on folio 28r, with the text starting just below. Page titles continue with DE SPHAERA on the recto of each folio and LIBER VNVS on the verso through folio 29 and change to ECCLESIASTICAS on each recto and COMPUTUS on each verso starting with 30. The copy in Indiana is missing leaves 47 and [54], and leaf 46 is numbered on the verso with running titles reversed.

Five early printings of this title are known. The 1591 and 1603 printings are unchanged from that of 1581 in spite of the fact that the Gregorian

---

74. G. Vacca, "Maurolycus, The First Discoverer of the Principle of Mathematical Induction," *Bulletin of the American Mathematical Association* XVI/2 (Nov. 1909): 70–73. Also, W. H. Bussey, "The Origin of Mathematical Induction," *The American Mathematical Monthly* XXIV/5 (May 1917): 199–207.

calendar, with a new rule for calculating the epact (the age of the moon at the beginning of the year), was then in effect.

Year	City	Notes
1575	Venice	Second item in *Opuscula Mathematica* (pages 26–47); see previous entry for holdings
1578	México	Item at hand; see preceding text for holdings
1581	Köln	Bound with *Theoricæ Novæ Planetarum* by Georg von Peurbach (1423–1461), *De Geographia* by Henricus Glareanus (1488–1563), and a preface by Albertus Hero. *Computus* is pp. 124–167. Held by Biblioteca Nacional de España, Biblioteca Nazionale Braidense, Biblioteca Nazionale Centrale di Roma, Biblioteca Universitaria Alessandrina, Biblioteca Vallicelliana, Bibliothèque National de France, Bibliothèque de l'Observatoire de Paris, Brown University, Burndy Library, Carleton College, Library of Congress, Linda Hall Library, New York Public Library, Universidad Complutense de Madrid, Universidade de Coimbra, University of Minnesota, University of Oklahoma, University of Oxford (two copies), University of Texas at Austin, and Yale University; there are at least five more copies in Germany
1591	Köln	Reprint of the same titles as in 1581; Bibliothèque National de France, Bibliothèque de l'Observatoire de Paris, Cornell University, Harvard University, New York Public Library, Princeton University, Swarthmore College, Universidad de Sevilla, Université Louis Pasteur, Universidad de Barcelona, Universidad de Sevilla, University of Glasgow, and University of Oxford, with at least five more copies in Germany
1603	Köln	Reprint of the same titles as in 1581; Albion College, Bayerische Staatsbibliothek, Eidgenössische Technische Hochschule Zürich, University of Michigan

This text treats all things related to the Julian calendar, including epact and golden number. These topics will be covered for the Gregorian calendar in the entry discussing the *Instrucción Náutica* of 1587. For a detailed account of how the Julian calendar evolved, see David Ewing Duncan's *Calendar* or J. L. Heilbron's *The Sun in the Church*. These accounts show how the Roman Church was drawn into astronomy and calendar issues because of the desire to have everyone in the Christian world celebrate Easter on the same day. Duncan writes throughout *Calendar* that *computus* was the term used in the Middle Ages for the intellectual pursuit of time reckoning, with the ultimate purpose of determining the true date of Easter.

The *Computus Ecclesiasticus* became obsolete only seven years after its first printing when the Gregorian calendar was decreed in 1582. For example, on page 35 the year still begins with "Martius."

There is an index at the end (folios 50v–54v) that covers all three titles, which suggests that the printer intended for these items to be bound together.

The authors Francisco de Toledo and Francisco Maurolico may be the only two on our list to have never visited the New World. (This is assuming that we are correct that the 1692 almanac generally attributed to John Partridge should actually be credited to Benjamin Harris; see Part 2.)

1583

Diego García de Palacio. 8

DIALOGOS / MILITARES, / DE LA FORMACION, / è informacion de Personas,Instrumen / tos, y cosas nescessarias para el / buen vso dela Guerra. / Compuesto por el Doctor Diego Garcia de Palacio,del / Consejo de su Magestad, y su Oydor en la / Real Audencia de Mexico. / CON LICENCIA, / En Mexico, en casa de Pedro Ocharte. / Año de 1583.

Pedro Ocharte, México.

Icazbalceta 90. Sabin 58268. Medina, *México*, 98. Wagner 126. Karpinski [2].

Originals located at: Biblioteca Nacional de España, Biblioteca National de México, Escorial, John Carter Brown Library.

This item has been microfilmed several times.

Paging: Four unnumbered leaves followed by 192 leaves, numbered except for 1. (Folio 192 is missing in the John Carter Brown copy.)

A facsimile edition of the copy in Madrid was published in 1944. A facsimile of Book 3 on muskets and artillery was produced in Seville in 1984. A reprint in modern Spanish with 65 pages of introductory material was

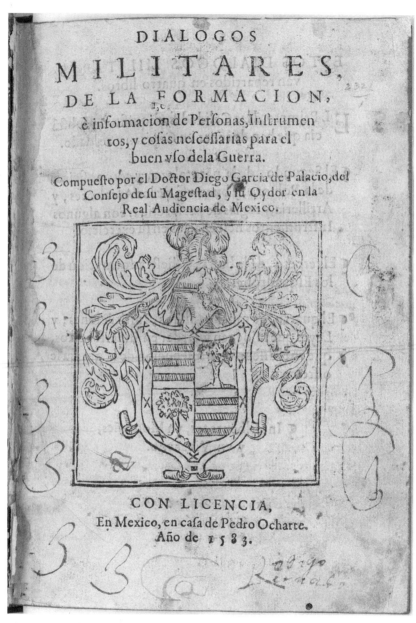

Figure 12. Title page of the *Dialogos Militares,* 1583. The John Carter Brown Library at Brown University, Providence, RI.

done in 2003. Elías Trabulse gives an excerpt in the second volume of his series on the history of science in Mexico.[75]

We note that García refers to his doctorate. Degrees awarded by the university at this time were Bachiller, Licenciado, Maestro, and Doctor. We will see other instances of these degrees on title pages from our list.

The biographical information about Diego García de Palacio, the author of this entry and the next, has been in flux. Manzano Baena says that he was born in the 1530s in Ambrosero, Cantabria, in northern Spain and that this information corrects some earlier sources.[76] He studied at Salamanca and became an administrator. "In April of 1572 . . . he was given the title of *procurador fiscal* and *promotor de justicia* of the *Audiencia* of Guatemala, and two months later, without having departed yet toward the American continent, he was promoted to *Oidor* of the *Audiencia* of Guatemala, where García de Palacio started to work in March of 1574."[77]

When Francis Drake entered the Pacific Ocean and began to raid Spanish holdings on the west coasts of the Americas, García was given a ship and sent to catch the English, but he failed to do so.[78] He was granted a doctorate in law in 1580, and from December 10, 1581, to December 10, 1582, he was rector of the Real y Pontifical Universidad de México (the university that Alonso de la Vera Cruz was involved in founding from 1551 to 1553, ancestor of today's Universidad Nacional Autónoma de México).[79]

---

75. Elías Trabulse's series, *Historia de la Ciencia en México*, will be referred to numerous times in our text. When we are, as here, noting that he includes an excerpt from one of the entries, we will refer to the first three volumes simply as *Historia, Siglo XVI*; *Historia, Siglo XVII*; and *Apéndices e Índices*. Note that this name (e.g., *Siglo XVII*) doesn't always agree with the century that the excerpted work first appeared.

76. Diego García de Palacio, *Diálogos Militares. Estudio Preliminar de Laura Manzano Baena*, Madrid, 2003: 17. Lamb and others had said that García was born in Arce. Ursula Lamb, "Nautical Scientists and Their Clients in Iberia (1508–1624): Science from Imperial Perspective," *Revista da Universidade de Coimbra* XXXII (1985): 56.

77. Ibid.: 18. According to E. G. Squier, in his notes to García's *Carta Dirijida*, a *Real Audencía* was a court of law consisting of a president, four judges (*oidores*), a prosecutor (*fiscal*), and their staff.

78. Squier writes that García wrote a letter to the king about Drake in 1579 but did not get his commission as Captain General until 1587. Both documents are in Seville, according to Squier.

79. Ursula Lamb, "Nautical Scientists and Their Clients in Iberia (1508–1624): Science from Imperial Perspective," *Revista da Universidade de Coimbra* XXXII (1985): 57.

In 1589, a decree imposed on García a large fine and a nine-year suspension from the office of *oidor*.[80] He was charged with seventy-two counts involving the misuse of his office.[81] García de Palacio died in 1595, and his widow and children were left in sufficiently difficult financial straits that they had to be granted assistance from the king.[82]

The list of contents on the reverse of the title page is not completely accurate. A discussion of perspective can be found in Book 3 and a treatment of the formation of squadrons in Book 4, instead of Books 2 and 3, respectively, as indicated on the contents page.

The book is written as a dialogue between two characters, Viscayno and Montañes. The dialogue starts on 7r, with the letters V and M indicating the two speakers. Then, beginning with Question 1 on 9r, Viscayno asks the questions and Montañes answers them. García says (in the dedication of the next entry on our list), "soy Montañes" (I am Montañes), and Bankston translates this as, "I am from Santander," indicating that this was supposed to mean that García was not a flatterer. It may be that García intended a double meaning, saying both that he identified with one of the two characters in the book and that he was from a province where people are known for their truthfulness. Laura Manzano Baena agrees that Montañes is an "alter ego" for the author.[83]

Santander is now the largest city in the autonomous region of Cantabria in Spain. Viscaya is the westernmost of the three subdivisions of the

---

80. Ibid.: 57.

81. Edmundo O'Gorman broke the news of this in his article of 1946, "Nuevos Datos sobre el Dr. Diego García de Palacio," noting that previously Joaquín García Icazbalceta had written that after the Drake incident "history doesn't return to make mention" of García de Palacio. O'Gorman calls García a "great rogue." He includes in his article the entire text of King Philip II's confirmation of García's sentencing. He also reproduces the signature of García de Palacio. Laura Manzano Baena adds, on p. 19 of her biography of García for the 2003 edition of *Dialogos*, that the Viceroy, Álvaro Manrique de Zúñiga, to whom García dedicated his *Instrucción Nautica* (the next entry), was also removed, and this left García without his chief protector.

82. Ursula Lamb (in "Nautical Scientists and Their Clients in Iberia," *Revista da Universidade de Coimbra* XXXII [1985]: 57) says he returned to Spain and died in Santander, but Bankston (in his English translation of the next entry, García's *Instrucción Nautica*) writes that he died in Mexico, based on the record of a funeral for him that occurred on November 15, 1595, in Mexico City. The most recent of García de Palacio's biographers, Laura Manzano Baena, in her introductory material in the 2003 reprint of *Dialogos*, doesn't say either way.

83. Diego García de Palacio, *Diálogos Militares. Estudio Preliminar de Laura Manzano Baena*, Madrid, 2003: 30.

modern-day autonomous region of País Vasco, or Basque Country, and it borders Cantabria. So it is easy to imagine that when García de Palacio was growing up, an encounter between a Montañes and a Viscayno would not have been unheard of.

Folios 135v to 139r include an explanation of the use of an instrument called the *escala altimentria*, which allows trigonometry problems to be solved via similar triangles, circumventing any mention of the trigonometric functions. The instrument is a square of metal $SVQT$. Sides VT and VS are marked off into 12 intervals, which are numbered 1 to 12 in the direction of V. A plumb line hangs from point Q so that a reading can be made on either side VT or side VS.

In an example problem, two points D and C are 285 feet apart on the ground. Sightings are made to a point A using the *escala altimentria*. The readings found on side VT are 8 at point D and 10 at point C. García says to divide these numbers into 12 and then find the difference of the results:

$$\frac{12}{8} - \frac{12}{10} = \frac{3}{10}.$$

Then divide 285 by this (he says multiply but he clearly means we should divide) and get 950 feet as the height of point A above the line DC. In many modern trigonometry textbooks, similar problems are posed, and the solution is that the height h can be expressed in terms of the distance x between the two sightings and the two angles of elevation, $\angle C$ and $\angle D$, by the equation

$$h = \frac{x}{\cot\angle D - \cot\angle C}.$$

The genius of the *escala altimentria* is that it allows one to read the trigonometric functions (actually twelve times the tangent) directly from the instrument without having to find angles and then look up the trigonometric functions in a table.

Propositions of Euclid are mentioned on folio 126v. There are also numerous examples of the use of a combination of arithmetic and geometry in the formation of squadrons. The first example given is how to arrange 81 soldiers in a square. A diagram is given demonstrating that it can be done with nine ranks and nine files. This paradigm of elementary arithmetic, with different numbers and shapes, is repeated extensively throughout the last section, Book 4, of this work. This obsession with the arithmetic of squadrons as a branch of applied military science is not unique to García. Among the works we discuss here, it can be found also in Pasamonte Atanasio Reatón's *Arte Menor* of 1649, Antonio de Heredia y Estupiñán's *Teórica, y Práctica* of 1660, and Benito Fernández de Belo's *Breve Aritmética* of 1675. One wonders whether captains of squadrons really used such books.

Manzano Baena's take on this is that these examples evoked the popular image of the conquistadors Cortés and Pizarro using small contingents of armed men to defeat the native hordes.[84] She says that "the lesser number of men [results in] the greater ease of their management."[85] For a systematic dismantling of this image, see Chapter 3 of Matthew Restall's *Seven Myths of the Spanish Conquest*.

1587

Diego García de Palacio.   9

INSTRVCION / NAVTHICA, PARA EL BVEN / Vso, y regimiento de las Naos, su traça,y / y [sic] gouierno conforme à la altura de Mexico. / Cōpuesta por el Doctor Diego garcia de / Palacio, del Cōsejo de su Magestad, / y su Oydor en la Real audiē- / cia de la dicha Ciudad. / Dirigido, al Excellētissimo Señor Don Aluaro Manrrique,de / çuñiga, Marques de Villa manrrique, Virrey, Gouer- / nador, y Capitan general destos Reynos. / Con licencia, EnMexico, En casa de Pedro / Ocharte. Año de 1587.

Pedro Ocharte, México.

Icazbalceta 97. Sabin 58269. Medina, *México*, 106. Wagner 137. Karpinski [4].

Originals located at: British Library, Biblioteca Cervantina, Escorial, Hispanic Society of America, Huntington Library, John Carter Brown Library, Library of Congress, Museo Naval, New York Public Library, Universidad de Salamanca, Yale University.

This item has been microfilmed several times.

Paging: Four unnumbered leaves, then 156 folios numbered except for 1, 16 to 23, and 78 to 87. The Huntington copy is missing twelve leaves. The Yale copy has leaves 51 and 52 bound between 54 and 55.

A facsimile edition of the copy at the Museo Naval was produced in 1944. (A note in the introduction says that "there are scarcely half a dozen copies in Spain." Since we have only listed three copies in Spanish

84. Ibid.: 52.
85. Ibid.

❊INSTRVCION❊

NAVTHICA, PARA EL BVEN

Vſo, y regimiento de las Naos, ſu traça, y
y gouierno conforme à la altura de Mexico.
Cópueſta por el Doctor Diego garcia de
Palacio, del Cóſejo de ſu Mageſtad,
y ſu Oydor enla Real audié-
cia de la dicha Ciudad.
Dirigido, al Excellétiſsimo Señor Don Aluaro Manrrique, de
çuñiga, Marques de Villa manrrique, Virrey, Gouernador, y Capitan general deſtos Reynos.

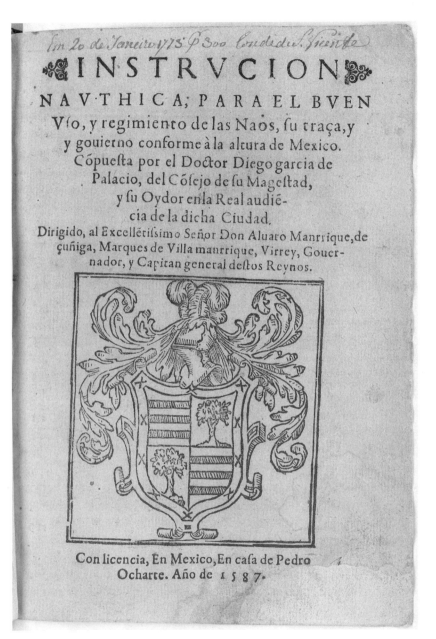

Con licencia, En Mexico, En caſa de Pedro
Ocharte. Año de 1587.

Figure 13. Title page of the *Instrucción Nautica*, 1587. The John Carter Brown Library at Brown University, Providence, RI.

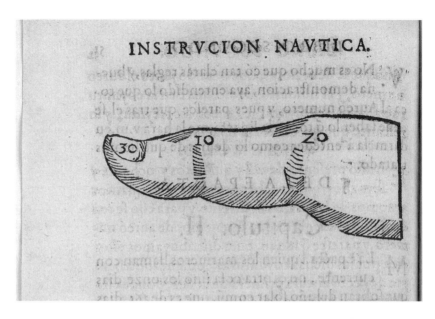

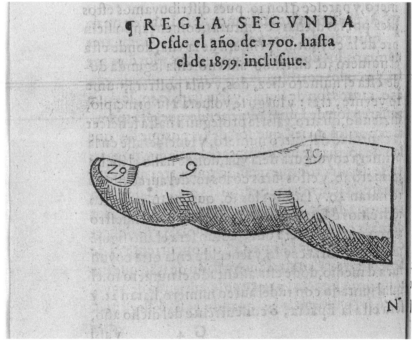

Figure 14. Garcia de Palacio's thumb diagrams for computing the epact from the golden number. Each diagram, after the first, works for the range of years in the heading. The John Carter Brown Library at Brown University, Providence, RI.

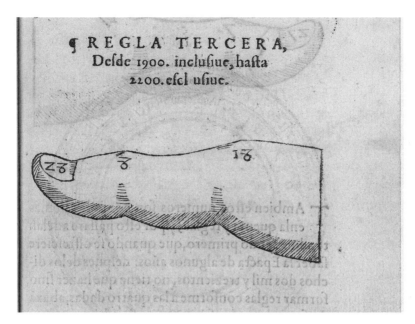

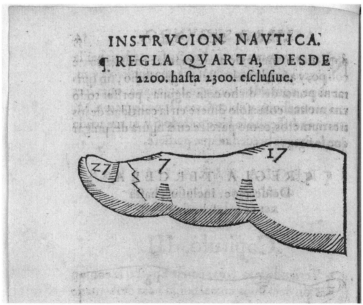

Figure 15. Garcia de Palacio's thumb diagrams for computing the epact from the golden number. Each diagram works for the range of years in the heading. The John Carter Brown Library at Brown University, Providence, RI.

libraries, this would suggest that our list of originals may not yet be complete.) The Museo Naval published a transcription of their copy, with modern spelling, in 1993. This edition has an introduction by Mariano Cuesta Domingo, with a very thorough bibliography. In 1998 the complete text was included in a compilation of thirty-two texts on navigation, compiled by González-Aller Hierro, and released on disk. An English translation by J. Bankston, with extensive notes by the translator, has had two editions, in 1986 and 1988. There is an excerpt included in Elías Trabulse's *Historia de la Ciencia en México, Estudios y Textos, Siglo XVII*. A book on nautical vocabulary was printed in 1722 in Sevilla under the title *Vocabulario marítimo, y explicación de los vocablos, que usa la gente de mar, en su exercicio del arte de marear*, and a copy of this is held at the University of Minnesota. It was attributed to García de Palacio by Martín Fernádez de Navarette and may be a reprint of the last section of the *Instrucción Náutica*.

See the preceding entry for some information about the author. This book follows the format of the *Dialogos Militares*, that of a dialogue between one Viscayno and one Montañes. Again, it is Viscayno that asks the questions and Montañes who answers them.

This title is a thorough exposition of sixteenth-century Spanish seafaring knowledge. The glossary of nautical terms at the end may have been particularly important for generating interest in this work among historians over the years. Bankston writes (in his English translation) that a modern shipwright should be able to build and outfit a period Spanish ship using García's descriptions alone. Ursula Lamb writes that the *Instrucción* was a "famous and popular text."[86]

It is appropriate here to mention an earlier work, *Ytinerario de Navegaçion de los mares y tierras oçidentales* by Juan de Escalante de Mendoza, printed in Seville in 1575. A manuscript copy (held at the Biblioteca Nacional de España, with the maps, according to Lamb, held by the Museo Naval) has copious corrections[87] as if it is in preparation for printing. It is in the form of a dialogue between "Tristan"[88] and "Piloto." The printed version is included on the disk of thirty-two maritime texts mentioned earlier, and

---

86. Ursula Lamb, "Cosmographers of Seville: Nautical Science and Social Experience," in *Cosmographers and Pilots of the Spanish Maritime Empire*, Aldershot, 1995: 682.

87. The original title, for instance, is *Libro Nombrado Regimiento de la Navegaçion de las Indias Oçidentales*.

88. Originally "Maestro."

the manuscript is the subject of a book produced by the Museo Naval in Madrid. The manuscript is presumed to have been written while the author traveled in the New World, even though it was later printed in Europe. Some of the same topics discussed in the following are present. Could García have been aware of this work? Julio Guillén suggests this in his introductions to the 1944 facsimile editions of García's two books.

A later manuscript that belongs to the same family is the *Arte de Navegar* by Juan Gallo de Miranda. Trabulse (*Historia de la Ciencia en México, Siglo XVI*) writes that in this Mexican manuscript of 1621, Gallo de Miranda attempts to solve the longitude problem using eclipses.

There are in *Instrucción Nautica* a number of instances of mathematics that are worthy of comment. At the beginning of the section on finding the golden number for a year, there is an example given with the division worked out. Bankston, in his first edition of his English translation says, "Also it should be noted that the translator, who in no way pretends to be so much as a novice in the science of mathematics, has been unable to work out the author's method of division, but neither has an accomplished mathematician, who shall go unnamed, now instructing in graduate mathematics at one of the better institutions of higher learning in the East." By the time of his second edition, however, he had found a collaborator who undertook the explanation. Bankston includes as his Appendix IV "The Galley Method of Division" by James E. Kelly, Jr. Since Bankston's original note suggests that not everyone interested in García de Palacio's works is familiar with this algorithm, a quick explanation is appropriate here.

The problem is to divide 1588 by 19 and find the remainder. García gives

$$\begin{array}{r} 31 \\ 761 \\ 1588 \quad |\underline{83} \\ 199 \\ 1 \end{array}$$

and says the problem is done. In view of the subsequent explanation, we suggest that the intended arrangement of numbers was

$$\begin{array}{r} 1 \\ 3 \\ 761 \\ 1588 \quad |\underline{83} \\ 199 \\ 1 \end{array}$$

and that perhaps the changes were made by the printer. So one first writes down 1588 and writes 19 below it, with the 1 of 19 below the 5 in 1588. Eight 190s can go into 1588, so we write 8 in the quotient box and take 8 times one away from 15, leaving 7. At this point we have

$$\begin{array}{r} 7 \\ 1\cancel{5}88 \quad |\underline{8} \\ 19 \end{array}$$

where we have crossed out the 15 since we do not need it any more. Then, 8 times 9 subtracted from 78 leaves 6. Thus

$$\begin{array}{r} \cancel{7}6 \\ 1\cancel{5}88 \quad |\underline{8} \\ 19 \end{array}$$

and one continues by writing another 19 below and finding that three 19s can go into 68. Three 1s from 6 leaves 3, and then three 9s are subtracted from 38. So the final array with used numbers crossed out is

$$\begin{array}{r} 1 \\ \cancel{3} \\ \cancel{7}61 \\ 1\cancel{5}88 \quad |\underline{83} \\ 19\cancel{9} \\ 1 \end{array}$$

and we see that the remainder is 11, the golden number for 1587. (See below for the explanation of the golden number of a year.)

If one is navigating close to shore, it might be useful to know whether the tide is in or out. There is herein a discussion of the epact as a way to calculate the phases of the moon, later motivated by its use in a treatment of the problem of predicting when the high and low tides will be. The overall outline of the discussion is that one finds the golden number to get the epact, the epact then lets one find the age of the moon, this gives the position of the moon, and that leads to the tides.[89]

The chief use of the epact in other contexts is to find the date of Easter. This issue is obscure enough that it will be prudent to review here the history of this calculation.

---

89. The high tide, or *pleamar*, is said to come when the moon is in the Southwest or Northeast, and the low tide, or *baxamar*, comes when the moon is in the Southeast or Northwest. We assume that this rule works for some part of the Gulf of Mexico, but not necessarily around the world.

The Council of Nicæa decreed in 325 that Easter should be celebrated on the Sunday following the first full moon after the vernal equinox. This was an attempt to reconcile the different Easter customs of various Christian communities. It left open the possibility, however, that people might disagree about when the vernal equinox was and when exactly the full moon was.

In 525 Dionysius Exiguus proposed a system that would make astronomical observations unnecessary. The equinox was assumed to fall always on the twenty-first of March, and the phases of the moon were approximated by noting that nineteen years was very close to a whole number of lunations. (A lunation is the time for the moon to run through all of its phases.) In principle, under this system, all communities in all regions would agree on the same date for Easter.

Centuries later, a new problem arose. The Julian calendar, with its leap day every fourth year, implied a time between vernal equinoxes of $365\frac{1}{4}$ days. That was just enough off from the true value that the vernal equinox was creeping earlier and earlier in the calendar year. People could see that if the trend continued, in a great many centuries Christmas would be in the spring and Easter would, under Dionysius's computation, fall in the summer.

In 1582 the Gregorian calendar, because of the work of Clavius and Lilius, was introduced to solve this problem. First of all, the calendar skipped ten days to make up for past drift. Thursday, October 4, 1582, was followed by Friday, October 15. After that, any year divisible by 100 but not divisible by 400 would not be a leap year. So 1700, 1800, and 1900 were not leap years, but 1600 and 2000 were. Finally, corrections were made to the epact to take into account better data for the length of a mean lunation.

Epact is generally defined to be the age of the moon (the number of days since the new moon) at the beginning of the calendar year.[90] It will be convenient here to use a mathematical scheme known as modular arithmetic to discuss the method of finding the epact, although the reader must keep in mind that it was not available to Dionysius Exiguus, Pope Gregory XIII, or Diego García de Palacio. In modern times, modular arithmetic is a cornerstone of number theory and has applications to coding and computing. In modular arithmetic, one has a whole number, called the modulus, and

---

90. See, for example, Åke Wallenquist, *Dictionary of Astronomical Terms*, Garden City, NY, 1966.

one reduces all numbers to whatever remainders they have after whole number division by this modulus. For example we can say that

$$3 + 4 \equiv 2 \pmod{5}$$

since 7 reduces to the remainder 2 after division by 5.

The Greek astronomer Meton is credited with the observation that nineteen years is almost exactly the same as 235 lunations. Thus, the nineteen-year cycle of the moon's behavior is called the Metonic cycle. The epact as implemented by Dionysius exploits the Metonic cycle. His epact increases by 11 every year (because when twelve $29\frac{1}{2}$-day lunations are taken away from 365 days, the remainder is 11), but when that produces a number over 30, that number is reduced by 30. Then, at the end of every nineteen years, the cycle of epact numbers starts over.

So, to implement the method of Dionysius, we find the so-called golden number[91] by taking the year plus 1 modulo 19. Take this answer to be in the set $\{1, 2, 3, \cdots, 19\}$. (Since zero as a counting number was not yet a concept in western culture at the time of Dionysius, a remainder of 0 was seen as a golden number of 19.) Then the epact is given for the Julian calendar by

$$\text{Epact} \equiv 11 \times \text{Golden Number} \pmod{30}$$

taking the answer in the set $\{1, 2, 3, \cdots, 30\}$.

When the Gregorian calendar took effect, the shift of ten days was reflected by a similar shift in the epact, so the Gregorian formula was

$$\text{Epact} \equiv 11 \times \text{Golden Number} - 10 \pmod{30}.$$

What's more, two other small corrections were introduced in the Gregorian epact. On the one hand, in every century year that was not a leap year, the epact was reduced by 1 from that year on. On the other, because of the fact that nineteen years is not exactly 235 lunations, a correction is needed that increases the epact by 1 in certain century years from that year on.

In the *Instrucción Nautica*, to keep the calculations as simple as possible, there is a description (with woodcut diagrams) of how to use three points

---

91. Laura Delbrugge, in the introduction to Andres de Li's *Reportorio*, says that this term has its origins in the custom of writing the number in gold ink on early calendars.

on one's thumb to do arithmetic modulo 30. García de Palacio's method for finding the epact (starting on folio 51v) starts as follows. Having found the golden number (as in the example we have discussed), García draws the numbers 10, 20, 30 on the thumb and invites the reader to count off the golden number, starting with the 30 and cycling through 10 and 20 to get back to 30. The number we stop on when we have counted off the golden number is the number we need.[92] This amounts to finding the golden number minus 1 modulus 3 and multiplying that by ten, taking this answer to be in the set {10, 20, 30}. Take this result and add it to the golden number modulus 30 to get the epact in the set {1, 2, 3, · · ·, 30}. One can see that this is equivalent to

$$\text{Epact} \equiv 11 \times \text{Golden Number} - 10 \pmod{30}$$

but García de Palacio is trying to appeal to the reader who might find even multiplication by eleven a challenge, and so he keeps the arithmetic as elementary as possible. This formula works for the years from the establishment of the Gregorian calendar to 1699. He has us correct the epact in later centuries by subtracting 1 for 1700–1899, 2 for 1900–2199, and 3 for 2200–2299. He accomplishes this by drawing three new "thumb diagrams" using the sets {9, 19, 29}, {8, 18, 28}, and {7, 17, 27}, respectively.

The rule for the epact is presented here as covering any year from the time of the Gregorian reform of the calendar (1582) up to 2299. Sure enough, if one follows García de Palacio's instructions for May of 2001, one calculates that the new moon falls on May 22, which is correct. Suffice it to say that García de Palacio has incorporated both the non-leap century year correction and the corrections to the Metonic cycle into his algorithm.[93]

---

92. We see this counting on the thumb also in Jerónimo de Chávez's almanac of 1572, although in that year the Julian calendar was still in use and so there was a ten day difference in the epact. Chávez has his reader start the count at the 10 and run toward 20 and 30.

93. If he were just compensating for the non-leap century years the directions would be to subtract 1 for 1700–1799, 2 for 1800–1899, 3 for 1900 to 2099, 4 for 2100–2199, and 5 for 2200–2299. However the correction to the Metonic cycle adds 1 in the 19th century and beyond, and again this will happen in the twenty-second century. This has the effect of gluing some intervals of years together in the correction to the epact. If we look beyond the years covered by García de Palacio for a bit, it is interesting to note that while the total epact correction for 2300–2399 will be 4, the year 2400 is due for a Metonic cycle correction but not a leap year correction, and so the total correction will go back to 3 for 2400–2499.

Since the Gregorian calendar had only been in place for a few years, García must have been among the first authors to explain the new system for finding the epact.[94]

Many sources discuss epact at greater length. The historical side is treated without excessive mathematics in *Calendar* by David Ewing Duncan. See also J. L. Heilbron's *The Sun in the Church* for another point of view. *Calendrical Calculations* by Nachum Dershowitz and Edward Reingold gives a computer algorithm approach to how epacts were calculated under the Julian calendar and how the method was modified for the Gregorian calendar.[95] The article "Calendar" in early editions of the *Encyclopedia Britannica* is also a very concise and highly algorithmic pre-computer treatment of epact, Easter, and more. And one of the earlier entries here, discussing the *Computus Ecclesiasticus* of 1578, also explains this for the Julian calendar.

One of the most interesting statements in Duncan's book is the following: "The history of science in the Middle Ages would have been very different if the bishops at Nicaea had decided simply to name a fixed date for Easter in the solar calendar. But they did not. Instead, in the wake of Nicaea, Christians developed what became a complex equation to determine the proper day, forcing time reckoners to return to something Caesar had dispensed with centuries earlier: a dependence on the moon."[96]

Here we will focus on García de Palacio's intent, which is to predict the phases of the moon so that he can predict the tides. But there will be more to say about epact and Easter in Part 2 in one of the entries for 1663.

We should stop and ask what this epact really measures. In a year with epact 30, for instance, García predicts that the first new moon will fall on January 29. (One takes the epact of 30, plus 1 for January, and subtracts this from 60 to get the date of the new moon.) Taking noon on the 29th as an average and regressing 29.5 days, we see the previous new moon falling in the middle of the night a full day before the beginning of the year. So an

---

94. According to Lamb, Alonso de Chávez corrected the 1584 edition of his son's *Chronographía o Repertorio de los Tiempos* to incorporate the Gregorian calendar.

95. Be advised, however, that the Gregorian-shifted epact defined on page 54 of *Calendrical Calculations* is the traditional Gregorian epact plus 6. Since Nachum Dershowitz and Edward Reingold are using the epact only as a way to calculate Easter, their epact is the number that one subtracts from April 19 to get a full moon.

96. David Ewing Duncan, *Calendar, Humanity's Epic Struggle to Determine a True and Accurate Year*, New York, 1999: 72.

epact of 30, i.e., 0, means that the moon is already a full day old when one is celebrating the New Year at the stroke of midnight. We will return to this point later.

The epact rule incorporates some approximations that should introduce small errors. For instance, using 29.5 days as the length of a lunation, one sees that twelve lunations are 354 days. A non-leap year is eleven days longer, so García de Palacio tells us that the epact advances eleven days every year. This ignores leap years, so his epact should acquire an error of one day every four years. But a mean lunation is actually almost three quarters of an hour longer than 29.5 days. This means that the year-to-year drift of eleven days is about nine hours too long. Over four years, this makes the epact too high by a day and a half, cancelling the previous error and then some.

Moreover, the rule has us doing arithmetic mod 30 instead of mod 29.5, and so this adds on an error of half a day every three years or so, approximately cancelling the net effect of the two errors above. Another inconsistency is that a year of the form $19n - 1$ gets an epact of 19, so the next year should get a 30. But years of the form $19n$ get an epact of 1, a discrepancy of one day every nineteen years.

A way of rationalizing these observations is provided by the article "Calendar" in early editions of the Encyclopedia Britannica. One can see the article online at www.1911encyclopedia.org/Calendar, although the tables and formulas are not reproduced well there.[97] Imagine a lunar calendar with months that alternate between twenty-nine and thirty days. Every nineteen years, add seven extra months to make a total of 235, the number of lunations in the Metonic cycle. One of these extra months goes at the very end of the cycle and has twenty-nine days. The other six have thirty days each, for a total of 6,935 days in nineteen years. The authors of the article do not specify where these six months go, but to get the answer we want, we will need calendar years with epacts in the range of 20 to 30 to either have a month of thirty days as their last full month or to get an extra month of thirty days within them, so we might as well put the six extra months into those positions. Finally, add one day to any lunar month that would contain February 29 on the civil calendar. This yields a total of $6,939 \frac{3}{4}$ days, on average, which is exactly nineteen Julian years.

---

97. This Web site says it is "based on" the 1911 edition of Britannica. By the way, in later editions of Britannica, the article on the calendar has been greatly simplified and the information we use here is not present.

The authors call this the ecclesiastical calendar. One can check that if the first day of the Metonic cycle (i.e., the first day of a year with golden number 1) is day one of a lunar month, then the epact formula will give the day of the lunar month that January 1 falls on each year (at least until the first correction to the epact is made in 1700).[98] This fact puts into perspective the inconsistencies we have discussed above. The missing leap days, the use of mod 30 arithmetic, and the jump of 1 at the end of the cycle are all put into a context where they make more sense. The inconsistencies don't go away, however. They still produce errors, not only with respect to what the actual moon is doing, but even with respect to a hypothetical lunar cycle that is exactly $\frac{1}{235}$ of nineteen Julian years.

Let's explore how these various errors sort themselves out over a period of seventy-six years sometime in the sixteenth and seventeenth centuries. We will assume, for the sake of argument, that all lunations last the same period of time, and that one Metonic cycle of nineteen Julian years of 365.25 days each is exactly 235 lunations.[99] Then one lunation is about twenty-nine days, 12 hours, 44 minutes, 25.5 seconds,[100] and the shortfall of twelve lunations from 365 days is about 10 days, 15 hours, 6 minutes, 53.5 seconds.[101] Let's start with a leap year that is divisible by 19 (say 1596) and assume that in the previous year the true epact was exactly 19.

In the following table, the first column is the golden number (aureo numero, or A. N.) for the year. The second column is the epact as calculated

---

98. 1700 is not a leap year in the Gregorian calendar, so if an extra day is not added somewhere the Metonic cycle lasting from 1691 to 1709 would be one day short. The rule that tells us to subtract 1 from the epact for 1700 is equivalent to adding a day to one of the lunar months wholly within 1699. A funny thing happens in 1710 when the epact 30 occurs for the first time ever. This is problematic because the nineteen-year cycle is supposed to end with a lunar month of twenty-nine days. We do not know what interpretation this was given in the ecclesiastical calendar.

99. This last assumption is justified by saying that we will correct for the Gregorian calendar and for the error in the Metonic cycle later, by subtracting something from the epact for later centuries.

100. The actual mean length of one lunation is 29 days, 12 hours, 44 minutes, and 2.8 seconds.

101. This is the amount by which the epact should advance from a non-leap year to the next year. The average amount of advance under these assumptions would therefore be 10 days, 21 hours, 6 minutes, 53.5 seconds, where we have added 6 hours per year for the one leap day that occurs every four years. The actual value for one tropical year minus twelve mean lunations is 10 days, 21 hours, and 12 seconds.

in *Instrucción Náutica*. The other columns are the hypothetical epact (subject to the assumptions already given) in days, hours, and minutes for four cycles of 19 years, calculated by successively adding 10 days 15 hours 6 minutes 53.5 seconds to the number above, adding one more day if the number given is for a leap year, and throwing away 29 days 12 hours 44 minutes 25.5 seconds whenever we surpass that quantity, with the first entry in the second column following the last entry in the first column and so on. Leap years are indicated by asterisks. The last column is the average of the four before it.

A. N.	Epact	D H:M	D H:M	D H:M	D H:M	Average
1	1	0 02:22*	0 08:22	0 14:22	0 20:22	0 11:22
2	12	11 17:29	10 23:29*	11 05:29	11 11:29	11 08:29
3	23	22 08:36	22 14:36	21 20:36*	22 02:36	22 05:36
4	4	3 10:59	3 16:59	3 22:59	3 04:59*	3 13:59
5	15	14 02:06*	14 08:06	14 14:06	14 20:06	14 11:06
6	26	25 17:13	24 23:13*	25 05:13	25 11:13	25 08:13
7	7	6 19:35	7 01:35	6 07:35*	6 13:35	6 16:35
8	18	17 10:42	17 16:42	17 22:42	17 04:42*	17 13:42
9	29	28 01:49*	28 07:49	28 13:49	28 19:49	28 10:49
10	10	10 04:11	9 10:11*	9 16:11	9 22:11	9 19:11
11	21	20 19:18	21 01:18	20 07:18*	20 13:18	20 16:18
12	2	1 21:41	2 03:41	2 09:41	1 15:41*	2 00:41
13	13	12 12:47*	12 18:47	13 00:47	13 06:47	12 21:47
14	24	24 03:54	23 09:54*	23 15:54	23 21:54	23 18:54
15	5	5 06:17	5 12:17	4 18:17*	5 00:17	5 03:17
16	16	15 21:24	16 03:24	16 09:24	15 15:24*	16 00:24
17	27	26 12:31*	26 18:31	27 00:31	27 06:31	26 21:31
18	8	8 14:53	7 20:53*	8 02:53	8 08:53	8 05:53
19	19	19 06:00	19 12:00	18 18:00*	19 00:00	19 03:00

This table is not meant to reflect what the actual moon is doing, but only to show how the small errors mentioned work themselves out over a cycle of 76 years (the four-year leap cycle times the nineteen-year Metonic cycle). By hypothesis, the numbers were set to be exact in years 0 and 76. In between, we see a maximum error in year 41 of 1 day, 3 hours, and

24 minutes. This epact algorithm appears to represent a reasonable compromise between accuracy and ease of computation.

The epacts as presented by García de Palacio should be compared to the actual motions of the moon. We will call these epacts Gregorian epacts below to distinguish them from the true epact, i.e., the actual age of the moon at the new year. At sunearth.gsfc.nasa.gov/eclipse/phase/phasecat.html the phases of the moon are listed for a 5,000-year span, courtesy of Fred Espenak. Using these dates, one can calculate an epact for a year as the time, at midnight on the morning of January 1, since the last new moon of the previous year. This has been done in the next table, adding 6 hours to the figures to account for the difference between Greenwich Mean Time and the time in Mexico City (Central Time in the United States). Gregorian epacts are lower by 2 than in the previous table since we are using the twentieth and twenty-first centuries. In the case where this produces a 30, a 0 is used to make comparison with the true epact easier.

Years	Epact	D H:M	D H:M	D H:M	D H:M
1976, 1995, 2014, 2033	29	29 05:10*	29 06:06	29 05:39	29 09:08
1977, 1996, 2015, 2034	10	11 03:52	10 03:38*	10 04:25	10 11:14
1978, 1997, 2016, 2035	21	21 12:28	21 13:05	20 19:31*	21 09:46
1979, 1998, 2017, 2036	2	2 10:25	2 13:04	2 23:07	2 15:29*
1980, 1999, 2018, 2037	13	12 21:37*	13 07:17	13 23:30	14 14:26
1981, 2000, 2019, 2038	24	24 15:26	24 07:28*	24 22:40	25 06:21
1982, 2001, 2020, 2039	5	5 19:50	6 12:37	6 00:46*	6 04:57
1983, 2002, 2021, 2040	16	16 20:41	17 09:12	17 13:43	16 13:29*
1984, 2003, 2022, 2041	27	27 17:34*	27 22:25	27 22:17	27 22:29
1985, 2004, 2023, 2042	8	9 18:12	8 20:16*	8 19:44	8 21:55
1986, 2005, 2024, 2043	19	20 05:05	20 04:32	19 06:29*	19 15:31
1987, 2006, 2025, 2044	0	1 02:50	1 02:49	1 07:34	0 20:12*
1988, 2007, 2026, 2045	11	11 11:35*	11 16:00	12 04:17	12 21:07
1989, 2008, 2027, 2046	22	23 00:25	22 12:20*	23 05:09	23 18:19
1990, 2009, 2028, 2047	3	4 02:41	4 17:22	4 09:48*	4 19:20
1991, 2010, 2029, 2048	14	15 01:38	15 17:58	16 03:54	15 06:22*
1992, 2011, 2030, 2049	25	26 02:04*	26 12:24	26 15:08	26 14:31
1993, 2012, 2031, 2050	6	8 05:17	7 11:53*	7 12:28	7 12:10
1994, 2013, 2032, 2051	17	18 20:33	18 21:19	17 20:56*	18 00:44

Note that in this table, in which the numbers come from the actual behavior of the moon, we see a significant variation from the advance of 10 days (11 for leap years), 15 hours, and 7 minutes that was used to generate the last table. This means that the length of one lunation varies enough from 29 days, 12 hours, and 44 minutes that twelve consecutive lunations can be several hours shorter or longer than the average value we used earlier.[102]

As stated, we believe that an epact of 0 should mean that the age of the moon at midnight on the morning of the first day of January is about 1, and this is approximately what we see in the row for 1987, 2006, 2025, and 2044. So if we compare the calculated true epacts to Gregorian epacts plus 1, the maximum error in Gregorian epacts is 1 day, 5 hours, and 17 minutes (in 1993). The maximum by which the true epacts precede Gregorian epact plus 1 is 1 day, 4 hours, and 29 minutes in 2016. (García de Palacio himself says that the error in calculating the date of the new moon by his method can never be more than a day.) Gregorian epacts with 1 added do appear to be the best whole-number approximations to the numbers in the four columns, taken from the actual behavior of the moon. We note that Dershowitz and Reingold state on page 53 of *Calendrical Calculations* that "A one-day bias was deliberately introduced in [the Gregorian epact's] initial sixteenth-century value ... to minimize coincidences of Easter and Passover ... as often as possible"; this may explain what we are seeing here.

After discussing the epact, García de Palacio describes how to find the age of the moon. One adds the epact, the day of the month, and a number for the month as follows: January = 1, February = 2, March = 1, April = 2, May = 3, June = 4, July = 5, August = 6, September = 7, October = 8, November = 9, and December = 10. If this number is greater than 60, one subtracts 59. Otherwise, if the number is greater than 30, subtract 29 for months with 30 days and subtract 30 for months with 31 days. (He leaves open what to do in February.) The result is the age of the moon.

This method overcompensates for the shortfall of 1 that we have noted. For on January 1 we should add to the epact 1 for the month and 1 for the date. This means that for a year with an epact of 30, the age of the moon on January 1 is 2. To be generous, we might allow that García de Palacio is saying that an age of 2 means that the moon is in its second day on January

---

102. According to Nachum Dershowitz and Edward Reingold, the difference between mean and true times of the new moon can be as much as the neighborhood of 14 hours (*Calendrical Calculations*, Cambridge, 1997: 152).

MATHEMATICAL WORKS PRINTED IN THE AMERICAS, 1554–1700

1. We note also that this discrepancy is not limited to García. In the *Cambridge Ephemeris* for 1687 by William Williams, listed in Part 2, there is a table of the moon's age for every day of the year. There, as here, the epact plus 2 gives the age of the moon on January 1. (This is subject to the fact that Williams's calendar runs from March to February. So look at the next year, 1688, and find the epact, 7, and compare it with the age of the moon, 9, listed by Williams for January 1, 1687/1688.) So it may be universal that the epact is off by 2 from the age of the moon at the beginning of the year, in spite of its definition.

This book also contains a primitive table of secants and tangents, in the form of the question, "how many leagues does one travel if one moves one degree of longitude at various angles to the latitude lines?" Seven angles are given and are referred to as the "Seven Quarters." In the following table the first column is the angle made to the latitude line based on the assumption that García de Palacio means to divide ninety degrees into eight equal parts. The next three columns are taken from the text. The second column appears to be the secant of the angle times $17\frac{1}{2}$ (the distance in leagues that one travels at that heading if the length of one degree of longitude is $17\frac{1}{2}$ leagues), the third column likewise appears to be $17\frac{1}{2}$ times the tangent, and the fourth column is just the tangent (with the number after the dot representing so many sixtieths—a descendent of the Babylonian system of fractions).

78.75°	$98\frac{1}{6}$	$87\frac{1}{2}$	5
67.5°	$45\frac{3}{5}$	42	2·24
56.25°	$13\frac{1}{2}$	$26\frac{1}{4}$	1·30
45°	$24\frac{3}{4}$	$17\frac{1}{2}$	1
33.75°	21	$11\frac{2}{3}$	·40
22.5°	19	$7\frac{1}{3}$	·25
11.25°	$17\frac{5}{6}$	$3\frac{1}{2}$	·12
0°	$17\frac{1}{2}$		

Some of these numbers are remarkably accurate. (The $13\frac{1}{2}$ in the third row must be a printer's mistake—$31\frac{1}{2}$ is better, and this correction has been written in the Salamanca copy.) There may be more going on here than a mere rounding off of the true values. It is well known to students of trigonometry that a triangle with sides 5, 12, and 13 is a right triangle; less well known is that the acute angles in this triangle are fairly close to 22.5° and 67.5° (they are about 22.6199° and 67.3801°, or 22°37'12" and

67°22′48″, respectively). If we use this triangle to estimate tan 22.5° and tan 67.5°, we come up with $\frac{5}{12}$ and $\frac{12}{5}$, which are exactly the $\frac{25}{60}$ and $2\frac{24}{60}$ that we see in García's table.

If we then bisect the angle whose tangent is $\frac{5}{12}$, we get exactly the smaller acute angle in the triangle with sides 1, 5, $\sqrt{26}$. This can be verified by using the half-angle formulas from trigonometry.

$$\tan\left(\frac{1}{2}\tan^{-1}\frac{5}{12}\right) = \frac{1 - \cos(\tan^{-1}\frac{5}{12})}{\sin(\tan^{-1}\frac{5}{12})}$$

$$= \frac{1 - \frac{12}{13}}{\frac{5}{13}}$$

$$= \frac{13 - 12}{5}$$

$$= \frac{1}{5}$$

This angle (about 11°18′36″) and its complement (about 78°41′24″) may be used to estimate tan 11.25° and tan 78.75°. The answers are $\frac{1}{5}$ and 5, which again match the table in the text exactly.

Finally, taking half the angle whose tangent is $\frac{12}{5}$, we get exactly the smaller of the acute angles in the triangle with sides 2, 3, $\sqrt{13}$. This angle and its complement (about 33°41′24″ and 56°18′36″, respectively) can be used to estimate tan 33.75° and tan 56.25°. The answers are $\frac{2}{3}$ and $\frac{3}{2}$, once again perfectly matching the table. So, in the six instances in the last column of the table in which the answer for the tangent is approximate, the answers coincide exactly with the answers from three triangles that approximate the correct angles, and these triangles are related to each other via the halving or doubling of angles. Whether García de Palacio realized this or was just rounding is food for speculation.

Moving left, the second to last column is exactly $17\frac{1}{2}$ times the last column, except for the entry $7\frac{1}{3}$, which should be $7\frac{7}{24}$. Results for García's first column, for $17\frac{1}{2}$ times the secant, are less consistent. One of the best numbers there is the entry for 45°: $24\frac{3}{4}$. The exact value of sec 45° is $\sqrt{2}$. The approximation of $24\frac{3}{4}$ for $17.5\sqrt{2}$ works because $\frac{99}{70}$ is a very good approximation to $\sqrt{2}$.

If we use the 5, 12, 13 triangle to estimate 17.5sec 22.5° and 17.5sec 67.5° we get $18\frac{23}{24}$ and $45\frac{1}{2}$, respectively. Rounding the $18\frac{23}{24}$ to 19 is very reasonable, but the $45\frac{3}{5}$ in the table is inexplicable until we realize that $45\frac{3}{5}/19 = 12/5$, i.e., the 19 and the $45\frac{3}{5}$ in the table, represent an increase by exactly the same factor $[\frac{456}{455}]$ over the values from the triangle, $18\frac{23}{24}$ and $45\frac{1}{2}$. It is true that $45\frac{3}{5}$ is closer to the true answer (17.5sec 67.5° is about 45.7297) but it is not that much closer. $45\frac{3}{4}$ would be a very good approximation.

Let's use the triangle with sides 2, 3, and $\sqrt{13}$ to estimate 17.5sec 33.75° and 17.5sec 56.25°. Calculating $\sqrt{13}$ gives us about 3.60555. Using 3.6 as an approximation we get exactly the 21 and the $31\frac{1}{2}$ that we see in the table (subject to interchanging the digits in the $13\frac{1}{2}$ as noted).

Lastly, we use the triangle with sides 1, 5, and $\sqrt{26}$ to estimate 17.5sec 11.25° and 17.5sec 78.75°. Calculation of $\sqrt{26}$ gives us about 5.09902. Using 5.1 as an approximation we get $17\frac{17}{20}$ and $89\frac{1}{4}$, respectively. In the first case, lowering the value to $17\frac{5}{6}$ is only a tiny change, and the true value is in between (17.5sec 11.25° is about 17.8428). In the second case lowering the value to $89\frac{1}{6}$ is very hard to understand when the true value of 17.5sec 78.75° is closer to 89.7020. The only explanation we can see is that $89\frac{1}{6} = 5 \times 17\frac{5}{6}$. The 1 to 5 ratio between the two sides has been exactly preserved even though they are both a little off. These numbers imply $5\frac{2}{21}$ as an approximation to $\sqrt{26}$.

There are tables of the declination of the sun on folios 15v to 23r for the four-year cycle (three ordinary years and a leap year), and tables of the full and new moons on folios 78r to 87r for the years 1586 through 1604. The latter table indicates, for each year, the solar cycle, the golden number, the Roman indiction, Ash Wednesday, Ascension, Corpus Christi, the dominical letter, the epact, Septuagesima, Easter, Pentecost, and Advent.

There are explanations of the use of various instruments, including a discussion of the astrolabe in Chapter 4 of Book 1.

There is a discussion of the variation of the compass needle from true north. Bankston seems to be overdoing the interpretive side of his role as a translator when he suggests in a footnote that García de Palacio believed that the variation of the compass needle could be used to determine longitude at sea.[103] The passage he offers as evidence follows a discussion of

---

103. According to Ursula Lamb, the Spanish crown proposed a prize in 1582 for the solution to the problem of how to calculate longitude at sea ("Nautical Scientists and Their Clients in Iberia (1508–1624)," *Revista da Universidade de Coimbra* XXXII [1985]: 50). Charles Cotter says 1598 (*A History of Nautical Astronomy*, London, 1968). Both dates may be right. For more about the problem, and its eventual solution by an English clockmaker, see the book *Longitude* by Dava Sobel.

how the variation should be recorded as one travels about so that when one returns to the same places it may be corrected for. We might translate the footnoted passage as, "For this reason you will see the distance and separation [from true north?] that is made in navigation—navigating accordingly the needles show it." It is far from clear that García is talking about longitude.

1597

Joán de Belveder. **10**

LIBRO GENERAL / DELAS REDVCIONES / DE PLATA, Y ORO DE DIFERENTES / leyes y pesos,de menor à mayor cantidad,y de sus / interesses à tanto por ciento , con otras re- / glas, y auisos muy necessarios para / estos Reynos del Piru. / COMPVESTO POR IOAN DE BELVEDER , NATV- / ral dela villa de Tahuste,enel Reyno de Aragon. / *DIRIGIDO AL DOCTOR IOAN RUYZ* / *de Prado,Inquisidor Apostolico destos Reynos del Piru &c.* / CON PRIVILEGIO. / En Lima por Antonio Ricardo. Año de / M. D. XCVII.

Antonio Ricardo, Lima.

Medina, *Lima*, 11. Karpinski [5]. Vargas Ugarte 17.

Originals located at Biblioteca Nacional de Chile, Biblioteca Nacional de España, Biblioteca Nacional del Perú, Duke University, Indiana University, New York Public Library, Real Biblioteca, Yale University.

Microfilm copies are available.

Paging: Nine preliminary leaves followed by 198 mostly numbered folios. Of the first nine leaves, 2 to 6 are numbered at the bottom with an errata page inserted between 3 and 4—this leaf is missing in the Duke University copy and 3 is missing in the New York copy. The tables start with folio 1. Folios 36, 49, 50, and 67 are unnumbered; 84 is misnumbered 72; 85 is unnumbered; 93 is misnumbered 92; 103 is misnumbered 102; 113 is misnumbered 114; 124 is misnumbered 122; 126 is misnumbered 138; 127 is misnumbered 139; 137 is misnumbered 149; 138 is misnumbered 150; 140 is misnumbered 152; 148, 151 are unnumbered; 152 is misnumbered 151; 158 and 161 are unnumbered; 169 is misnumbered 157; 192 is misnumbered 162; and 198 is unnumbered. Folio 198, which has a coat of arms recto and verso blank, is missing in the copies at Duke and Indiana.

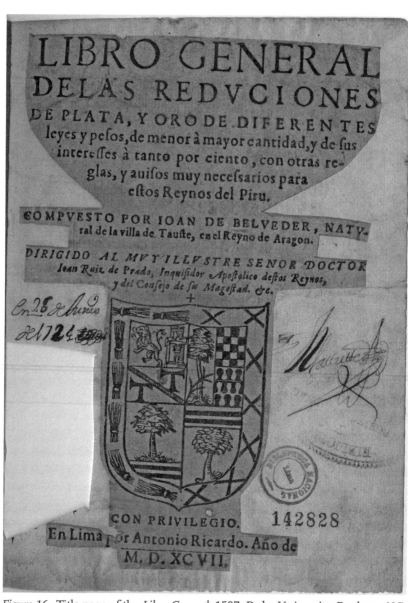

Figure 16. Title page of the *Libro General*, 1597. Duke University, Durham, NC.

Tabla breue, para resumir de pesos corrientes a ensayados, por multiplicacion.

A 140 ps, t, Por 100	Multiplica por	71 ps, 3 t. 5 g.
A 140 ps, 2 t, Por 100	Multiplica por	71 ps, 2 t. 5 g.
A 140 ps, 4 t, Por 100	Multiplica por	71 ps, 1 t. 5 g.
A 140 ps, 6 t, Por 100	Multiplica por	71 ps, t. 5 g.
A 141 ps, t, Por 100	Multiplica por	70 ps, 7 t. 4 g.
A 141 ps, 2 t, Por 100	Multiplica por	70 ps, 6 t. 4 g.
A 141 ps, 4 t, Por 100	Multiplica por	70 ps, 5 t. 4 g.
A 141 ps, 6 t, Por 100	Multiplica por	70 ps, 4 t. 4 g.
A 142 ps, t, Por 100	Multiplica por	70 ps, 3 t. 4 g.
A 142 ps, 2 t, Por 100	Multiplica por	70 ps, 2 t. 4 g.
A 142 ps, 4 t, Por 100	Multiplica por	70 ps, t. 5 g.
A 142 ps, 6 t, Por 100	Multiplica por	70 ps, t. 5 g.
A 143 ps, t, Por 100	Multiplica por	69 ps, 7 t. 5 g.
A 143 ps, 2 t, Por 100	Multiplica por	69 ps, 6 t. 5 g.
A 143 ps, 4 t, Por 100	Multiplica por	69 ps, 5 t. 6 g.
A 143 ps, 6 t, Por 100	Multiplica por	69 ps, 4 t. 6 g.
A 144 ps, t, Por 100	Multiplica por	69 ps, 3 t. 6 g.
A 144 ps, 2 t, Por 100	Multiplica por	69 ps, 2 t. 7 g.
A 144 ps, 4 t, Por 100	Multiplica por	69 ps, 1 t. 7 g.
A 144 ps, 6 t, Por 100	Multiplica por	69 ps, t. 8 g.
A 145 ps, t, Por 100	Multiplica por	68 ps, 7 t. 8 g.
A 145 ps, 2 t, Por 100	Multiplica por	68 ps, 6 t. 9 g.
A 145 ps, 4 t, Por 100	Multiplica por	68 ps, 5 t. 10 g.
A 145 ps, 6 t, Por 100	Multiplica por	68 ps, 4 t. 10 g.
A 146 ps, t, Por 100	Multiplica por	68 ps, 3 t. 11 g.
A 146 ps, 2 t, Por 100	Multiplica por	68 ps, 3 t. g.
A 146 ps, 4 t, Por 100	Multiplica por	68 ps, 2 t. 1 g.
A 146 ps, 6 t, Por 100	Multiplica por	68 ps, 1 t. 1 g.
A 147 ps, t, Por 100	Multiplica por	68 ps, t. 2 g.

Aqui va la Figura del Exemplo, y Prueua Real desta Regla.

Cambio 15800 ps.   Prueua Real 11010 ps. 3 t. 7 g. 3 m. 3 q.
Exemplo 69 ps. 5 t. 6 g. 9   143 ps. 4 t.
```
   142200                64              33030
    94800                                44040
     7900                                11010
     1975                                 5505
      790                                  35.7
      158                                  17.7.2 q.
       22. 1 m 2 q                          7.1.2 q.
    11010(45. 1 m 2 q                       4
                                        15800(00
```

Figure 17. Joán de Belveder shows how dividing by a percentage can be converted into multiplication. Note the use of currency units to express fractions. Duke University, Durham, NC.

The copies at Duke, Indiana, New York Public, Real Biblioteca, and Yale have the following variations on the title page: ral de la villa de Tauste, en el Reyno de Aragon. / *DIRIGIDO AL MVY ILLVSTRE SENOR DOCTOR* / *Ioan Ruiz de Prado, Inquisidor Apostolico destos Reynos,* / *y del Consejo de su Magestad. &c.* / CON PRIVILEGIO. / En Lima por Antonio Ricardo. Año de / M. D. XC VII.

In the *Declaración del Valor* we find, ". . . por aver muchas diferencias entre contadores sobre dezir si el tomin de plata ensayada es doze granos, o doze granos y medio, y digo que el peso de plata ensayada vale 450 maravedis, que son ciē granos, o centavos . . ." This may refer to the inconsistency (sometimes 12 *granos* to the *tomín*, sometimes $12\frac{1}{2}$) that we pointed out earlier in the discussion of Juan Diez Freyle's *Sumario Compendioso*. Here $4\frac{1}{2}$ *maravedis* is 1 *grano*, $12\frac{1}{2}$ *granos* is 1 *tomín*, and 8 *tomines* is 1 *peso*. Hence, a *grano* is equal to a *centavo*.

We note that, in contrast to the *Sumario Compendioso*, there are no Roman numerals in the tables. In the tables, an elongated U for 1,000 is used in lieu of a comma. This symbol is called the *calderón*. According to Florian Cajori, it was used in Mexican manuscripts as late as 1855. In his *The Early Mathematical Sciences in North and South America* of 1928, he reports seeing it used with both Roman and Hindu-Arabic numerals after 1600, but with Roman numerals only before 1600, which means that this book of 1597 had not been seen by Cajori at the time of that writing. In his *A History of Mathematical Notations*, also published in 1928, he amends his previous remarks to say that the *calderón* is used "more commonly" with Roman numerals before 1600, and that "[Louis] Karpinski has pointed out that it is used with the Hindu-Arabic numerals as early as 1519, in the accounts of the Magellan voyages." It does appear that this book of 1597 represents a very early use of the *calderón* with Hindu-Arabic numerals. See our discussion of the *Sumario Compendioso* for the earliest use of the *calderón* (with Roman numerals) in the works on our list.

Some terms used here were introduced in our discussion of the *Sumario Compendioso* of 1556, and so we recommend that this section should be read after that discussion.

Chapter 1, starting on the ninth preliminary leaf, contains the first table (folios 1–36), which has values of silver for 1,000 to 2,400 *de ley*. At 2,200 *de ley*, 1 *marco* of silver is 4 *pesos*, 7 *tomines*, 1 *grano*, $1\frac{3}{4}$ *maravedis*. This is exactly the value that Diez Freyle gives in 1556, given that 1 *grano* for Belveder is $4\frac{1}{2}$ *maravedis*.

Chapter [2], starting on 36v, contains the second and third tables. The second table (folios 37–49) expands this information for the concentration 2,380. The third table (folio 50) gives the same information in a different form.

Chapter 3, the fourth table (folios 51–67), is a table of percentages for $r = 8\%$ to 50%. *Pesos* and *tomines* are multiplied by $1 + r$ to give *pesos, tomines, granos, maravedis,* and *quartillos*. Chapter 4, the fifth table (folios 68–83), divides by $1 + r$, as in a similar table in the *Sumario Compendioso*. There is a strange feature in the labeling of the last two tables. Multiplying by $1 + r$ is labeled as converting from *pesos de plata ensayada* to *pesos corrientes de a nueve reales*. Dividing by $1 + r$ is labeled as converting back, i.e., *pesos de plata corriente de a nueve reales* to *pesos ensayados de a quatrocientos y cinquenta maravedis*. We believe that these conversions include a tax, explaining the variable percentage. At the going rate given earlier of 34 *maravedis* to the *real*, the conversion of *pesos de minas* to *pesos de plata corriente*, without tax, should be

$$1 + r = \frac{450}{9 \cdot 34} \approx 1.4706 \approx 147\%$$

So any percentage below $r = 47\%$ represents a tax or fee applied to assayed silver before it is converted to silver currency. See our discussion of percentage tables in the *Sumario Compendioso* for reasons why there might be such a fee with a range of values.

Clarence Haring points out that a *peso corriente* was already known in 1531, and that it was valued at $\frac{2}{3}$ of the *peso de minas*.[104] This would correspond to the $r = 50\%$ in the tables at hand. He writes shortly after that this *peso corriente* "may also help to explain the statements of some seventeenth-century writers that there was an imaginary unit called the peso ensayado of nine reals (306 maravedis)."[105] The term, *pesos corrientes de a nueve reales*, used here, supports Haring's conjecture that the *peso* of 1531, valued at roughly 300 *maravedis*, became the *peso* of 306 *maravedis* used later.

Folio 84 (numbered 72) explains how to do the previous divisions by changing them to multiplication. Belveder gives a table of the

---

104. Clarence H. Haring, "American Gold and Silver Production in the First Half of the Sixteenth Century," *The Quarterly Journal of Economics* XXIX (May 1915): 477.

105. Ibid.: 478.

necessary multiplying factors on 84v for the range of $1 + r = 140\%$ to 147%.[106]

For instance, if one wishes to divide by $1 + r$ when $r = 40\%$, we might say that this is the same as multiplying by $\frac{5}{7}$. Belveder wishes to express this as multiplication by a percentage—today we would be comfortable with $71\frac{3}{7}\%$ or $71.\overline{428571}\%$. Belveder favors a notation based on currency, something he can assume that his readers are intimately familiar with. He says multiply by 71 *pesos*, 3 *tomines*, 5 *granos*, and $\frac{1}{6}$ of a *grano*.[107] In fact, $71\frac{3}{7}$ works out to 71 *pesos*, 3 *tomines*, $5\frac{5}{14}$ *granos*, so the $\frac{1}{6}$ is not best.

In another example, $r = 43\frac{1}{2}\%$, which Belveder calls dividing by 143 *pesos*, 4 *tomines por* 100. He says just multiply by 69 *pesos*, 5 *tomines*, 6 *granos*, and $\frac{9}{64}$ of a *grano*.[108] This is equivalent to multiplying by 0.6968640625 when the actual reciprocal of 1.435 is about 0.6968641115, a relative error of 7×10^{-8}.

It is interesting to look carefully at Belveder's worked example of the multiplication. We see at the bottom of 84v:

15800 ps.	11010 ps. 3 t. 7 g. 3 m. 3 q.
69 ps. 5 t. 6 g. $\frac{9}{64}$	143 ps. 4 t.
142200	33030
94800	44040
7900	11010
1975	5505
790	35.7
158	17.7.2 q.
22. 1 m. 2 q.	7.1.2 q.
	4
1101045. 1 m. 2 q.	1580000

106. In Francisco Juan Garreguilla's work of 1607 (see later entry), there is a table similar to the fourth table here, but with a range of just 140% to 144%. In 1620, Juan Vázquez de Serna's book, printed in Spain, gives a table of such rates going from 143% to $143\frac{1}{2}\%$ by ninths. He explains that when gold and silver is brought from Peru to Puerto Belo, there one gets only 143 to $143\frac{2}{3}$ *pesos* of 9 *reales* for 100 *pesos de minas*.

107. Because of a printer's error, in the table on 84v all these fractions (of a *grano*) have been written with the denominators only. In the copies at the Biblioteca Nacional de España, Duke, Indiana, New York, Real Biblioteca, and Yale, someone has written in numerators over all the denominators in spite of this information not being listed on the errata page. Hardly any of the fractions thus produced are close to what they should be. The "corrected" fractions are consistent from copy to copy, so there may have once been another errata, now lost.

108. Since this is his illustrative example on 84r, we can confirm that the anonymous editor has correctly changed 64 to $\frac{9}{64}$ on 84v. (See previous footnote.)

The left column is the computation for dividing by $143\frac{1}{2}$%, i.e., Belveder has multiplied his amount, 15,800 *pesos*, by 69 *pesos*, 5 *tomines*, $6\frac{9}{64}$ of a *grano*. The first two lines below the bar are the result of multiplying by the 69 *pesos*. The 7,900 is the result of multiplying by 4 *tomines* and then he adds a quarter more, 1,975, for the fifth *tomín*. The 790 is caused by multiplying by 5 *granos* and the 158 is for the other *grano*. The penultimate number is supposed to be the $\frac{9}{64}$% times the 15,800, and Belveder says this is 22 *granos*, 1 *maravedi*, and 2 *quartillos*. Actually the product is very close to 22 *granos*, 1 *maravedi*, so the 2 *quartillos* in the sum at the bottom are under suspicion.

In that sum, 11,010 45. 1 m. 2 q., the 11,010 are *pesos* and the 45 must be regarded as *granos*, because we should have been multiplying by a number around 69%, not 69. The 45 *granos* are 3 *tomines*, 7 *granos*, 2 *maravedis*, and 1 *quartillo*, so Belveder's answer is 11,010 *pesos*, 3 *tomines*, 7 *granos*, 3 *maravedis*, and 3 *quartillos*. This answer is closer than it should be (the correct answer should end with about 2.33 *quartillos*) because Belveder's error with the 2 *quartillos* compensates for the slight inaccuracy of the $\frac{9}{64}$.

The right-hand column above is the check on the work. Belveder multiplies his answer by 143 *pesos*, 4 *tomines* to see if he gets back the 15,800. So the first three lines under the bar are the results of multiplying just the 11,010 by the 143. The next line is the 11,010 by the 4 *tomines*. In the next line he is taking 143 *pesos*, 4 *tomines* by 2 *tomines*, i.e., by $\frac{1}{18}$. He gets 35 *pesos* and 7 *tomines*. He adds in half of that—17 *pesos*, 7 *tomines*, and 2 *quartillos*—and now he has multiplied in the full 3 *tomines* from the top number. He next should multiply the $143\frac{1}{2}$ by the 7 *granos*, but instead he multiplies by 5 *granos* since that is easier. Now he must wrap this up by taking the remaining 2 *granos*, 3 *maravedis*, and 3 *quartillos* times the 143 *pesos*, 4 *tomines*. He says this is 4. That's actually not too bad—the number above it is farther off.

Most of the fractions (of a *grano*) given in this table are not even good approximations of the right ones. We have not been able to guess at Belveder's method for calculating these fractions. Only the last example of $\frac{9}{64}$ makes some sense. The correct fraction is $\frac{81}{574}$, and adding 2 to its denominator produces a fraction that reduces to $\frac{9}{64}$.

Folios 85 through 99, found in Chapter 5, are a table of values of gold. Weights are in *pesos* and *tomines* (with an entry for 6 *granos*, which appears to actually be for half a *tomín*) and values are in *pesos*, *tomines*, *granos*, *maravedis*, and *quartos*. The standard is still $22\frac{1}{2}$ carats, called *buen oro*. In fact, the labeling indicates that conversion to money *is* converting to $22\frac{1}{2}$ carats.

Chapter 6 contains several tables of conversions for units of weight. We find here that 1 *quintal* = 4 *arrobas*, 1 *arroba* = 25 *libras*, 1 *libra* = 16 *onças*, 1 *adarme* = $\frac{1}{16}$ *onça*, and 1 *grano* = $\frac{1}{32}$ *adarme*. A *libra* should be 2 *marcos* since a *marco* is 8 ounces. A *libra* of $22\frac{1}{2}$ carat gold is listed here with the value of 100 *pesos de oro*, implying once again a ratio of exactly 10 to 1 for the values of gold and silver.

By 1597, the year of publication of this work, Peru had had a mint, off and on, for 29 years. (The Lima mint, which struck its first coins in 1568, was closed in 1571. When it reopened in 1577 there was already a mint in Potosí, where the great silver mines were. The Lima mint closed again in 1588 and stayed closed for almost a century.) So it is not surprising that more space is given to conversions between different minted coins than we see in the *Sumario Compendioso*. The rest of the tables serve this purpose.

Many tables use percents for a variable exchange rate, i.e., tax, as in the third and fourth tables. The name *patacones* for *pesos* of 8 *reales* is introduced in Chapter 11. Chapter 22 gives a table for computation of taxes. The output is called the *almojarifazgo*. (An *almojarife* is the king's tax collector. Haring translates *almojarifazgo* as "customs dues."[109]) In this table a given rate of 40% means multiplying by 0.14, as if the 40% is a surcharge on a 10% tax.

Chapter 23 explains how the *quinto* on silver is calculated. If one has silver at 2,380, take the weight in *marcos* times 2,380 to get *maravedis*. Take out 1% for mercy and $\frac{1}{2}$% for the assayer, and the king gets one fifth of what is left. This means that the total assessment for a *quinto* when Belveder was writing was $0.015 + 0.985/5 = 21.2\%$.

Chapter 24 deals with other metals, including mercury. Chapter 25 shows that the *quinto* on mercury is a straight 20%.

As in the *Sumario Compendioso*, there are word problems in the back of the book. One example is, "A man buys a certain heirloom for 56 ducats and wants to turn it around such that he can make a profit of 20%. It is asked to what price will he raise the said heirloom so that he can profit the 20%." Using the rule of three, he finds the answer to be $67\frac{1}{5}$.

In one he asks, "Three fifths of what number will be three quarters?" In the next sentence he changes the problem, because his answer is that "thus you will say that three fifths are three quarters of twelve fifteenths."

---

109. Clarence H. Haring, "American Gold and Silver Production in the First Half of the Sixteenth Century," *The Quarterly Journal of Economics* XXIX (May 1915): 477.

Another problem is to give two numbers with their sum equal to their product and "that so much be the third of one as the half of the other." He notes that 6 is the product of 2 and 3, and the sum of these numbers is 5. So divide this 5 by 3 and divide 5 by 2 to get the answers, $1\frac{2}{3}$ and $2\frac{1}{2}$.

His last problem is given without solution, saying it is "ordinary for any average accountant." Three merchants have moneys such that the second has double the wealth minus a quarter of the first, the third has $8\frac{2}{3}$ more than the second, and the three of them together have $43\frac{2}{3}$ ducats. In present-day algebra, we would say that this is solved by setting up three equations in three unknowns. Belveder says that one should use "the first false position of the rule of the *cosa*." (The answers are $7\frac{7}{9}$, $13\frac{11}{8}$, and $22\frac{5}{18}$.)

1603

Felipe de Echagoyan.   11

TABLAS / DE REDV- / CIONES DE MONEDAS, / *y del valor de todo genero de plata y / oro, y del modo de hazer las cuẽtas del, / y delos derechos q̃ se deuẽ a su Mag. / enel quintar la plata, y delos intereses / de uno hasta diez por ciento, y delos / censos desde 14. hasta 20U el millar, y / de lo que se ha de pagar en las aualia- / ciones, y de otras cosas necessarias y / conuenientes para las cuentas del trato / y contrato de estos Reynos. /* Hechas por Philippe de Echa- / goyan vezino de Mexico. / DIRIGIDAS / *A Juan de Tellaeche Çauala, Caualle- / ro hijodalgo de Solar conoscido de ca- / sas Infanzonadas enel Señorio / de Viscaya-* / Impressas con licencia. / *Por Henrico Martinez. Año* 1603.

Enrico Martínez, México.

Medina, *México*, 208.

Original located at: John Carter Brown Library.

Paging: 126 leaves. The pattern is twelve unnumbered leaves, thirty-six numbered leaves, and the first twelve leaves repeated; numbering resumes with 37 through 102 with 72 misnumbered 62, 82 misnumbered 28, and 93 and 95 unnumbered.

José Toribio Medina spells the author's name as Echegoyan.

There is a note to the reader from the printer Enrico Martínez on the sixth and seventh unnumbered leaves. If Joaquín García Icazbalceta is right about the *Sumario Compendioso* being produced for use in Peru, then this could be the first book of tables printed for use exclusively in Mexico.

A table of contents exists on the eighth to eleventh preliminary leaves, but it is incomplete, covering only the material up to folio 89v. Another list of contents appears on 92r and covers the rest of the book. These later folios give tables for the value of silver.

Echagoyan uses the *calderón* with his Hindu-Arabic numerals (see the discussion within our entry for 1597).

Some terms used here were introduced in our discussion of the *Sumario Compendioso* of 1556, and so this section should be read after that discussion.

There is an explanation of the purity of silver on 14r to 14v that says that 12 *dineros* times 24 *granos* per *dinero* times $8\frac{1}{3}$ *maravedis* per *grano*[110] makes 2,400 *maravedis* for pure silver. This is the first book on our list to give this second meaning for *maravedi*, an increment of $\frac{1}{2400}$ in the purity of silver.

Echagoyan prefers transactions in *pesos de tipuzque* as opposed to the use of *pesos de minas* that we see in the 1556 *Sumario Compendioso*, the 1597 *Libro General*, and the 1607 *Libro de Plata*. As he puts it, "La mayor parte de las contrataciones en esta tierra es en pesos de tipuzque." Seventeenths come up often, because 272, the number of *maravedis* in a *peso de tipuzque*, factors as $272 = 16 \times 17$. For instance, one *marco* of silver at the 2,200 level is worth 8 *pesos* and $8\frac{8}{17}$ *granos*,[111] and these *pesos* are *pesos de tipuzque*, not *pesos de minas*. This change of currency exactly accounts for the discrepancy between this and the value of 4 *pesos*, 7 *tomines*, and $6\frac{1}{4}$ *maravedis* that we see in the *Sumario Compendioso*, the *Libro General*, and the *Libro de Plata*, and thus the price of silver does not vary at all during the time spanned by these four texts. Since *pesos de tipuzque* are only 272 *maravedis*, the corresponding *tomines* and *granos* (at 96 *granos* to the peso) are only 34 *maravedis* and $2\frac{5}{6}$ *maravedis*, respectively.

The author's list of conversions among the important coins includes: 1 *escudo* = 400 *maravedis*, 1 *ducado* = 375 *maravedis*, 1 *peso de minas* = 450 *maravedis*, 1 *peso de plata corriente* = 306 *maravedis*, 1 *peso de tipuzque* = 272 *maravedis*, 1 *real* = 34 *maravedis*, 1 *maravedi* = 2 *blancas*, 1 *blanca* = 3 *cornados*, and therefore 1 *grano* [$=2\frac{5}{6}$ *maravedis*] = 5 *blancas* and 2 *cornados*.

In addition to the usual tables for silver and gold, we see tables for the value of quantities of wine and *cacao*, tables for the computation of salaries,

---

110. He notes that silversmiths use $8\frac{1}{4}$ *maravedis* to the *grano*.

111. The author writes 1.17 *avos* to mean $\frac{1}{17}$, so this is not yet an example of decimal fractions (which first appear in their modern form among the works on our list in the *Libra Astronómica* of 1690).

and many other things. There is a table of distance measurements that includes the following examples:

$$100 \text{ anas de Flandes} = 81 \text{ varas castellanas}$$
$$100 \text{ canas de Italia} = 200 \text{ varas castellanas}$$
$$100 \text{ cobodos de Portugal} = 80 \text{ varas castellanas}$$

One interesting feature is the discussion of gold from China. In that context we see two new units of weight, the *mae* and the *tae*. Conversions are

$$1 \text{ marco} = 6 \text{ taes } 14\tfrac{1}{2} \text{ maes} = 110\tfrac{1}{2} \text{ maes}$$
$$1 \text{ tae} = 16 \text{ maes}.$$

The very first table for silver incorporates some rounding. The value for 2,400 should be 8 *pesos*, 6 *tomines*, $7\tfrac{1}{17}$ *granos*, but appears instead as 8 *pesos*, 6 *tomines*, 6 *granos*. This is thematic for Echagoyan as we see by comparing the next two tables. One of these gives values for *plata del rescate* at 6 *pesos*, 3 *tomines*, $5\tfrac{13}{17}$ *granos* to the *marco*, and the next rounds this down to 6 *pesos*, 2 *tomines*. The author writes on folio 16v that, "no dan por el marco enesta ciudad mas de 6. pesos y 2. tomines q̃ es vn real y 16. m r̃s y vn tercio de marauedi menos de lo q̃ vale, y enlas minas quita 16. m r̃s y vn tercio por que dan 6. pesos 3. tomines y no mas—" He gives in his tables what is actually paid and not some hypothetical value. The first value for *plata del rescate* reflects a reduction of 20.8% from the value of 2,210 *de ley*, just like what Juan Diez Freyle calls *plata del quinto*. The second value amounts to a reduction by $23\tfrac{1}{13}$%, just like a similar table that Diez Freyle gave for *plata del rescate*.

There is a single table for *plata del diezmo* that is 10.9% lower than 2,210. This puts the *marco* at 7 *pesos*, 1 *tomín*, 11 *granos*, or $1,969\tfrac{1}{6}$ *maravedis*. The text makes some rather contradictory statements about this table, saying that the true value is $1970\tfrac{1}{10}$ *maravedis* and that this is 7 *pesos*, $11\tfrac{2}{17}$ *granos* (when it is actually 7 *pesos*, 1 *tomín*, $11\tfrac{5.6}{17}$ *granos*), but that they give one third of a *maravedi* less than this.

The author says that the value of *plata del quinto* is 2,211 *maravedis* per *marco*, but the table rounds this down to 8 *pesos*, 1 *tomín*. This is the value that Juan Diez Freyle gave for *plata quintada*. For the next table we are told that *plata del Peru* is valued at 2,380 but that they give only 2,376 or 8 *pesos*, 5 *tomines*, $10\tfrac{10}{17}$ *granos*. In fact the table gives the full 2,380, which is 8 *pesos*, 6 *tomines*.

There is a table for what *pesos* will buy in *plata del rescate*. For example, 7 *pesos* correspond to 1 *marco* and 6 *tomines* in change (*sobras*). There are similar tables for *diezmo*, *quinto*, and *Peru*.

The silver tables at the very end of the book are more like those at the beginning of the *Sumario Compendioso*, giving values of silver in a range of 2,200 to 2,360 *de ley*.

There are two different tables for values of gold. In either table there are 16 *reales* in a *castellano* and 50 *castellanos* in a *marco*. Because *castellanos* are a unit of weight for this text, they are not mentioned among the conversions of currency discussed earlier, and they do not appear until the section on gold. (But note that the term *peso de oro* is synonymous with *castellano* for Echagoyan.)

The first table for gold is labeled 20 *maravedis* to the carat, and the second is labeled $24\frac{8}{11}$ *maravedis* to the carat. The meaning of this is that this number, 20 or $24\frac{8}{11}$, respectively, is to be multiplied by the number of carats to get the value of a *castellano* in *maravedis*. If the first table gives the base value of the gold, the second table represents an additional tax of $23\frac{7}{11}\%$. The text suggests that this tax includes a component that is 15%, but it does not explain what this 15% is or what the rest of the tax is.

In the first table we see that at $22\frac{1}{2}$ carats, 1 *marco* of gold is 22,500 *maravedis*, 1 *castellano* is 450 *maravedis*, and 1 *real* is $28\frac{2}{16}$ *maravedis*. This is the same value for gold given by Juan Diez Freyle in 1556. In the second table, one example is that at 22 carats, 1 *marco* of gold is 27,200 *maravedis*, 1 *castellano* is 544 *maravedis*, and 1 *real* is 34 *maravedis*.

Since Echagoyan uses *marcos* as a measure of weight for gold, it is possible to confirm that gold is still worth exactly ten times its weight in silver. The *pesos* used by Juan Diez Freyle as a unit of weight are probably the *pesos de oro*, or *castellanos*, used here for one fiftieth of a *marco*.

1606

Enrico Martínez. **12**

REPORTORIO: / DE LOS TIEM- / POS, Y HISTORIA NATVRAL / DESTA NVEVA ESPANA. / *Compuesto por Henrrico Martinez Cosmographo de su Ma- / gestad è Interprete del Sancto Officio deste Reyno. / Dirigido al Excellentissimo / Señor Don Iuan de Mendoça y Luna Marques de / Montesclaros, Virrey, Gouernador, Presidente y Cappi- / tan General por el Rey nuestro Señor en esta Nueua España &c.* / CON LICENCIA Y PRIVILEGIO. / En Mexico. / *En la Emprenta del mesmo [sic] autor año de 1606.*

Enrico Martínez, México.
Sabin 44952. Medina, *México*, 228.

Originals located at: Baylor University, Biblioteca Cervantina, Biblioteca Nacional de España, Biblioteca National de México, Biblioteca Palafoxiana, British Library, Brown University, Centro de Estudios de Historia de México Condumex, Hispanic Society of America, John Carter Brown Library, Library of Congress, New York Historical Society, New York Public Library, Tulane University, Universidad de Salamanca, University of Texas at Austin (two copies), Wellcome Library, Yale University (two copies).

Microform copies are available.

During one visit to the British Library we examined two different originals, one partial. As of 2005, copy 2 is listed as "destroyed."

Paging: 199 leaves numbered irregularly as follows: twelve unnumbered leaves, the last six of which are the table of contents; eighteen leaves numbered front and back as 1–36, with 1 unnumbered, 9 misnumbered 8, 19–23 misnumbered 22–26 and 24–28 unnumbered; one unnumbered leaf containing a volvelle recto with verso blank; eight leaves numbered front and back as 37–52 with 38 reversed to 83; twenty-three leaves numbered on recto as 53–75 with 54 unnumbered; twenty-one unnumbered leaves; four leaves numbered front and back as 77–84; four leaves numbered front and back as 75–82; fifty-two leaves numbered front and back as 93–196 with 95 unnumbered; fifty-six leaves numbered front and back as 167–277 with page 185 misnumbered 183 in some copies, 232 and 233 unnumbered, 270 misnumbered 278 in most copies, 274 misnumbered 74 in some copies, and 278 unnumbered. The table of contents appears to be consistent with this numbering. Some copies have had new page numbers inked in with such thoroughness that it is difficult to determine what the original numbering was.

There are two versions of the original. Version 1 has the title page given here and has the errors in paging noted for pages 185 and 274. Version 2 has the page numbers corrected for 185 and 274 and has the following variant title page: REPORTORIO / DE LOS TIEM- / POS, Y HISTORIA NATVRAL / DESTA NVEVA ESPANA. / *Compuesto por Henrrico Martinez Cosmographo de su Magestad è Inter- / prete del Sancto Officio deste Reyno.* / Dirigido al Excellentissimo / Señor Don Iuan de Mendoça y Luna Marques de / Montesclaros, Virrey,Gouernador, Presidente y Cappitã / *General por el Rey nr̃o Señor en esta Nueua España.*

Figure 18. Title page of the *Reportorio de los Tiempos*, 1606. The John Carter Brown Library at Brown University, Providence, RI.

&c. / CON LICENCIA Y PRIVILEGIO. / En Mexico. / *En la Emprenta del mesmo autor año de* 1606.

Another difference between the two versions is the design used to fill the page at the end of the *Prologo*. Later we will quote four paragraphs starting from folio 53 and will note some differences there. There are so many changes in Version 2, and so few of them could be called corrections, that it seems reasonable to conclude that the type had been dismantled and had to be reset for certain pages.

Version 1 is represented by Biblioteca Nacional de México, British Library, Centro de Estudios de Historia de México Condumex, Hispanic Society of America, John Carter Brown, New York Historical Society, New York Public, Tulane University, Universidad de Salamanca, University of Texas at Austin copy 2, Wellcome Library, and Yale copy 2 with the exceptions that page 274 is correct at the Hispanic Society and New York Public and page 270 is numbered correctly at the New York Historical Society. Version 2 is Baylor University, Biblioteca Nacional de España, Biblioteca Palafoxiana, Brown University, Library of Congress, University of Texas at Austin copy 1, Yale copy 1. Copy 2 in Austin actually has a title page made from newer paper that doesn't quite match either version. The Library of Congress copy has a title page reset in modern type. The Hispanic Society copy is missing preliminary leaves [6] and [7] and the New York Historical Society copy is missing the last four leaves.

A reprint was done in 1948, with an appendix by Francisco González de Cossío that lists all the works that came from the press of Martínez. This version was again reprinted in 1991. Grupo Condumex published a facsimile edition in 1980, and the Centro de Estudios de Historia de México Condumex reprinted this in 1981. An excerpt appears in Elías Trabulse's *Historia de la Ciencia en México, Siglo XVII*. A survey of Martínez's work as a printer, including facsimiles of several pages from the *Reportorio*, was published in 1996. The entire text of the Biblioteca Nacional de México copy is available online at www.cervantesvirtual.com.

In Francisco de la Maza's *Enrico Martínez, Cosmógrafo e Impresor*, we see that Martínez's origins have been assigned to Mexico, Spain, France, Germany, the Netherlands, and Flanders by various authors. José Toribio Medina quotes Vicente de Paula Andrade, who cites the argument made by Nuñez Ortega that Martínez was French. The online version of the Catholic Encyclopedia at www.newadvent.org/cathen/ is one reference that still adheres to this proposition in its article, "Martín, Enrico" by Camillus

Crivelli.[112] The sources that cite de la Maza, however, seem to be convinced by his claims to the contrary.

According to de la Maza, Enrico Martínez (c. 1555–1632) was born Heinrich Martin in Hamburg. His family moved to Seville when he was eight and there converted to Catholicism. He travelled Europe at age 19 and studied mathematics in France. In 1589 he went to New Spain and served the Inquisition there as an interpreter for Flemish and German. Apparently he acquired his print shop by receiving from the Inquisition the holdings of a Dutch prisoner. He held the post of Royal Cosmographer and was placed in charge of the difficult project of drainage of the Valley of Mexico.

According to Jonathan Kandell's *La Capital*, the drainage project, which involved digging a canal and tunnel through the side of the Valley of Mexico so that floodwaters would drain out, was extremely unpopular with Mexican Spanish society, both because of the amount of tax money that it consumed and the demand for native labor (470,000 native workers were used for six months, according to de la Maza[113]). When it was completed in 1608, Martínez found it difficult to get the colony to fund the maintenance of the system, and after 1623, when all maintenance stopped, the tunnel and canal became blocked by debris.

In 1629 a flood began that was to last for 5 years. The city was under six feet of water. In the search for a scapegoat, the Viceroy briefly imprisoned Martínez, but soon he was put to work to find a solution to the disastrous flood. He did not live to see the waters recede in 1634, as he had died near Mexico City in Cuautitlán in 1632, but in the long term, his plans for the water management of the Valley of Mexico were put into effect, since no one wanted a repeat of the Great Flood of 1629 to 1634.

Edmundo O'Gorman gives Henrrico or Henrico Martín, Henrrico or Henrico Martínez, Enrico Martínez, and Enrique Martínez as forms of his name that he used. We will use the name Enrico Martínez when referring to him here. This is the form used on the street that bears his name in modern-day Mexico City.

---

112. For this Web site the 1913 edition of the Catholic Encyclopedia was used because it was in the public domain. In the 1967 edition of the New Catholic Encyclopedia, the article "Martínez, Enrico" was authored by de la Maza and describes Martínez as a German scientist.

113. Francisco de la Maza, "Enrico Martínez," New Catholic Encyclopedia, Washington, 1967.

Martínez printed this work on his own press. Juan Pascoe quotes de la Maza as saying that "the *Reportorio* is the unique work badly printed by Martínez . . . and it causes surprise to see how much care he put into the books of others and neglected so much his own." Pascoe goes on to disagree, saying, "The anarchic foliation and eccentric signaturation are the result of the constraint with which it was made."

The title could be borrowed from any of several well-known almanacs published in Spain. A *Reportorio de los Tiempos* by Andrés de Li, for example, is known to have been printed in Zaragoza in 1492. A complete copy of a 1495 edition was used for a reprint in 1999. De la Maza gives a list of other "Reportorios" printed in Spain in the sixteenth century and notes that *Historia Natural* as well was used as a title by José de Acosta in 1590. *Reportorio* appears to be an obsolete spelling of *repertorio* (repertoire).

José Miguel Quintana, in his *La Astrología en la Nueva España, en el Siglo XVII*, espouses the theory that Martínez was heavily influenced by the *Chronographía o Reportorio de los Tiempos* of Jerónimo de Chávez. This was first published in 1561 and had six editions.[114] Quintana gives a chapter-by-chapter comparison of the two texts (using the 1588 edition of Chávez). We will note some of the similarities that Quintana brings to light. Chávez was the translator into Spanish of the *Sphere* of John of Sacrobosco, which we have mentioned in the entries for 1557 and 1578. Quintana thinks that the *Cronographía* was Chávez's own amplification of the *Sphere*, and so the latter work may also be an influence on Martínez. In his biography of Martínez, de la Maza reconstructs what titles may have been in Martínez's library. The Chávez *Chronographía* is on this list and so is Sacrobosco's *Sphere*.

This is chiefly an astrological work, coordinating events in the history of Mexico and the rest of the world with phenomena in the sky. Pages 225 through 276 are a history of the world from 1520 to 1590 and include a short biography of Mary, Queen of Scots, starting on page 273. There is one paragraph about the Spanish Armada on page 276. The note on the bottom of that page says that a second volume (apparently never published) will update this history to 1605.

114. Ursula Lamb, "The Teaching of Pilots and the *Chronographía o Reportorio de los Tiempos*," in *Cosmographers and Pilots of the Spanish Maritime Empire*. Aldershot, 1995: 12. Jerónimo de Chávez died in 1574 but apparently the editions kept coming out, including a 1584 edition corrected for the Gregorian calendar, under the editorship of his father, Alonso de Chávez.

There is a list of the Aztec kings of Mexico and a list of the viceroys of colonial Mexico.

Treatise 5, beginning on page 201, is probably identical to the book *Discurso sobre la magna conjuncion de los planetas Jupiter y Saturno, acaecida in 24 de Diciembre de 1603 en el 9 g. de Sagitario*, which Martínez authored and published in 1604, of which no copies are known to have survived. Since no mathematics are in this version of it, this work is not listed separately here. Within this treatise on the "magna conjunción," pages 212 through 224 are a treatment of the history of the wars between the Christian kingdoms and the Turks. Naturally enough, the author predicts the "fall and destruction" of the Turkish Empire.

The volvelle on the leaf following page 36 can be turned so that the time of the night meets the day of the year, and this puts a sign of the zodiac under the curved fixed arm.[115] This was intended as a tool for determining the rising sign of the subject, the sign (one of the twelve divisions of the ecliptic) that is crossing the eastern horizon at the time of birth.[116] The volvelle is designed with a pattern one sees on an astrolabe, with a smaller circle off center in a larger circle. The curve in the arm is important, and so is the fact that the wheel of zodiac signs is off center, to incorporate the fact that the "signs of slow ascension"—Cancer to Sagittarius—take longer to rise than the "signs of fast ascension"—Capricorn to Gemini.

---

115. The copies at Tulane University and at the British Library have both a movable wheel and the fixed arm over it. The copy at Salamanca and both copies at the University of Texas have the wheel and arm, but the arm is only attached at the middle; in copy 2 the wheel doesn't turn. The copy at the John Carter Brown Library has just the wheel, and the copies at Baylor, Brown, the Hispanic Society, the Library of Congress, the New York Historical Society (with the leaf following p. 44), the Wellcome Library, and both copies at Yale are missing both the wheel and the arm. The copies at the Biblioteca Nacional de México and the New York Public Library have the wheel and arm replaced with new paper. The copy at the Biblioteca Nacional de España is missing the whole page. The first Condumex facsimile has the wheel and it turns, but in the second facsimile it doesn't turn in some copies.

116. Martínez refers to this only as *"en que signo nació"* ("in what sign one was born"). In today's popular usage that phrase would describe the sun sign, the sign that the sun is in at the time of birth. In astrological practice, however, the rising sign has a more primary role than that of the sun sign. The sign, and degree of the sign, where the ecliptic meets the eastern half of the horizon at the time of birth of the subject, also known as the *ascendent*, or *horoscope*, is the cusp of the so-called "first house"; thus it forms the starting point for the building of a natal chart for the subject.

LA letra M. que está junto a la ♓ denota el mes de Março, y la
A. el mes de Abril, y la otra M. el mes de Mayo, y la I. Iunio
y la otra I. Iulio y lo mismo se ade entender de las demas letras que cada vna significa su mes, y va la sucessión dellos de la mano derecha azia
la izquierda.∽∽∽

Figure 19. The volvelle from the *Reportorio de los Tiempos,* 1606. The inner disk turns so that the date along its outer edge can be lined up with the time of day. Then the rising sign can be found on the offset circle. The circle of signs is offset so that the signs of fast ascension cross the arm faster than the signs of slow ascension. Tulane University Latin American Library, New Orleans, LA.

This entry is probably the first book printed in the New World to mention trigonometric functions by name.[117] Martínez wrote a *Tratado de Trigonometría*, now lost, according to Trabulse (*Apéndices e Índices*). We have found three examples of applied trigonometry in the *Reportorio*.

Chapter 34 of Treatise 1 is a treatment of the Gregorian calendar reform. The next chapter asks, in so many words, "How could they know that the calendar was ten days off?" The answer offered by Martínez involves solving a spherical triangle.

Here are the last four paragraphs of Chapter 35 of Treatise One as they appear in the text, starting with folio 53r. We have preserved the peculiarities of spelling, accent marks, capitalization, and the use of periods after all numbers. Only a few typographical changes have been introduced here. We have written each s the same way, without the use of the elongated s. Also, one sees in the third paragraph the use of U for 1,000 in lieu of a comma.[118] The actual U used in the text is somewhat more elongated than can be reproduced here. In the copy held at the John Carter Brown Library, some corrections have been made in ink, and some of these have been noted here in the footnotes. Versions 1 and 2 have numerous differences from each other in these paragraphs. Some typical differences are noted below in footnotes.

PARA saber agora[119] mediante la dicha operacion, que tanto tiempo à que[120] passò ò falta para llegar el Sol al dicho equinocio, se multiplicarà el seno recto del arco del apartamiẽto del Sol dela equinocial, por el seno total, y el producto diuidirlo por el seno recto de la maxima declinaciõ del Sol, y el quociento[121] serà el seno del arco de la eclipta, que el Sol entonces estarà apartado del principio de Aries, el qual arco diuidido por el mouimiento diario del Sol, vendrà el tiempo que a[122] passado el Equinocio,[123] si el Sol se hallare de la

---

117. See *Instrucción Nauthica* for a table of secants and tangents without the names, and see *Diálogos Militares* for the solution of trigonometric problems using only similar triangles (without the use of angle measure).
118. This is the *calderón*. For more information, see our discussion of its use in 1597.
119. The 1948 transcription into modern Spanish gives this as *ahora*.
120. q̃ in Version 2.
121. *quociẽto* in Version 2.
122. *à* in Version 2.
123. *Equinociò* in Version 2.

Equinocial al Norte, ò el tiempo que falta, si estuuiere apartado della al Sur, mas para que esto mejor se entienda pondrè vn breue exemplo.

ANO De[124] 1604. a[125] treinta dias del mes de Março en la ciudad de Mexico, cuyo[126] eleuacion de polo Artico es diez y nueue grados y quinze minutos, tome[127] la altura del Sol al punto de medio dia[128] y halle[129] le sobre el Orizonte[130] setenta y quatro grados y treinta y nueue minutos, que restados de nouenta que ay desdel Orizonte al Zenith, quedan quinze[131] grados y 21. minutos, està pues apartado la Equinocial del Zenith al Sur otro tanto quanto es la eleuacion del Polo, que ya se dixo ser 19. grados y quinze minut.[132] de los quales restados los dichos quinze[133] grados y 21. minut.[134] quedo[135] siendo la differencia entre el Sol y la equinocial tres grados y cinquenta y quatro minutos, y estos declinaua el Sol de la linea al norte,[136] en el dicho instante.

Esto hecho, busque[137] el seno recto destos tres grados y cinquenta[138] y quatro minutos y halle ser 4080. los quales multipliquè por el seno total, que[139] enesta operaciõ fueron[140] sesenta mil, y vinieron al producto 244. quentos y 800. U. estos parti por 23. U. 924. que[141] es seno recto de veinte y tres grados y medio (que en este exemplo se tomaron por maxima declinacion del Sol) y salieron al quociente 10 U. 232. cuyo

---

124. *de* in Version 2.
125. *à* in Version 2.
126. This has been changed to *cuya* in the John Carter Brown copy and in the 1948 edition.
127. John Carter Brown: *tomè*.
128. A comma has been inserted here in the John Carter Brown copy.
129. John Carter Brown: *hallè*.
130. John Carter Brown: insert comma here.
131. 15. in Version 2.
132. 15. *minutos* in Version 2.
133. 15. in Version 2.
134. *minutos* in Version 2.
135. *quedò* in Version 2. Inked in at John Carter Brown.
136. *Norte* in Version 2.
137. John Carter Brown : *busquè*.
138. *cinquẽta* in Version 2.
139. *q̃* in Version 2.
140. *fuerõ* in Version 2.
141. *q̃* in Version 2.

arco es de nueue grados y cincuenta min.[142] escassos[143] y tanto[144] dirè que[145] estuuo el Sol al medio dia a[146] los treinta de Març o enel dicho año apartado del principio de Aries segun secession[147] de signos, enel meridiano de Mexico, cuya longitud es 267. grados 12 minut.[148] respecto el meridiano fixo antiguo.[149]

Estos 9. grad. y 50. minutos los reduzgo todos à minutos y hazen[150] 590. mi. los quales parto por 59 min.[151] (que es casi el mouimiento diario del Sol en dicho tiempo) y vienen al quociento diez, que son dias, estos 10. dias quitados de los 30. de Março, quedan 20. y assi digo, que el Equinocio del Verano en el referido año sucediò en el meridiano de Mexico, à los 20. de Março cerca de las 12. del dia. Digo pues, que este es vno de los modos por donde se supo el tiempo que los Equinocios se auian anticipado, por que haziendo esta diligencia a[152] los 21. de Março del año de 1580. hallaron que auian passados mas de diez dias, que el Sol auia llegado al principio de Aries, que es el Equinocio del Verano.

Here is a translation. In the passage above we have noted the points where the anonymous editor of the John Carter Brown Library copy has added accent marks to certain verbs, changing the tense from present subjunctive to preterite. We have elected to make similar changes in tense throughout the second and third paragraphs below.

To know now, by means of the said operation, that so much time passed since or lacks until the arrival of the sun at the said equinox, one will multiply the sine of the arc of the separation of the sun from the equinoctial [celestial equator] by the total sine [sin 90°], and the product divided by the sine of the maximum declination of the sun,

142. 50. *minutos* in Version 2.
143. Comma in Version 2.
144. *tāto* in Version 2.
145. *q̃* in Version 2.
146. *à* in Version 2.
147. *sucession* in Version 2.
148. *mi.* in Version 2.
149. No dash in Version 2.
150. *hazẽ* in Version 2.
151. *mi.* in Version 2.
152. *à* in Version 2.

and the quotient will be the sine of the arc of the ecliptic that the sun then will be parted from the beginning of Aries, which arc divided by the daily movement of the sun, will accommodate the time that it has passed the equinox, if the sun will be situated to the North of the equinoctial, or the time that lacks, if it be parted to the South of the equinoctial, moreover for this to be understood better I set a brief example.

March 30, 1604, in Mexico City, whose elevation of the [celestial] North Pole is 19°15', I took the height of the sun at the midday point, and I found it 74°39' above the horizon, that deducted from 90 that there is from the horizon to the zenith, left 15°21'. The equinoctial is then separated from the zenith to the south another amount as much as is the elevation of the pole, which already was said to be 19°15' from which deducted the said 15°21', 3°54' was left over being the difference between the sun and the equinoctial, and this declined the sun by the line to the north, in the said instant.

This done, I sought the sine of these 3°54' and I found it to be 4,080, which I multiplied by the total sine, that in this operation was 60,000, and it came out to the product 244,800,000, this I divided by 23,924 that is the sine of 23°30' (that in this example was taken for the maximum declination of the sun) and it came out to the quotient 10,232, whose arc is just 9°50' and so much I will say that the sun was, at midday on March 30th of said year, separated from the beginning of Aries according to succession of signs, in the meridian of Mexico, whose longitude is 267°12' respect to the ancient fixed meridian.

These 9°50' I convert them all to minutes and they make 590 minutes which I divide by 59 minutes (that is almost the daily movement of the sun in said time) and they come out to the quotient 10, which are days, these 10 days taken from the 30th of March, leave 20. And thus I say, that the equinox of summer in the reported year happened, in the meridian of Mexico, on the 20th of March about 12 noon. I say then, that this was one of the ways whereby they knew the time that the equinoxes have advanced themselves, because making this diligence on March 21, 1580, they found that more than 10 days had passed, that the sun had left the beginning of Aries, which is the equinox of summer.

Thus the author makes use of a spherical triangle with one corner at the vernal equinox (the point in the sky where the sun passes from south to north through the celestial equator) and with the opposite side created by dropping a perpendicular from the sun's position to the celestial equator. The angle A at the vernal equinox is 23.5°, the axial tilt of the earth. The opposite side a is the arc 3°54'. We are to find the hypotenuse c, which is the arc along the ecliptic that the sun has traversed since the day it crossed the equator (the first day of Aries). By the Law of Sines for spherical triangles,

$$\frac{\sin A}{\sin a} = \frac{\sin C}{\sin c},$$

we have

$$c = \sin^{-1}\frac{\sin a \sin C}{\sin A}$$

$$= \sin^{-1}\frac{\sin 3°54' \sin 90°}{\sin 23°30'}$$

$$= 9°50'.$$

Note that the scale is $\sin 90° = 60{,}000$, the *seno total*.[153]

On page 82 (the first one) Martínez addresses the problem of calculating the length of one degree of longitude at various latitudes and works an example. In contrast to the calculation already shown, here he uses $\sin 90° = 100{,}000$. One might conclude from this that Martínez had at his disposal two different sets of trigonometry tables. In his reconstruction of Martínez's library, de la Maza suggests that Martínez may have had access to the works of Regiomontanus, and he mentions elsewhere that a cosmographer would have been familiar with Ptolemy's *Almagest*.

There is more trigonometry in Treatise 3, Chapter 4, pages 164 to 166. Again Martínez uses $\sin 90° = 100{,}000$. He calculates the length of the day (hours from sunrise to sunset) using the sun's declination on a given day. As with the other uses of trigonometry, he gives the calculation without explaining why it works.

---

153. It wasn't until Leonard Euler's *Introductio in Analysin Infinitorum* in 1748 that the sine of 90°, the total sine, was fixed at the value 1. In his Chapter VIII he writes, "*Ponamus ergo Radium Circuli seu Sinum totum esse* = 1 . . ." V. Frederick Rickey, in his notes to the *Introductio*, says that "this is essentially the last use of the phrase 'total sine' for the largest sine."

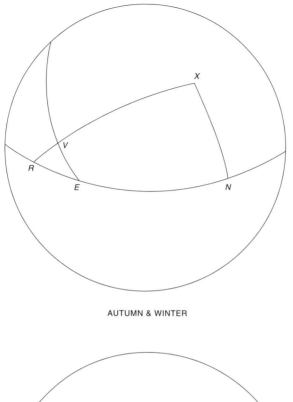

AUTUMN & WINTER

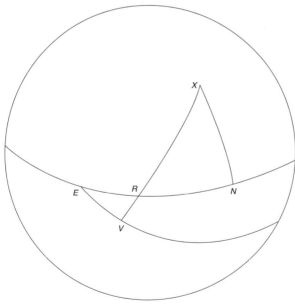

SPRING & SUMMER

Figure 20. The method that Martínez uses to calculate the length of the period of daylight makes use of right triangles on the celestial sphere. The time difference between sunrise and 6 A.M. is proportional to the length of arc EV.

The method makes use of two spherical triangles, one above and one below the horizon. Let X be the celestial North Pole, let E and N be the eastern and northern points on the horizon, let R be the point on the horizon where the sun rises, and let V be the point where the arc XR meets the celestial equator. V will be below the horizon in spring and summer and above the horizon in autumn and winter. Martínez calls $\angle XRN$ the "angle of the horizon with a circle of declination." He can find it because arc XN is the latitude, arc XR is 90° minus the declination of the sun, and angle RNX is 90°, and so, by the Law of Sines applied to triangle XRN,

$$\frac{\sin \angle XRN}{\sin (\text{Latitude})} = \frac{\sin 90°}{\sin (90° - \text{Declination})}.$$

Then he can use this answer in the other triangle, since $\angle XRN = \angle ERV$. He calls arc EV the "ascensional difference." (For the purpose of the next equation, the ascensional difference will be negative when the sun's declination is negative, i.e., during autumn and winter.) Arc RV is the declination and angle REV is 90° minus the latitude. So, using the Law of Sines for triangle ERV

$$\frac{\sin \angle XRN}{\sin (\text{Ascensional Diffence})} = \frac{\sin (90° - \text{Latitude})}{\sin (\text{Declination})}.$$

Now that he knows the ascensional difference he can divide it by 15 (the number of degrees which the celestial equator appears to move through in an hour), add 6, and double the result to get the length of the day in hours.

In this same chapter he gives the declination of the sun at solstice (which equals the axial tilt of the earth) as 23°28′.

Folios 56 through the twentieth unnumbered leaf are a lunario for the years 1606 to 1620, with an explanation of their use on 55r. This may give us a sense of what Martínez's lost almanac of 1604 was like. Chávez's *Chronographía o Reportorio*, which Quintana thinks was the model for the *Reportorio* of Martínez, gives a lunario for the years 1588 to 1606.

Following the lunario on pages 77 to 80 is a table of differences in local time between México and other cities in the world. Chapter 41 discusses conjunctions. Chapter 43 covers the determination of longitude by eclipses. Chapter 44 (p. 84) is on the cause of eclipses. Pages 75 to 82 and 93 to 99 are an ephemeris of eclipses for 1606 through 1615. (Chávez gave a list of eclipses for 1560 to 1605.) Martínez promises to take this to 1640 in the second volume (which he apparently never got around to doing).

In the words of de la Maza, Martínez died "old, wise, and sad" in 1632.

1607

Francisco Juan Garreguilla.   13

LIBRO DE / PLATA REDVZIDA / QVE TRATA DE LEYES BAIAS / desde 20. Marcos, hasta 120. Con sus Abe- / zedarios al margen. Con vna ta- / bla general a la postre. / FECHO POR EL CONTADOR FRANCIS- / co Iuan Garreguilla natural de la Ciudad de / Valencia en Espana. / DIRIGIDO ALOS SEÑORES PRESIDENTE / y Oydores dela Real Audencia y Chancillares desta / Ciudad de los Reyes. / CON LICENCIA. / Impresso en Lima por Francisco del Canto. / Año. M. DC. VII.

Francisco del Canto, Lima.

Medina, *Lima*, 37. Karpinski [6]. Vargas Ugarte 49.

Originals located at: Biblioteca Nacional del Perú, John Carter Brown Library.

A microfilm copy has been made of the John Carter Brown original.

Paging: 300 leaves. This book contains tables, which are indexed by tabs cut into the margins of the pages. There is no page or folio numbering. There are eight leaves of preliminary material.

The tables start on the verso of the ninth unnumbered leaf. Starting with the ninth leaf the signatures go: [no signature], A2, A3, [no signature], B, B2, B3, [no signature], C, C2, C3, [no signature], and so on, skipping J, and ending with N4 (N3v and N4 are blank except for marginal tab "120" in lower right N4r). So the first table has fifty-two leaves.

The second table uses signatures Aa2, Aa3 and so on, taking another fifty-two leaves. A third table goes for another fifty-two leaves using signatures with the pattern Aaa.

Next are four unnumbered leaves. The fourth of these leaves is blank.

Then there are eight more leaves, with a new title page on the first one. The new title is LIBRO DE / PLATA REDVZIDA / QVE TRATA DESDE TREYNTA / Marcos, hasta ciento y veynte y nueue de toda ley, / de dos mil trezientos y ochenta. Con su / Abezedario al margen. Con tres / tablas ala postre. Much of the material after the second title page is identical to what is found in the first eight leaves. This repeated part consists of a *tassa*, a license covering two pages, three approbations, a dedication, and three sonnets. What is changed is the title, the *al lector* (which covers two pages instead of four), and a table of contents (which is omitted). The extra space is taken up by three blank pages.

The next table has signatures in the pattern [no signature], 2, 3, 4, b, 2, 3, 4, and so on, skipping j, skipping u, and skipping w, and after z there is aa, and so on, ending on cc[1]. There are then text pages continuing this pattern through ee4 with many of the signatures missing. So this table and text add up to 112 leaves.

There are then four leaves with no signatures. The last eight leaves have signatures in the pattern [no signature], 2, 3, [no signature], A, A2, A3, [no signature].

As the title would indicate, there are no tables for gold in this text.

Some terms used in this entry were introduced in our discussion of the *Sumario Compendioso* of 1556, and so this section should be read after that discussion.

Tables 1 through 3 give the values of silver in *pesos de minas* for weights from 20 *marcos* to 120 *marcos* and concentrations from 2,100 to 2,370 *de ley*. (Table 1 is 2,100–2,190, Table 2 is 2,200–2,280, and Table 3 is 2,290–2,370.) It would seem that the price of silver has changed not at all since Diez Freyle's *Sumario Compendioso* in 1556, since at 2,200 *de ley* the increment in price for an increase of 1 *marco* is 4 *pesos*, 7 *tomines*, and 2 *granos*, as opposed to 4 *pesos*, 7 *tomines*, $6\frac{1}{2}$ *maravedis* for Diez Freyle. These increments, by the *marco* or ounce, are very inconsistent, however.

Table 4 gives percentages from 140% to 144%. This must be the *tabla general* of the first title. The input is in *pesos, tomines,* and *granos,* and the output is given in both *pesos* of 9 *reales* (the coin that Joán de Belveder and Felipe de Echagoyan called the *peso de plata corriente*) and *pesos* of 8 *reales*. The latter *pesos* are called *patacones* here, as they were by Belveder in 1597, instead of the term *pesos de tipuzque,* which has been used in other sources. A typical entry is that at 140% a 50 *peso* input gives 70 *pesos* of 9 *reales*. As stated in earlier entries, we believe that this sort of table incorporates a tax on top of a conversion. (The rate for converting *pesos de minas* to *pesos de plata corriente* without the tax would be 147%.)

Table 5, after the second title, takes *marcos* of silver at 2,380 *de ley*, converts these to both *maravedis* and *pesos de minas*, and then, as in Table 4, gives the values in both *pesos* of 9 *reales* and *pesos* of 8 *reales* at rates of 140% to 144%. Weights run from 30 *marcos* to 129 *marcos*, $7\frac{1}{2}$ ounces.

Table 6, on the verso of the folio with signature cc, supplements Table 5 by showing how many *reales* to add for different weights and for increments of one eighth of a percent. There is an explanation of this table on the recto of the next folio. Garreguilla uses *tomines* for eighths of a percent, similar to what we see Belveder doing in 1597.

Table 7, on the four folios with no signaturation, is a table of interests from $\frac{1}{8}$% to 5%. We are told that these are the "interests paid in Potosí." Percents are given as *patacones* and *tomines* per 100 *patacones*.

Table 8, on the last eight folios, converts *marcos* of silver at 2,380 *de ley* to *patacones*. The table incorporates a rate of $12\frac{1}{2}$ *reales* to each *peso de minas*, instead of the true value of $13\frac{4}{17}$. This means there is a tax here of $5\frac{5}{9}$%.

1610

Pedro de Aguilar Gordillo.   **14**

ALIVIO DE MERCADERES: / Y TODO GENERO DE GENTE, PARA FACILIDAD / de las quentas que se an de hazer de las Platas que al presente corren / en esta Nueua España. Con las reduciones de Plata à reales, con / forme à la ley del genero que se reduxere, desde media on- / ça hasta dos mil marcos. Con la Tabla de pesos de mi- / nas, reduzidos à pesos de ocho reales, desde vn / peso hasta diez mil. Con el orden de hazer / la quenta de la plata que tiene oro. / POR PEDRO DE AGVILAR GORDILLO, / *Mercader vecino de Mexico*. / Al Doctor Antonio de Morga, del Consejo de su / Magestad, y su Alcalde de Corte en esta Audencia y Chancille- / ria Real de Mexico. / *Con Preuilegio* [sic] *por ocho Años*. / En Mexico, en la Emprenta de Geronymo Balli. / Año, M. DC. X.

Jerónimo Balli, México.

Medina, *México*, 252.

Original located at: Biblioteca Nacional de España.

Paging: Four unnumbered leaves, sixty numbered folios with 47 misnumbered 35.

Some terms used here were introduced in the discussion of the *Sumario Compendioso* of 1556, and so this section should be read after that discussion.

The first three tables are labeled *plata del diezmo*, *plata del rescate*, and *plata quintada*. We believe that these are related to taxes, as in the similar but much shorter tables in the *Sumario Compendioso*. *Plata del rescate* here is the *plata de quinto* there.

The first table starts on the verso of the fourth leaf. Values of silver are given for finenesses of 2,210 to 2,380 in *pesos de tipuzque, reales*, and *granos*. On 1v through 19r, the same range of finenesses is represented with a greater range of weights. We see that 1 *marco* of silver at 2,250 is worth 7 *pesos*, 2 *reales*, 11 *granos* $\frac{9}{17}$ y $\frac{1}{2}$. The fraction at the end appears to stand for

$$\frac{9\frac{1}{2}}{17}$$

of a *grano*, as this makes the total exactly 2,004.75 *maravedis*—exactly 2,250 reduced by 10.9%. The rate of 10.9% can be confirmed from other entries that do not use fractions, e.g., 1 *marco* at 2,210 is 7 *pesos*, 1 *real*, 11 *granos*.

Plata del rescate is covered for 1 ounce and 1 *marco* on 20r and for more weights on 20v through 38r. Finenesses are again 2,210 to 2,380. One *marco* at 2,330 is 6 *pesos*, 6 *reales*, 3 *granos* $\frac{5}{17}$. This is $1,845\frac{1}{3}$ *maravedis*, and that is very close to a reduction by 20.8%. The name *rescate* was used for something else in the *Sumario Compendioso*, but here, *plata del rescate* is what the *Sumario* called *plata del quinto*.

Recto 39 gives the *plata quintada*, this time for 2,200 to 2,380. The meaning seems to be the same as for the *Sumario*—values of silver already taxed at the time of weighing. The value of 1 *marco* at 2,200 is 8 *pesos*, $8\frac{8}{17}$ *granos*, exactly the value that Echagoyan gave in 1603 when the currency was *pesos de tipuzque*. Verso 39 through 58r expands to more weights.

The text on 58v to 59r is a table for converting *pesos de minas* to *pesos de tipuzque*. This is the first use of these terms in this book, as the *pesos de tipuzque* in the earlier tables were simply called *pesos*.

Finally, 59v to 60v deals with silver that contains some gold, with example problems.

Jerónimo Valera.   15
COMMENTARII / AC QVAESTIONES / IN VNIVERSAM ARISTOTE- / LIS AC SVBTILISSIMI DOCTORIS IHOAN- / NIS DVNS SCOTI LOGI- CAM. / TOTVM HOC OPVS IN DVAS PARTES DISTRIBVTVM / offertur:prima continet breve[154] quoddam Logicæ Compendium quod vulgo solet Summa seu Sum- / mulæ Dialecticæ nuncupari Quæstiones pro lego menales , prædicabilia Porphirij , & / Aristotelis Anteprædicamenta,

---

154. The Chile, Harvard, and Indiana copies have "breue."

Prædicamenta & post Prædicamenta. / SECVNDA PARS LIBROS PERIHERMENIARVM SEV / de interpretatione, libros priorum,Posteriorum, Topicorum & / Elenchorum comprehendit. / AVCTORE R. P. F. HYERONIMO UALERA PERVANO ORDINIS MI- / norum Regularis obseruantiæ,Prouinciæ duodecim Apostolorum,Sacræ Theologiæ Lectore iubilato / & in Celeberrimo Limensi Conuentu S. Francisci Guardiano. / CVM PRIVILEGIO. / Limæ Apud Franciscum à Canto. Anno. M. DC. X.

Francisco del Canto, Lima.

Medina, *Lima*, 43. Vargas Ugarte 57.

Originals located at: Biblioteca Nacional de Chile, Biblioteca Nacional del Perú, Harvard University, Indiana University, John Carter Brown Library, Saint Bonaventure University. The California State Library has a copy listed in their catalog, but it could not be found when we visited.

Paging: 432 pages as follows: six unnumbered leaves; thirty-six numbered pages with 23 to 30 misnumbered 7 to 14 (in the Chile and Harvard copies, 7 is printed over with 21 and 14 with 30); and 384 numbered pages with 60 misnumbered 62, 194 misnumbered 192 (except in the Chile and Harvard copies), 196 misnumbered 194, 197 misnumbered 198, 203 unnumbered, 218 misnumbered 162, 330 misnumbered 340, 369 misnumbered 371, and 370 misnumbered 372.

The title page is printed in red and black. In the John Carter Brown copy this has been replaced with a facsimile.

José Toribio Medina lists this entry under 1609, as does Ruben Vargas Ugarte. The main title refers to a date of 1610, but the printer's colophon on page 384 says 1609.

This is the first logic book out of Lima, and, therefore, out of South America. Unlike the previous logic books, this one has no figures. In addition to Aristotle, the title mentions the Franciscan John Duns Scotus (1265/1266–1308) an important figure in medieval philosophy.

Valera (c. 1569–1625) was a Franciscan who became Provincial in Peru in 1614. He was born in the Chachapoyas district in Peru. It is not known whether his background was *criollo* or native.

1615

Álvaro de Fuentes y de la Cerda.   16

LIBRO / De quentas, y / REDVCCIONES / de plata y oro , y otras ta- / blas faciles y prouechosas, para / la contratacion desta Prouin- / cia de

Guathemala . / *PARA CVYO EFECTO* / *fueron hechas por Don Aluaro* / *de Fuentes y de la Çerda* / *vezino della* . / *Con licencia y priuilegio* . / EN MEXICO. / *En la Emprenta de Ioan Ruyz*. / Año de M. DC. XV.
Juan Ruiz, México.
Medina, *México*, 294. Karpinski [7]. González de Cossío 137.

Elías Trabulse includes a copy of the title page on page 66 of *Historia, Siglo XVI*. Francisco González de Cossío says that his reference comes from Bibliografía Americana, núm. 5, de Porrúa Hnos. y Cía., núm. 6883, which also shows a title page.

1621

Pedro de Paz. 17
Declaracion de los puntos convenientes y necesarios para repartir con exactitud las Rentas eclesiasticas en las catedrales de la Nueva España. Por D. Pedro Paz, contador de diezmos de la Iglesia Metropolitana de Mexico. Mexico, 1621.
México.
Medina, *México*, 335.

1623

Pedro de Paz. 18
Arte para / aprender Todo el / menor del Arith- / metica, sin Maestro. / Dirigido al Doctor / Don Diego de Gueuara y Estrada, Chantre / de la sancta Iglesia Metropolitana / de Mexico. / Hecho por Pedro de Paz Conta- / dor de la dicha sancta Iglesia. / Con Privilegio. / Impresso en Mexico, por Ioan Ruyz. / Año de 1623.
Juan Ruiz, México.
Sabin 59336. Medina, *México*, 352. Karpinski [8].

Original located at: University of California Berkeley, University of Oxford.

Paging: 195 leaves as follows: eight preliminary leaves, 181 numbered folios with 1 unnumbered, 107 misnumbered 106, and 109 misnumbered 108, and finally six unnumbered leaves for a table of contents.

The Oxford copy has preliminary leaves [4] and [7] and all of the content pages; it is missing leaf [4] of the table.

The copy at Berkeley is very far from complete, but is accompanied by photocopies of most of the missing pages. Original leaves remaining are 25 to 41, 43 to 45, 48 to 71, about half of 72, 80 to 89, 91 to 94, 96 to 178, and the second leaf of the table, bound in backwards. At the time of acquisition (1927), another copy was owned by Dr. Nicolás León, a Mexican physician and book collector, and this copy has been used to supply in facsimile the remaining pages except for preliminary leaves [1] to [3], preliminary leaves [6] to [8], folios 2 to 7, 42, 47, 73, and the last three leaves of the table. Some of the contents of the missing preliminary leaves are supplied in typewritten form, probably taken from José Toribio Medina, who indicates that Andrade saw an incomplete copy.

A survey of this book was done by Florian Cajori in his 1927 paper.[155] We will just give the highlights of this paper.

Cajori reports that Paz borrowed heavily from a European work, the 1562 *Arithmetica practiva, y especulative* of Juan Pérez de Moya. Many of the same examples are used, with the same numbers, and sometimes with the same words.

One interesting piece of notation that Paz picked up from Pérez de Moya is illustrated by the following example (in Cajori's words) from the chapter on "The Rule of Three for Fractions":

For example (fol. 131), if $\frac{2}{3}$ yd. of cloth is worth $\frac{4}{7}$ ducados, how much is $\frac{1}{3}$ worth? Write

$$\begin{matrix} 2 \\ 3 \end{matrix} \times \begin{matrix} 4 - 1 \\ 7 - 3 \end{matrix}$$

Multiply together the numbers connected by dashes: $3 \times 4 \times 1 = 12$, $2 \times 7 \times 3 = 42$. Answer: $\frac{12}{42}$ or $\frac{2}{7}$.

This use of the X can be seen in many places in this work. It is used in adding and subtracting fractions (chapters VII and VIII, respectively), and we will give an illustration of that in the next entry.

Cajori notes that Paz makes no use of the *calderón* (see our entry for 1597) and that, like Perez de Moya, Paz uses *cuento* for millions. Paz's method of division is the galley method (see our discussion of this in our entry for 1587).

155. Florian Cajori, "The Earliest Arithmetic Printed in America," *Isis* IX 3/31 (1927): 391–401.

There is a trick for mental multiplication of numbers between 10 and 20. Cajori gives 14 × 17 as an example. 14 + 7 = 21, and this times 10 is 210. 4 × 7 = 28 and 210 + 28 = 238, the product of 14 and 17.

One interesting way to manipulate fractions is not mentioned by Cajori. On 112r and 112v, Paz shows an example of what he announces to be a new operation on fractions in addition to the four operations of addition, subtraction, multiplication, and division. The example is to divide $17\frac{7}{13}$ by 6. He does this by first dividing 17 by 6 and getting $2\frac{5}{6}$. Then he performs what amounts to the calculation

$$\frac{5 \times 13 + 7}{6 \times 13} = \frac{72}{78} = \frac{12}{13}.$$

He says that the Italians call this to *infilzar* the fractions, probably from an Italian verb meaning "to pierce." The answer is then $2\frac{12}{13}$. What he did to the fractions is similar to combining fractions with a common denominator, but here, by failing to multiply the 7 by 6, he makes up for the fact that the $\frac{7}{13}$ was never divided by 6.

Paz promised his readers a second book on algebra. It is not known to have appeared.

1649

Atanasio Reatón, Pasamonte.   19

ARTE / MENOR / DE / ARISMETICA, / Y / MODO DE FORMAR / CAMPOS. / TRATA LAS CVENTAS QVE SE PVEDEN / OFRECER EN LOS REYNOS DE SV / MAGESTAD: / POR ESTILO MVY CLARO, Y BREVE PARA / QVE SE APRENDAN SIN MAESTRO. / *A PEDRO DE SOTO LOPEZ,* / *Contador del Tribunal del S. Oficio de la Inquisicion* / *desta Nueua Espana , Prior de la Vniuersidad de los* / *Mercaderes deste Reyno , Iuez administrador* / *de las Reales Alcaualas.* / POR ATANASIO REATON, PASAMONTE, / vezino desta Ciudad. / *CON LICENCIA EN MEXICO.* / Por la Viuda de Bernardo Calderon, Año de 1649.

Widow of Bernardo Calderón, México.

Medina, *México*, 690. Karpinski [9].

Originals located at: Biblioteca Nacional de Chile, University of Texas at Austin.

Microfilm copies are available.

Paging: Four preliminary leaves, folios numbered 1 to 78.

The text is divided into twenty-four chapters.

This book was the subject of a 1926 article by Louis Karpinski, based on a talk given in 1925.[156] The article appeared in the year just before a copy of the previous entry was acquired by the Bancroft Library in Berkeley and became available to Florian Cajori for his article with a similar title.

In this work another use of the X in arithmetic appears. The problem is to subtract fractions, $\frac{3}{4}$ from $\frac{4}{5}$. The fractions are written before and after the X and numbers along the dashes are multiplied, as in the previous entry. The products along the dashes are written above the fractions, and the product of the denominators is written below the X. Thus we have

$$\begin{matrix} 15 & 16 \\ \frac{3}{4} \diagdown \diagup \frac{4}{5} \\ \diagup \diagdown \\ 20 \end{matrix}.$$

The difference of the products at the top, 16 − 15, over the product at the bottom, 20, is the result, i.e.:

$$\frac{4}{5} - \frac{3}{4} = \frac{1}{20}.$$

Squadrons are discussed in Chapter 23, 52r to 68r, the longest chapter.

The rule of three, with and without time, is treated. An example of the rule of three with time is the following: If with 30 *reales* in 4 months I gain 44, with 50 *reales* in 3 months I gain how much? The calculation is

$$\frac{50 \times 3 \times 44}{30 \times 4} = 55.$$

The problem reduces to the usual rule of three, using the numbers $30 \times 4 = 120$, 44, and $50 \times 3 = 150$. Note that for this to work, simple interest is being assumed.

1650

Pedro Diez de Atienza. **20**

Exmo S^{or}. / El Bachiller Pedro Diez de Atiença presbytero, dize, que por quanto en dias / passados presentò ante el señor Marques de

---

156. Louis C. Karpinski, "The Earliest Known American Arithmetic," *Science* 63 (Feb. 19, 1926): 193–195.

Mancera, Virrei q' fue destos Rei- / nos del Peru, ciertos memoriales, que, estàn en el oficio de gouierno, sobre la reduc- / cion de monedas, i otros puntos, etc.

Lima.

Medina, *Lima*, 318. Vargas Ugarte 396.

Medina lists a copy at the Archivo General de Indias, Sevilla.

Paging: Two leaves.

This is the first of six entries on currency reform through 1691. The status of these items as members of this list hangs upon what we see in the two that are available on microfilm, one in 1657 and the other in 1691.

Francisco de Villegas.   21

Repvesta / a la obiecion opvesta por / Migvel de Roxas al informe pre- / sentado a / los Señores de la junta, en orden a representar los inconuenien- / tes, que se han de seguir a los comerciantes deste Reyno, y de Es- / paña, no corriendo la quenta de las barras por mareuedis, / y ajustado a decenas, como hasta aqui / se á acostumbrado. / Discvrrida por el mismo Francisco / de Villegas, Ensayador, y Contraste publico desta / ciudad de los Reyes. / A pedimeeto [sic] del Prior, y Consvles / del comercio deste Reyno.

Lima.

Medina, *Lima*, 333. Vargas Ugarte 432.

Medina lists a copy at the Archivo General de Indias, Sevilla, with number 109-8-17, tomo 17. When we visited the Archivo General de Indias, we were told that this number had never been used.

Paging: Four leaves.

1652

Pedro Diez de Atienza.   22

Paracer, qve dio el Lic. Pedro Diez de / Atiença, presbitero, sobre el cumplimiento de la cedula, que tuuo / de su Magestad, cerca de la reformacion de la moneda, el Exce- / lentissimo Señor Conde de Saluatierra, Virrey del Peru.

Lima.

Medina, *Lima*, 345. Vargas Ugarte 465.

Medina lists a copy at the Archivo General de Indias, Sevilla.

Paging: Five leaves.

1657

Luis Enríquez de Guzmán.   **23**
Don Lvis Enriqvez de Gvzman Conde / de Alua de Aliste, y de Villaflor, &c. ... / ... Por quanto cumpliendose en esta Ciudad la vltima prorrogacion que se concedio, para que pu- / diesse correr la moneda resellada de reales de a ocho, y de a quatro, a / siete reales y medio los de a ocho, y a la mitad los de a quatro: mandè / etc.
Lima.
Medina, *Lima*, 378. Vargas Ugarte 510.
Medina lists a copy at the Archivo General de Indias, Sevilla.
Paging: Two leaves.
Enriquez de Guzmán was sent to Mexico as Viceroy in 1649 and on to Peru as Viceroy in 1655.

Francisco de Villegas.   **24**
REPAROS / QVE PROPONE AL EXC.^{MO}S.^{OR} / CONDE DE ALVA DE ALISTE / Virrey deste Reyno, en orden a la mo- / neda resellada. / A PEDIMIENTO DEL PRIOR, Y / *Consules desta Ciudad de los Reyes.* / FRANCISCO DE VILLEGAS ENSAYADOR APRO- / bado,y Contraste publico della.
Lima.
Medina, *Lima*, 383. Vargas Ugarte 514.
Microfilm copies are available. The image shows the stamp of the Biblioteca Nacional de Chile.
Paging: Eight leaves. Text starts on title page. There are forty numbered paragraphs and four unnumbered, two at the beginning and two at the end.

1660

Juan de Figueroa.   **25**
OPVSCVLO / DE ASTROLOGIA EN / MEDICINA, Y DE LOS TERMINOS, / Y PARTES DE LA ASTRONOMIA / NECESSARIAS PARA EL VSO DELLA: / COMPVESTOS POR IOAN DE FIGVEROA, FAMI- / liar del Santo Oficio de la Inquisiciõ, Regidor,y Tesorero de la / Casa de la Moneda de la ciudad de los Reyes,veintiquatro, / Ensayador,y fundador

mayor de Potosi. / DIRIGIDOS AL EXC.^{MO} S.^{OR} DON LVIS / Henriqvez de Gvzman Conde de Alva de Aliste , y Villaflor, / Grande de España, Virrey ,Gouernador,y Capitan general de / los Reynos del Peru,Tierrafirme, y Chile / Con licencia En Lima, Año de 1660.

Lima.

Medina, *Lima*, 401. Vargas Ugarte 536.

Originals located at: Biblioteca Nacional de Chile, Biblioteca Nacional de España, British Library, Real Instituto y Observatorio, Universidad Complutense de Madrid, Williams College, Yale University.

Microfilm copies are available.

Paging: Total of 357 leaves. Fourteen unnumbered leaves with a table of contents on the last five, then folios 1 to 349 with the following irregularities: unnumbered are 8, 16, 212, 258; misnumbered 80 as 79, 287 as 278, 321 as 322; skipped are 165, 166, 195, 196, 273, 274. The copy at Yale is missing leaf 160. Leaf 76 at Williams is inverted.

The work starts, as so many of its contemporaries do, with a description of the sphere.

Chapter 34 of Opúsculo I is "De la equacion del tiempo por la inequalidad de los dias" with a table on 65r and 65v. "Equation" is used here not in the current mathematical sense but the sense of balancing. The "equation of time" is the term that must be added to or subtracted from clock time to get solar time. This varies throughout the year owing to two causes. One is the eccentricity of the Earth's orbit, which leads to variation in the rate of the Sun's apparent motion through the ecliptic. The other is the varying slope of the ecliptic with respect to the celestial equator. This means that when the Sun's motion around the ecliptic is translated into a rate of change of right ascension, the rate of change will be greatest at the solstices and somewhat less the rest of the year.

The author states on 64r that the Primer Semicirculo, Aries to Virgo, takes 186 days 8 hours, and the Segundo Semicirculo, Libra to Pisces, takes 178 days 23 hours. The first figure is lower than we see now. The years from 1992 to 2001 saw the time between the Vernal and Autumnal equinoxes vary from 186 days, 9 hours, 33 minutes to 186 days, 10 hours, 1 minute, with an average of 186 days, 9 hours, 50 minutes, according to aa.usno.navy.mil/data/docs/EarthSeasons.php. The time for the combined seasons of spring and summer should have been longer in the seventeenth century, not shorter than now, because the perihelion was closer to the winter solstice then. Likewise, the span just given for the combined seasons of autumn and winter is too long. For the ten years ending with the vernal equinox of 2002, this averaged only 178 days, 20 hours, 1 minute. We will return to this issue in Part 2 when we discuss the 1678 almanac by Thomas Brattle.

Folios 55r through 57r have a table of latitude and longitude for American cities and places from Labrador to the Straits of Magellan.

Folio 211v has a table of declinations of each degree of the ecliptic. The table on 212r incorporates the latitude of a planet from the ecliptic. The axial tilt of the Earth implied by these tables is 23°31′30″.

All of Opúsculo V discusses eclipses, with explicit calculations for many examples.

Chapter 22 of Opúsculo VI is on comets. Figueroa quotes the opinions of many experts. (See the entry for 1690.)

Antonio de Heredia y Estupiñán. 26
TEORICA, Y / PRACTICA / DE ESQVADRONES, / *DEDVCIDA* / DEL TESORO MILITAR / QVE TIENE ORDENADO PARA / DAR A LA PRENSA / EL CAPITAN Y SARGENTO MAYOR / DON ANTONIO DE HEREDIA / Y Estvpiñan. / *DEDICALA* / AL EXC.^{MO} S.^{OR} CONDE / de Alva de Aliste , y de Villaflor, / Virrey,Gouernador, y Capitan ge- / neral destos Reinos del Peru, / Tierrafirme,y Chile,&c. / CON LICENCIA, / En Lima,por Ioseph de Contreras,Año de 1660.

Joseph de Contreras, Lima.

Medina, *Lima*, 402. Vargas Ugarte 537.

Originals located at: Biblioteca Nacional de Chile, Biblioteca Nacional de España.

Microfilm copies are available.

Paging: Eight unnumbered leaves, fifty-four numbered folios, two leaves for a plate, and four leaves for the table of contents. There are seven chapters with numbered paragraphs within each chapter.

This is the third book on our list to treat squadrons. On 32r the following problem is given: 1029 soldiers are to stand in *terreno* proportion, i.e., so that the ranks and files will be two and one third to one. Thus: $1029 \times 2\frac{1}{3} = 2401$ and $\sqrt{2401} = 49$. So the answer is 21 by 49.

The author speculates about who invented algebra on folio 27v. "[Some say] that the inventor was an Arab named Geber, and that from this name was derived 'algebra.'" Geber is the Latinized version of Jabir ibn Aflah (c.1100–c.1160), who was a native of Moorish Spain, and also of Jabir ibn Hayyan (ninth century). The currently accepted etymology of the word "algebra" is that it was derived from the word *al-jabr* in the title of al-Khwārizmī's *Al-kitāb al-mu<u>h</u>taṣar fī hisāb al-jabr wa-l-muqābala* (*The Condensed Book on the Operations of Restoring and Comparing*). *Al-jabr* or "restoring" refers to the act of adding the same quantity to both sides of an equation to simplify and solve it, as in

$$5x = 14 - 2x$$
$$7x = 14.$$

$14 - 2x$ is regarded as a "deficient" 14, and adding the $2x$ "restores" it to a whole 14. *Al-muqābala* or "comparing" refers to the act of subtracting equal quantities. Smith quotes a title containing the words "Algebra et Almucabala" for an early European (Paris, 1577) incarnation of al-Khwārizmī's terms.[157] Sometime later, "almucabala" was dropped and the subject became simply "algebra."

1668

Juan de Castañeda.   27

REFORMACION / DE LAS TABLAS, / Y QVENTAS DE PLATA, Y DE / LA QVE TIENE ORO. / VAN AÑADIDAS TRES REGLAS BREUES / generales , para que cada vno sepa con solo multi- / plicar lo que se deve á su Magestad , de la Plata y / Oro, que se fuere á quintar. Con tres Tablas para / lo mismo. Con dos Tablas para el quinzavo , y / consumido de la Plata, que los Mineros v an á / marcar. con la Tabla de pesos de Minas / reduzidos á pesos de Tipuzque. / Por Iuan de Castañeda , natural de / San Iuan de Pineda en Cataluña. / Con Licencia, / Del EXCELLENTISSIMO S^{or} / Marques de MANCERA, Virrey desta Nueva / España, &c. / En MEXICO: Por Francisco Rodriguez Luperci o / Mercader de libros en la puente de Palacio. 1668

Francisco Rodríguez Lupercio, México.
Medina, *México*, 990.

---

157. Juan Diez Freyle, *Sumario Compendioso*, Mexico, 1556: 15.

Original located at: British Library. The California State Library has a copy listed in their catalog, but it could not be found when we visited.

Paging: Total of sixty-eight leaves. Three unnumbered leaves followed by folios 1 to 64, with 9 and 64 unnumbered and two 28s. There are two blank leaves inserted between folios 39 and 40 in the British Library copy.

The copy at the British Library is bound with a similar book by Juan Vázquez de Serna published in Cádiz in 1620.

Some terms used in this entry were introduced in our discussion of the *Sumario Compendioso* of 1556, and so this section should be read after that discussion.

The first table, on 2v to 20r, gives values of *plata del diezmo* that are very close to those of Pedro de Aguilar Gordillo's *Alivio de Mercaderes* of 1610. Values are given in *pesos*, *tomines*, and *granos*, and these *pesos* are *pesos de tipuzque*, as in the *Alivio*. One difference is the presentation of the fractions. Aguilar Gordillo wrote

$$\frac{9}{17} \; y \; \frac{1}{2}$$

for nine and a half seventeenths, where Castañeda is not averse to writing $\frac{19}{34}$ and indeed uses much higher denominators. At a fineness of 2,250, one *marco* is 7 *pesos*, 2 *tomines*, $11\frac{19}{34}$ *granos*, exactly agreeing with Aguilar Gordillo. At 2,210, the value for one *marco* is 7 *pesos*, 1 *tomín*, $10\frac{49}{50}$ *granos*, where Aguilar rounds off to 11 *granos*. At 2,310, Castañeda gives 7 *pesos*, 4 *tomines*, $6\frac{363}{850}$ *granos*. These numbers are the exact answers that result from using 10.9% for the diezmo.

Folios 20v and 21r give values based on one *marco* being exactly 7 *pesos*, 6 *tomines*, 4 *granos*. This is an approximation to the 2,380 level in the previous table, where the value of 1 *marco* is 7 *pesos*, 6 *tomines*, $4\frac{11}{25}$ *granos*.

Folios 22v to 39r are for *plata del rescate*. For 1 *marco* the value is 6 *pesos*, 6 *tomines*, $3\frac{129}{425}$ *granos*. This is an exact answer for the rate of 20.8%.

Here there are two blank leaves on which a previous owner of the only known copy has worked out a problem. Then folios 40v to 58r give *plata quintada*. The denominators never have to go above 17 here because each 10 *maravedis* adds $3\frac{9}{17}$ *granos* to the value of a *marco*.

On 58v and 59r is a table of *quinzavos*, or fifteenths, that miners pay. Input is in *marcos* and output is in *marcos*, ounces, and *ochavas*. On 59v and 60r is a table of twentieths.

The next four tables give the cost of mercury used in processing silver. It is assumed that one *quintal* of mercury costs 60 *pesos*. The four tables make different assumptions about how much silver a quintal of mercury is good for.

On 64v is a table for converting *pesos de minas* to *pesos de tipuzque*. This is the first mention of *minas* in the book and the first use of the word *tipuzque* since the prologue.

Between 58r and 58v there should have been more material, according to the title and the introduction, which treated gold and three rules for taxes.

The *calderón* symbol, for one thousand, is used throughout the text.

1672

John Eliot. 28

THE / *Logick Primer*. / Some Logical Notions to initiate / the *INDIANS* in the know- / ledge of the Rule of Reason ; / and to know how to make / use thereof. / Especially for the Instruction of / such as are Teachers / among them. / Composed by *J. E.* for the / use of the *Praying Indians.* / The use of this Iron Key is to / open the rich Treasury of / the holy Scriptures. / Prov. 1. 4. *To give subtilty to the / simple ; to the young man know- / ledge and discretion.* / Printed by *M. J* 1672.

Marmaduke Johnson, Cambridge.

Sabin 22163. Evans 166. Wing E518.

Original located at: British Library. The British Library copy was incorrectly shelved after it returned from having been rebound and could not be located as of 2006.

Readex microprint and microform copies are available. The complete text of the British Library copy is available online at Evans Digital Edition.

Paging: Thirty-nine leaves.

A reprint was done in 1904. It has an introduction by Wilberforce Eames, which says that six photocopies were made in 1889 from the British Library copy.

Most of the text is in Massachusett, a member of the Algonquian language family, with an interlinear word-by-word translation into English. Inherent to this style of translation, sometimes the English is very awkward because the Algonquian word order is imposed upon it. The last $14\frac{1}{2}$ pages (18 in the reprint) are in Algonquian only.

It is recorded that 1,000 copies were printed. Johnson was paid £6 according to Eames and to Trumbull.

John Eliot was born in 1604 and died May 21, 1690. His work as the "Apostle to the Indians" led the establishment of fourteen villages of Christianized natives. Chief among these was Natick, which is still a town in the greater Boston area.

A corporation in England raised money for Eliot's work and arranged for the printer Marmaduke Johnson to travel to Cambridge, Massachusetts, to work with the one printer already there. Eliot produced a Bible, translated into Massachusett and printed by Samuel Green and Marmaduke Johnson in 1663. (The scientist, Robert Boyle, was head of the corporation at this time and made the presentation of Eliot's Bible to King Charles II in 1664.) This was followed by *The Indian Grammar* begun in 1666, *The Indian Primer* in 1669, and *Indian Dialogues* in 1671, all printed by Johnson. The logic is clearly part of a program that tried to assimilate the Massachusett tribe into the Puritan worldview.

With King Philip's war in 1675 and 1676, the Massachusett tribe began to disappear. Even though they had collaborated with the English, they were not trusted. Many were killed, and the few remaining were eventually assimilated into other tribes. The Eliot Bible is now known in some sources as "the book that no one can read."

In the text at hand, syllogisms are discussed and classified into types. Eliot gives examples of each type, and frequently this is done by citing Bible verses. His first example of a syllogism, on 23r to v, may be paraphrased as the following:

> Everyone to whom belongeth the promise may be baptised.
> The promise belongeth to the infants of believers.
> ∴ The infants of believers may be baptised.

A later example of a syllogism is an application of the law of *modus ponens*. On 29v we see

> If unbelief driveth us from God then we must beware of it.
> Unbelief driveth us from God.
> ∴ We must beware of it.

We have put into Appendix C our argument that such a work belongs in a list of mathematical texts. In particular, we may cite here the classification of arguments into types based on their structure as an example of what mathematics does.

For the fan of interesting words, we note that the Algonquian word for logic is *anomayag* and that the word for syllogism is *oggusanukoowaonk*.

1675

Benito Fernández de Belo.   **29**

BREVE / ARITMETICA, / POR EL MAS SVCINTO MODO, QVE / hasta oy se ha visto: / Trata en las quentas que se pueden ofrecer para formar / Campos, y Esquadrones. / *COMPVESTA POR* D. BENITO FERNANDEZ / DE BELO , *Natural del Lugar de Arce del Valle de Piela- / gos montañas de Burgos, en España : soldado que fue en Car- / tagena de las Indias el año de 54. Y en el de 55 se halló de sol- / dado en la ocasion que se derrotò al enemigo Ingles en la campa- / ña de la Isla de S. Domingo: y aviendo dexado la campaña el ene / migo, con perdida de mas de seis mil hombres , y seis vanderas. / Saliò con licencia del señor Presidente de dicha Isla , para los / Reynos de España , en cuya ocasion fue prisionero de dichos In- / gleses , y llevado al Castillo de Londres , en Inglaterra , donde / passò à los Estados de Flandes , y bolviò à continuar el servir à / su Magestad de soldado en la Armada Real,en la Real Capita- / na en la Compañia del Excelentissimo Señor / Duque de Beragua.* / DEDICALA, / A DON DIEGO DE SIERRA, OSSORIO , Y FLORES, / Alcalde Mayor, y Capitan Aguerra, Theniente General / de la Provincia de Guachinango , y costas de Barlovento / por su Magestad. / CON LICENCIA. / EN MEXICO , por la Viuda de Bernardo Calderon , Año de 1675.

Widow of Bernardo Calderón, México.
Medina, *México*, 1125. Karpinski [10].
Originals located at: Biblioteca Nacional de España, John Carter Brown Library.
A microfilm copy has been made of the John Carter Brown original.
Paging: four unnumbered leaves followed by eleven numbered folios and one folded plate of figures. The Madrid copy has a blank leaf before the plate.
The copy in Madrid has variations on the title page after the author's name as follows: *Natural del Lugar de Arce del Valle de Piela- / gos montañas de Burgos, en España: Soldado que fue en Cartagena de / las Indias el*

Año de 54. Y en el de 55. se halló de Soldado en la ocasion / que se derrotò al enemigo Ingles en la campaña de la Isla de Santo Do- / mingo : y aviendo dexado la campaña el enemigo, con perdida de mas / de seis mil hombres, y seis vanderas: Saliò con licencia del señor Pre- / sidente de dicha Isla, para los Reynos de España, en cuya ocasion fue / prisionero de dichos Ingleses, y llevado al Castillo de Londres, en Ingla- / terra, de donde passò à los Estados de Flandes, y bolviò à continuar el / servir à su Magestad de Soldado en la Armada Real, en la Real Capi- / tana, en la Compañia del Excelentissimo Señor Duque de Beraguas. / DEDICALA, / A Don Diego De Sierra, Ossorio, y Florez, / *Hijo primogenito de D. Diego de Sierra, Ossorio, y / Florèz, Señor de la Casa de Sierra, vna de las mas ilus- / tres del Principado de Asturias, y de quien en este Rey- / no descienden de ella ilustres Familias: Alcalde Mayor, / y Capitan-Aguerra, Theniente General de la Provincia / de Guachinango, y costas de Barlovento / por su / Magestad. /* CON LICENCIA. / En Mexico, *por la Viuda de Bernardo Calderon, Año de 1675.*

The entire text is available at www.southernct.edu/~sandifer/Ed/History/Breve%20aritmetica/breve.htm by courtesy of Ed Sandifer. An English translation was done as a thesis by April Wyler.

Fernández de Belo gives a short autobiography on the title page. He is from the village of Arce in the valley of Pielagos [in the] mountains of Burgos.[158] While soldiering in the Indies in 1655, he was captured by the English and taken to London as a prisoner. From there he somehow made his way to Flanders, then still part of the Spanish Empire, and thus returned himself to service in the Royal Navy of Spain.

This work treats basic arithmetic, with a section on squadrons on folios 4r through 11r. Pasamonte Atanasio Reatón's book from 1649 seems to be an influence, based on the similarity of the figures.

1681

Eusebio Francisco Kino. 30

EXPOSICION / ASTRONOMICA / DE EL COMETA, / *Que el Año de 1680. por los meses de / Noviembre, y Diziembre, y este Año de 1681. por los meses / deEnero y Febrero,se ha visto en todo el mundo, / y le ha observado en la Ciudad de Cadiz, /* EL P. EUSEBIO FRANCISCO KINO /

---

158. There is a city of Pielagos in Spain, but it is much closer to Santander than to Burgos.

De la Compañia de JESVS. / Con LICENCIA, en Mexico por Francisco Rodriguez Lupercio. 1681.
Francisco Rodríguez Lupercio, México.
Sabin 37936. Medina, *México*, 1228.

Originals located at: Biblioteca Cervantina, Biblioteca Nacional de México (two copies), Biblioteca Palafoxiana, British Library, California Historical Society, Centro de Estudios de Historia de México Condumex, Huntington Library, John Carter Brown Library, Universidad de Santiago de Compostela, University of Michigan, University of Texas at Austin (two copies), Yale University.

Paging: Eight unnumbered leaves, one folded plate insert, and folios numbered 1 through 28.

There is an excerpt in Elías Trabulse's *Historia de la Ciencia en México... Siglo XVII*. The stellar map showing the path of the comet was reproduced there on page 140 and also in *Kino and the Cartography of Northwestern New Spain* by Ernest J. Burrus in 1965. Microfilm copies are available. The entire text of one of the Biblioteca Nacional de México copies is available at www.cervantesvirtual.com.

The copy at the British Library is bound with Gabriel Lopez de Bonilla's *Discurso, y Relación* of 1654, José de Escobar's *Discurso Cometológico* of 1681, and Gaspar Juan Evelino's *Especulación Astrológica* of 1682, making this tome a one-volume anthology of works favoring the dire significance of comets. A slightly more balanced compilation exists at the Biblioteca Nacional de México, where we find within one binding Carlos de Sigüenza y Góngora's *Libra Astronómica* of 1690, Kino's title under discussion, López de Bonilla's *Discurso, y Relación* (title page missing), Diego Rodríguez's *Discurso Etheorológico* of 1652, and Evelino's *Especulación Astrológica*. (See Appendix B for more about comet titles.) Copy 1 at the University of Texas has two copies of folio 1. Copy 2 is preliminary leaves 2 through 8 only and is bound with works by Sigüenza. The plate is detached and held separately at the John Carter Brown Library, is replaced by a facsimile at the University of Michigan, and is missing from copy 2 at the Biblioteca Nacional de México.

Kino wrote to the Duchess of Abeiro y Arcos that he was sending her one hundred copies of this book, from which he asked her to send six to Seville and six to Rome for various contacts of his.[159]

---

159. Irving A. Leonard, *Don Carlos de Sigüenza y Góngora*, Berkeley, CA, 1929: 72.

Figure 21. Title page of the *Exposición Astronomica*, 1681. The John Carter Brown Library at Brown University, Providence, RI.

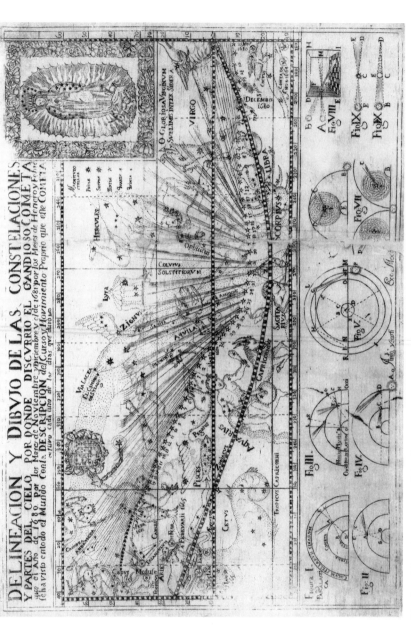

Figure 22. Eusebio Kino's map of the heavens, from the *Exposition Astronómica*, showing the path of the comet of 1680-1681. The lower strip contains the ten figures to which Kino refers within the text. The John Carter Brown Library at Brown University.

The book is divided into ten chapters, with Chapter 10 divided into five sections. Kino discusses parallax and at one point explicitly invokes a proposition of Euclid to support a step in his argument. He includes an insert showing a map of the sky with the path of the comet shown. Along the lower strip of this plate are Figures I through X, which are referred to in the text and which illustrate Kino's remarks about parallax and his other geometrical assertions.

Kino's story is intertwined with that of Sigüenza y Góngora for the purpose of the present study, so please refer to the discussion of the *Libra Astronómica* of 1690 for further information about the role played by Kino and his *Exposición Astronómica*.

1682

Martín de Echagaray. **31**
DECLARACION / DEL QVADRANTE / DE LAS CATHEDRALES DE / LAS INDIAS. / Con vna nueva Regla para facilitar sus / quentas, y otras. / *Por el H. MARTIN DE ECHAGARAY* / de la Compañia de JESVS. / DEDICALA / Al Señor Alferez IVAN HORTIZ DE ZA- / RATE, SAENZ DE MATVRANA, / Y TORREALDE. / Año de 1682. / CON PRIVILEGIO. / *En Mexico*: *por Francisco Rodriguez Lupercio. Mercader de libros en / la puente de Palacio.*

Francisco Rodríguez Lupercio, México.
Sabin 21767. Medina, *México*, 1243.
Originals located at: Biblioteca Nacional de Chile, Biblioteca Nacional de España, Indiana University.
Microfilm copies are available.
Paging: Eight preliminary leaves followed by twenty-six numbered folios, with 26 misnumbered 25. The title page is missing in the copy in Madrid and the first and last leaves are missing in Indiana.
The title page and dedication use the spelling Martín de Echagaray, but Medina gives Echegaray. The text is divided into thirty-two sections, with Part 1 starting on 1r.
The *quadrante* of the title seems to refer to a division of money into four unequal parts (*El Rey, fabrica, hospital*, and *mesa capitular*). On 7v, portions are created according to the fractions

$$\frac{2}{9}, \frac{1\frac{1}{2}}{9}, \frac{1\frac{1}{2}}{9}, \frac{4}{9}.$$

Some terms used here were introduced in the discussion of the *Sumario Compendioso* of 1556, and so this section should be read after that discussion.

The author uses 12 *granos* to the *tomín* (for instance, saying that 24 masses said times 3 *pesos* 2 *tomines* 3 *granos* each is 78 *pesos* 6 *tomines*) but at the same time treats a *peso* as approximately 100 *granos*, i.e., $8 \times 12\frac{1}{2}$.[160] We see this on folio 19r, where he shows how to divide 156 *pesos* among fifteen people. First one adds two zeros to make 15,600. Then divide this by the 15 to get 1,040 (using the galley method of division, which we have already treated in our discussion of the entry for 1587). He writes this as 10.40. to show that we must remove the factor of 100 that was put in.[161] The 40, he says, is 10 *granos* less than 50, which is 4 *tomines* (at 100 *granos* to 8 *tomines*, i.e., $12\frac{1}{2}$ *granos* to a *tomín*). So the answer is 10 *pesos* 3 *tomines* 2 *granos* (at 12 *granos* to the *tomín*). This method he calls the "rule of aggregation."[162] The division actually should give 10 *pesos* 3 *tomines* 2.4 *granos*.

Echagaray tries to explain his conflicting numbers for *granos* in a *peso* on 19r: "Aqui no se atiene a que el peso tiene 96. granos, sino a que la cosa entera vale 100. Ya que tanto vale la mitad de 100. como la mitad de 96. en razon de quebrados pues $\frac{50}{100}$ abos es $\frac{1}{2}$ como $\frac{48}{96}$ abos abreviados."

A similar example is worked on 19v and 20r. One is to divide 20 *pesos* 6 *tomines* 9 granos among 12 people. He converts the amount to 2,084 *granos* (which assumes $12\frac{1}{2}$ *granos* to the *tomín*). He then divides by 12 and writes 1.73, which ignores a whole $\frac{2}{3}$ of a grano. Then since 73 is 75 − 2, the answer must be 1 *peso* 5 *tomines* 10 *granos*. His answer times 12 is 20 *pesos* 6 *granos*, so he might as well have ignored the 9 *granos* in the question. The answer should have been 1 *peso* 5 *tomines* 10.75 *granos*.

In an example on 20v, we are to divide 256 *pesos* 6 *tomines* 9 *granos* by $12\frac{1}{2}$. He divides 2,568,400 by 1,250 and writes 20.54, again truncating the fractional part: 0.72 *granos*. He gives the answer as 20 *pesos* 4 *tomines* 4 *granos*. It should have been 20 *pesos* 4 *tomines* 4.56 *granos*. Echagaray's tendency to round down in every case has given him numbers not accurate to the nearest *grano* in these last two examples.

---

160. See our discussion of this point for the *Sumario Compendioso* of 1556 versus the *Libro General* of 1597.

161. The reader can decide whether this is an example of a decimal fraction or is a cousin to modern decimal notation. We claim below that the first examples of decimal fractions in print in the New World occur in Carlos de Sigüenza y Góngora's entry for 1690.

162. Echagaray begins his discussion of the rule of agregation in Part 25 and gives the example we have cited in Part 27.

Echagaray also made a map of the Gulf of Mexico in 1686. A photostat is part of the Karpinski map collection at the University of Michigan.

1690

Carlos de Sigüenza y Góngora. 32
LIBRA / ASTRONOMICA, / Y PHILOSOPHICA / EN QUE / D. *Carlos de Siguenza y Gongora / Cosmographo, y Mathematico Regio en la / Academia Mexicana,* / EXAMINA / no solo lo que à su MANIFIESTO PHILOSOPHICO / contra los Cometas opuso / el R. P. EUSEBIO FRANCISCO KINO de la Compañia de / JESUS; sino lo que el mismo R. P. opinò, y pretendio haver / demostrado en su EXPOSICION ASTRONOMICA / del Cometa del año de 1681. / *Sacala à luz D. SEBASTIAN DE GVZMAN Y CORDOVA, / Fator, Veedor, Proveedor, Iuez Oficial de la Real Hazienda / de su Magestad en la Caxa desta Corte.* / En Mexico: por los Herederos de la Viuda de BernardoCal deron[163] / IXI. DC. XC.

Heirs of the widow of Bernardo Calderón, México.

Sabin 80976. Medina, *México*, 1484.

Originals located at: Biblioteca Cervantina, Biblioteca Nacional de España, Biblioteca Nacional de México (two copies), Biblioteca Palafoxiana, Hispanic Society of America, Huntington Library, Indiana University, John Carter Brown Library, New York Public Library, Southern Methodist University, University of California Berkeley, University of Arizona, University of Chicago, University of Texas at Austin, Yale University.

Paging: Twelve unnumbered leaves followed by 188 pages, numbered except for 1. Pages 1 to 188 contain 395 numbered paragraphs.

Some copies include an extra page before the title page as follows: LIBRA / ASTRONOMICA / Y / PHILOSOPHICA / Si in defensionem mei aliqua scrip- / sero,in te culpa sit qui me provocasti, / non in me, qui respondere compul- / sus sum. *D. Hieronymus Epist.* 14. / *ad D. Augustinum.pag.* 704. (The copy at the John Carter Brown Library has this bound in as the third leaf.)

Copy 2 at the Biblioteca Nacional de México is bound with Eusebio Francisco Kino's *Exposición Astronómica*, Gabriel López de Bonilla's *Discurso, y Relación* of 1654 (title page missing), Diego Rodríguez's *Discurso Etheorológico* of 1652, and Gaspar Juan Evelino's *Especulación Astrológica* of 1682.

---

163. Most copies have "Bernardo Calderon" here.

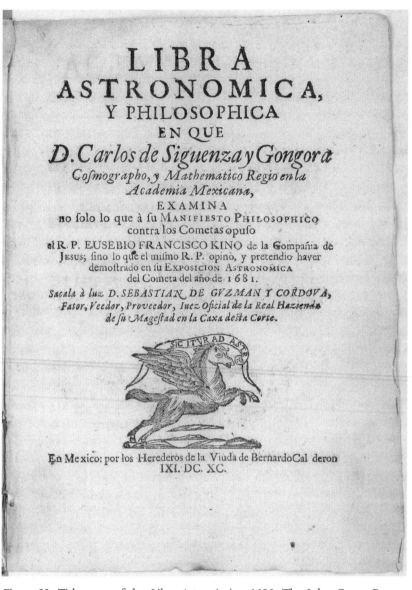

Figure 23. Title page of the *Libra Astronómica*, 1690. The John Carter Brown Library at Brown University, Providence, RI.

A reprint was done in 1959 and reissued in 1984. Also in 1984, an anthology *Seis Obras* included this work with five others by Sigüenza. There is now a 2001 facsimile edition with a prologue by Elías Trabulse. He had earlier included an excerpt in *Historia de la Ciencia en México, Estudios y Textos, Siglo XVII*. Microfilm copies are available. There is a reference to an inclusion of the *Libra Astronómica* in *Tres Obras de Sigüenza y Góngora* printed in Morelia in 1886.[164] The entire text of one of the Biblioteca Nacional de México copies is available online at www.cervantesvirtual.com.

Eusebio Francisco Kino (1644–1711) and Carlos de Sigüenza y Góngora (1645–1700) are possibly the two most written about authors on this list. There is an abundance of secondary material in both English and Spanish. Their principal biographers in the English language are Herbert Eugene Bolton[165] for Kino, and Bolton's student[166] Irving A. Leonard[167] for Sigüenza. We will tell here the story, as told by these two biographers and others, of how Kino and Sigüenza came to meet in Mexico, and how the comet of 1680 and 1681 caused them to author books in opposition to one another. Previous commentators may not have felt comfortable discussing the mathematics that appears in these books. This entry will make a small contribution in that direction here and fit it into the greater story.

The comet is perhaps a third character in the story, and it has its biographer, too. James Howard Robinson's book *The Great Comet of 1680* is a study in the history of cometary writings from antiquity to the seventeenth century and traces the gradual displacement of superstitious ideas in these writings by a more rational enlightened viewpoint. He focuses on developments in Germany, Switzerland, England and the English colonies, France, and the Netherlands, and he has very little to say about Spain or its colonies, so his book complements what has been written about Kino's and Sigüenza's interest in the comet. We will review here some of the aspects of seventeenth-century "comet fever" that Robinson reports, leading up to 1680 and the appearance of our third character, the "Great Comet." A coun-

---

164. Augustin and Aloys de Backer, *Bibliothèque de la Compagnie de Jésus*, new edition by Carlos Sommervogel, Bruxelles, 1898; entry for Kino, Eusebio Francisco.

165. Herbert Eugene Bolton, *The Padre on Horseback*, San Francisco, 1932, and *Rim of Christendom*, New York, 1936.

166. According to the dedication of *Books of the Brave*.

167. Irving A. Leonard, *Don Carlos*, Berkeley, 1929, and *Baroque Times*, Ann Arbor, 1959. See also Leonard's *Ensayo Bibliográfico de Don Carlos*, México, 1929, and "Sigüenza y Góngora and the Chaplaincy of the Hospital del Amor de Dios," *The Hispanic American Historical Review* XXXIV/4 (Nov. 1956): 580–587.

terpoint to Robinson's views is provided by Sara Schechner Genuth's *Comets, Popular Culture, and the Birth of Modern Cosmology*. She suggests that superstition about comets did not so much die out as transform itself. If people claimed less often that comets were portents, the causes of this were not limited to a supposed decline in superstition, but "the decline of prognostication in the seventeenth century was connected in part with the withdrawal of the elite from socially discredited and vulgar practices."[168]

A fourth character plays a minor role: a very famous nun who knew both Sigüenza and Kino. We will discuss her later and offer a poem of hers that touches upon the story at hand.

There was a wealth of spectacular comets in the seventeenth century and a minor industry of books about them.[169] Two of these titles have made our list of mathematical works, Kino's of 1681 and Sigüenza's of 1690. In what Florian Cajori calls "the earliest instance of a clash of intellects in public print on a scientific question that occurred in America," we see the former author for and the latter against the claim that comets were warning signs of impending doom. Some other works that were part of the same debate were not judged to be sufficiently mathematical to be included in the main list.[170] The entries for Kino and Sigüenza, however, are distinguished by a higher level of mathematics in their arguments.

---

168. Sara Schechner Genuth, *Comets, Popular Culture, and the Birth of Modern Cosmology*, Princeton, NJ, 1997: 13.

169. See in particular our Appendix B, where we have collected the titles of all the comet books we know of that were printed in the New World up to 1690.

170. In José de Escobar's *Discurso Cometológico* of 1681 we see a claim of a mathematical proof that comets are not made of fire. Some of these books mention parallax and throw some numbers around—Gabriel López de Bonilla, in his *Discurso, y Relación Cometographia*, states that the comet in question is 3,649 leagues above the Earth, and the sphere of fire is 3,278 leagues from the Earth, leaving the comet 371 leagues within the sphere of fire (folio 5v)—but this is the extent of the meager mathematical content of the books that we chose to leave out of the mathematical list. Samuel Danforth's *An Astronomical Description of the Late Comet* in 1665 mentions parallax briefly and then moves on. Likewise, Increase Mather's *Kometographia* of 1683 quotes "learned men" on the topic of parallax—he doesn't delve into the details himself. We have seen a copy at the Biblioteca Nacional de México of Diego Rodríguez's *Discurso Eteorológico* of 1652. (An excerpt is also given by Trabulse.) In spite of Rodríguez's reputation for mathematical accomplishment in his manuscripts, we did not find his one printed work to be very mathematical. In Juan Ruiz's *Discurso* (a copy of which is held at Berkeley, and which is also excerpted by Trabulse) he himself says that he is doing astrology, not astronomy. We have not seen the 1681 *Manifiesta Cristiano* of Martín de la Torre, nor Evelino's second comet book, *Disertación*, but we are guessing that they are no more mathematical than the other works mentioned here.

Robinson's book makes it clear that these books published in the New World were part of a worldwide phenomenon. He also lays out for us the context of the debate by tracing cometary theory back to one of the ancient venerable philosophers.

Aristotle believed that comets were atmospheric phenomena. "He taught that they were exhalations mounting to the upper regions of the air, where, in the diurnal movements of the upper atmosphere, these exhalations unite and condense. Here they take fire, possibly by the force of movement or from the neighboring fiery region or from the action of the stars and sun."[171]

This was generally accepted until Tycho Brahe measured the parallax of a comet in 1572 and showed that this comet was above the moon.[172] Parallax is the term for the process of finding the distance to an object by making two observations of the object from slightly different positions. The apparent shift of the object when the two observations are compared will be greater or lesser as the object is closer to or farther from the observer.

Johannes Kepler, in spite of his use of ellipses in his publication of 1609 to describe the path of planets, wrote that comets moved in straight lines. Both of these scientists, Brahe and Kepler, were also practicing astrologers.

Robinson notes that in Robert Grant's list of truly remarkable comets in history, the seventeenth century has more than its share: the list includes the comets of 1607, 1618, 1661, 1680, 1682, and 1689. The years 1607 and 1682 were two appearances of Halley's Comet. In 1618 there were actually three comets, with the third being the most remarkable. Later writers were to point out that this year marked the start of the Thirty Years' War. A trio of comets (1661, 1664, and 1665) appeared again in this century.

Gottfried Kirch is credited by Robinson for discovering the comet of 1680 early in the morning of November 14. The comet passed its perihelion on December 24 and thereafter appeared in the evening. The comet was distinguished, even among the great comets, for its spectacular tail.

---

171. James Howard Robinson, *The Great Comet of 1680*, Cleveland, OH, 1919/1986: 3.

172. We note in passing that by 1668 the mathematics necessary for such a calculation was available in English in a little book by Richard Holland, *Notes Shewing How to Get the Angle of Parallax of a Comet*, published at Oxford.

According to Robinson, "Kirch and several others did not identify the comet seen in November in the morning with the one visible after Christmas in the evening." Both Kino and Sigüenza made this identification, however, in the works we are considering here.

On December 2, so a contemporary story goes, an egg was laid in Rome with an image of the comet. Reports of this event caused such a stir in France that the Academy of Sciences reported on it in their journal, saying that the egg was actually marked with several stars instead of a single comet. In late January of 1681 a play entitled "La Comète" was performed in Paris. Robinson says that a popular Paris weekly "advised all who wished to be healed of the fear of the comet to see this comedy." In this play one character expresses unease at the prospect of eating an "omelette of comets."

Throughout history, appearances of comets had often stimulated a literary response. Robinson claims that this effect peaked with the comet of 1680. Over one hundred cometary pamphlets were produced in what is now Germany and German Switzerland alone. Robinson attributes the rise to the increasing number of doubters in the rationalist ranks. Toward the end of his book Robinson focuses on those who argued against superstition, devoting an entire chapter to one of them, Pierre Bayle. He would have had no shortage of material, however, had he chosen to use instead Carlos de Sigüenza y Góngora as his standard-bearer for this position.

Carlos de Sigüenza y Góngora was born in Mexico. His father had been tutor to a prince before emigrating from Spain, and the young Carlos was educated at home. At the age of 15 he was accepted as a novice in the Society of Jesus, and soon afterward he was a student at the College of the Holy Spirit in Puebla. It was here that shortly before his twenty-third birthday he was expelled from the order for indiscretions that seem to have involved the nightlife of the city. "He escaped from the dormitory to taste the forbidden fruit of nocturnal rambles about the city streets."[173]

Sigüenza made several requests to be readmitted to the order, but, while he was formally absolved of his breach of discipline, he was not welcomed back. He was left to make his way in the secular world, without affiliation with a religious order.

---

173. Irving A. Leonard, "Sigüenza y Góngora and the Chaplaincy of the Hospital del Amor de Dios," *The Hispanic American Historical Review* XXXIV/4 (Nov. 1956): 580–587.

Back in Mexico City, he was made Professor of Mathematics and Astrology at the Royal University of México in 1672.[174] He was well known to be critical of astrology, and Leonard suggests that his frequent extended absences from his lecturing duties may have been because astrology was a more popular topic at the university than mathematics was. He acquired other titles: Chief Cosmographer of the Realm, General Examiner of Gunners, University Accountant, Chief Almoner of the Archbishop, Corrector of the Inquisition, Chaplain of the Hospital del Amor de Dios. The latter title reflected the fact that he was still an ordained priest, even though he was no longer in the Jesuit order, and it included accommodations, a happy convenience since his salary from the University was meager. He held that position from 1682 until his death.

He began publishing almanacs in 1671 and continued until his death. We will have more to say about this in Part 2.

After the great comet appeared in the autumn of 1680, Sigüenza put out on January 13, 1681,[175] a pamphlet designed to ease the fears that this comet aroused.[176] It was the *Manifiesto Filosófico contra los cometas despojados del imperio que tenían sobre los tímidos*.[177]

174. Sigüenza had to compete for this position with two other applicants, Juan de Saucedo and José de Escobar Salmerón y Castro. (These two were both authors of rival almanacs to those of Sigüenza, and so we will see their names again in Part 2. Escobar Salmerón y Castro was also the author of a rival comet book, as we will see later in this entry and in Appendix B.) The latter claimed that he was the only one qualified for the position since he was the only one who had a degree from the university (his thesis in philosophy and sacred theology was published in 1667 by Juan Ruiz), but all three were admitted as candidates. The competition involved drawing a text at random and then preparing a discourse on the text within 24 hours. Sigüenza won this event on July 20, 1672. (See Part 2 under 1673 for the first almanac by Saucedo and 1678 for Escobar.)

175. Date given in paragraph 28 of *Libra Astronómica*.

176. This was the comet before the one that was to be named after Edmund Halley. Halley observed this comet while travelling through France and had trouble reconciling its motion with the prevailing view (from Kepler) that comets moved in straight lines. This experience prepared him to make more careful observations of "his" comet in 1682. According to Bolton this comet's orbit has been computed. It has a period of about 575 years and may have been the comet visible at Julius Caesar's death. It returned in 531 and 1106 and should be seen again in 2256, or thereabouts. In fact Isaac Newton computed some of the orbital data of this comet and published his conclusions in his *Principia* of 1687. Because of this, some sources refer to the comet of 1680 to 1681 as "Newton's comet." A curious false start in Newton's thinking is traced on p. 136 of Schechner Genuth's *Comets*. "In the spring of 1681, Newton began a letter to Flamsteed in which he discussed the non-linear path a comet would describe under the influence of a magnetic force seated in the sun.... But he suppressed this passage from the letter he eventually sent... Publicly, he dug in his heels, insisting that the comets of November and December were distinct."

177. Some sources give *Exposito Philosophica aversus Cometas*.

A response to the *Manifiesto Filosófico* came out immediately. *Manifiesta Cristiano en favor de los cometas mantenidos en su natural significación* by Martín de la Torre reasserted the prevailing view that comets were signs from God of dire events to come. Not much else is known about de la Torre. He was from Campeche, a part of the Yucatán peninsula, and there exists a map that he drew of the city of San Francisco de Campeche.[178]

Sigüenza replied with his *Belerofonte Mathemático contra la quimera astrológica de Martín de la Torre*. The title alludes to the story of the hero, Bellerophon, who rode the winged horse, Pegasus, to fight the creature, Chimera, which was part lion, part snake, part goat. However, it appears that the *Belerofonte* was never printed. Medina only lists this title within his entry for de la Torre. It is described as a pamphlet by Leonard in his *Baroque Times in Old Mexico* and in *Don Carlos* (page 59), but in the appendix to *Don Carlos* and in "Ensayo Bibliográfico" he lists it as a manuscript. Bolton says it was published. Joseph Sabin says it was unpublished. Sebastián de Guzmán y Córdova mentions it in his prologue to the *Libra Astronómica*, saying it "perished on the reef of his [Sigüenza's] carclessness." Because of this, Leonard says that the *Belerofonte* was lost by 1690.[179] Reading Guzmán more carefully, though, and contemplating his metaphorical allusion to reefs and shipwrecks, he seems to be saying it was never published at all.

Here is a longer translation of the relevant passage. It is Guzmán's version of what happened to the *Belerofonte* and how the *Libra* escaped the same fate.

". . . I don't know if he is speedier in imagining and forming a book as in forgetting it . . .

"The *Belerofonte Mathemático* . . . experienced this fortune, where is found such fine things and subtleties as using trigonometry in the investigation of the parallaxes and refractions, and the theory of the movements of comets, either by means of a straight-line trajectory in the hypothesis of Copernicus, or by conic spirals in the Cartesian vortices.

"This other most singular treatise perished (even though he got angry) on the reef of his carelessness, where by admirably easy methods, never used by any author, the eclipses of the sun were computed in the nonagesimal degreees from the ascendant in all their terms . . .

---

178. A photostat is in the Karpinski map collection at the University of Michigan. De la Torre is described on it as "alferez y ingeniero militar."

179. Irving A. Leonard, *Don Carlos*, México, 1929: 206.

"The present *Libra Astronómica, y Philosóphica*'s having escaped this shipwreck is due to me. Because I had it written at my request, and that of other friends, at the end of the year 1681, and had then secured, the following year, the licences for publishing it . . . [Sigüenza] carried it to me at my home, where, without fear of it being lost, it was saved until now, that it seems to me convenient that it come out in public."

Paragraph 318 of the *Libra Astronómica* also states that the *Belerofonte Mathemático* had already been written by the time Sigüenza was working on the *Libra Astronómica* at the end of 1681 and suggests that paragraphs 320 through 395 of the latter work may be an excerpt from the *Belerofonte*. (This is supported by William G. Bryant, in his notes to the *Libra* for *Seis Obras*.)[180] As we will see, this section contains some analysis of spherical triangles and uses both decimal fractions and logarithms. So had the *Belerofonte* been printed when it was written in 1681, it would change our date for the first appearance of decimal fractions and logarithms in print in the New World.

Then came José de Escobar Salmerón y Castro's *Discurso Cometologico, y Relacion del Nuevo Cometa: visto en aqueste Hemispherio Mexicano, y generalmente en todo el Mundo: el Año de 1680; Y extinguido en este de 81: Observado, y Regulado en este mismo Horizonte de Mexico*, still in 1681. Commentators often repeat that this work proposes that comets are composed of the exhalations of dead bodies and human perspiration. This belief is in accord with Aristotle's writings on comets, as stated earlier.[181]

Sure enough, we find on folio 7v of the *Discurso* the statement that "water, land, all living bodies, plants, and even the very dead bodies buried in the earth" are the material cause of comets. We must be careful of the word *exhalación*—it can mean either exhalation, or shooting star. It does appear, however, that Escobar was saying that evaporations from the earth cause the comets. On 7v and 8r he argues that the humors and spirits of the human body are known to be the cause of rain, and if that is so, why can't they also be the cause of comets?

---

180. Carlos de Sigüenza y Góngora, *Seis Obras*, Caracas, Venezuela, 1984: 409, note 74.

181. We see this view also expressed by Juan Ruiz in 1653: ". . . cometa no es otra cosa, que una copiosa cantidad de exhalaciones calientes . . ." One early example of the opposing view in the Americas is Samuel Danforth's work of 1665, which opens, "This comet is no sublunary Meteor or sulphureous Exhalation, but a Celestial Luminary, moving in the starry Heavens." Likewise, Increase Mather in 1683 starts his book with a chapter heading, "That Comets are not in the Air, but in the Starry Heaven."

Sigüenza states in paragraph 28 of the *Libra Astronómica* that he did not regard a response to this work of Escobar Salmerón y Castro as worth his time.[182] Trabulse, in a footnote to the passage quoted from the *Discurso*, says that "this idea of the exhalations of dead bodies as producers of the comet" brought about the fact that Sigüenza would "neglect to consider as serious the response of Salmerón. We believe however that, with the exception of this idea, [this book] merited a response from Don Carlos."[183]

Still in the same year of 1681, Kino's *Exposición Astronómica* was published. Kino, originally Chino or Chini, was born August 15, 1645, in Segno in what is now Italy, although at the time it was part of the Holy Roman Empire.[184] He became a Jesuit on November 20, 1665, and that entailed studies at several European colleges, culminating in three years spent studying Philosophy at the University of Ingolstadt. (Some of the rancor that Sigüenza would come to feel toward Kino may have been a result of the fact that the former had been expelled from his training to prepare him for the Jesuit order.)

Kino dreamed of doing missionary work in China. When his assignment finally came, he and another Tyrolese Jesuit, Antonio Kerschpamer, were told that one should be sent to the Philippines and the other to Mexico, and that they should decide the matter between them. They drew lots, and thus it was by chance that Kino was sent to Mexico.

On June 12, 1678, he set out from Genoa among a company of Jesuits bound for the New World, some of whom were to remain in Mexico and others to continue to the Philippines. They missed their connection in Cádiz by a few days because of delays at sea. Kino spent part of the next

---

182. ". . . a quien jamás pienso responder, por no ser digno de ello su extraordinario escrito y la espantosa proposición de haberse formado este cometa de lo exhalable de cuerpos difuntos y del sudor humano."

183. Elías Trabulse, *Historia de la Ciencia en México, Estudios y Textos, Siglo XVII*, México, 1984.

184. Bolton says in *Rim of Christendom* that Kino referred to the band of Jesuits he travelled with as "we Germans," and he gives the following excerpt from a letter to the Duchess of Aveiro y Arcos: "I am from Trent in the Tyrol, but I am in doubt whether I should call myself an Italian or a German. The city of Trent uses almost entirely the language, customs, and laws of the Italians. However, it is in the very edge of the Tyrol, and the Tyrol belongs to Germany. Besides, the college at Trent is a college of the province of Upper Germany, although we were instructed in our classes and talked together in Italian. Still, for the last eighteen years of my life I have lived almost in the center of Germany."

year in Seville. On July 7 of 1679 they boarded a ship that was part of a fleet heading out from Cádiz, but their ship ran aground on the 11th and had to be abandoned. Some of the Jesuits found passage on other ships, but Kino and the others had to return to Cádiz once more.

Thus it was that while staying at the Jesuit College in Cádiz two winters later, Kino had the opportunity to observe the comet of 1680 while waiting for his ship. He stated in a letter of December 28, 1680, that he didn't observe the comet himself until December 23, but later in his book he gave positions of the comet from late November on.

On January 27, Kino and others finally departed Cádiz for good. They arrived in Veracruz on May 3 and followed the mule trail to Mexico City, arriving by early June. Here Kino and Sigüenza were to meet.

At first they got along cordially, as Sigüenza invited Kino to his home to discuss their scholarly interests. Sigüenza loaned Kino some maps to help him prepare for his journey to the north, maps that Sigüenza later bitterly complained that he had to retrieve through a third party.

In mid-October of 1681, when Kino was about to depart for his missionary work in Baja California, he paid a visit to Sigüenza and bestowed upon him a copy of his freshly printed *Exposición Astronómica*.

Kino was familiar with the computation by Tycho Brahe of the parallax of a comet. His book contains a lengthy explanation of what parallax is and how to find it, with numerous geometrical figures and an explicit reference to Euclid at one point.[185] He includes a wonderfully drawn stellar map plotting the position of the comet; the direction, length, and width of its tail; and even its stellar magnitude over a period of time from late 1680 into early 1681. A suspicious feature of this map, however, is that the path of the comet is shown intersecting the ecliptic only once, where our modern understanding of the geometry of orbital motion entails that the comet should be observed crossing the ecliptic twice.

His position with regard to Aristotle's teaching that comets were atmospheric was that some comets were sublunar and others were not, and that those that were beyond the moon were too far away to be made of the same elements (earth, air, fire, and water) as the familiar terrestrial objects, and therefore they must be celestial, ethereal. He estimated that this particular comet was moving within the sphere of the sun.[186]

---

185. Euclid's *Elements*, Book 4, Proposition 15, is utilized on folio 9r.
186. ". . . paisano del sol y con vezino de Mercurio . . ." on 8v.

Kino puts the comet, and therefore the sun, at a distance of 1,153,000 leagues from the Earth.[187] Using the figure, from other texts on our list, that $17\frac{1}{2}$ Spanish leagues was one degree of latitude or 60 nautical miles, we find that the distance to the sun was assumed to be 3,953,000 nautical miles, or 4,549,000 miles, or 7,321,000 kilometers, close to 5% of the correct distance.[188] Kino shows us in his Figure V that the apparent length of the comet's tail is 60° and so the tail is also 1,153,000 leagues long.

He observes on 10r that the comet moves 4° per day, which is 10′ per hour. To these ten minutes he assigns 3,352 leagues, which is consistent with the distance to the comet given, if one uses $\pi = 3.14$. Kino goes on to say on 10v that when one includes the daily rotation of the sky the comet actually moves 356° in one day. He calculates that this corresponds to a speed 7,458 times that of a cannonball.

Kino's book is a short but significant addition to the body of cometary writing. It not only has a beautifully executed map of the comet's path, but (of greatest interest to us here) it has geometry and arithmetic woven throughout its arguments.

Sigüenza took the work to be a personal attack upon him, even though he was not mentioned. "No one knows better where the shoe pinches than the one who wears it."[189] He responded to Kino's efforts with some degree of sarcasm in his *Libra Astronómica*, written toward the end of 1681. This work was not printed, however, until 1690, when a friend came up with sufficient funds for the publication.[190] The friend was Guzmán, and he wrote the introduction and prologue that take up most of the preliminary leaves of the *Libra Astronómica*. This prologue is the source of the quote we have used above, where Guzmán writes that while the *Belerofonte Matemático*

---

187. Earlier, according to Bolton, while Kino was still in Spain, he wrote in a letter that the comet was only 3,000 leagues away.

188. Kino writes on 9r that "Tico Bray" (Tycho Brahe) had put the sun at 150 Earth radii away, but he corrects this to 1,150 on the next page, 9v. The number should be 23,481, the number of mean Earth radii in an astronomical unit.

189. Irving A. Leonard's translation in *Don Carlos de Sigüenza y Góngora*, Berkeley, 1929: 64.

190. In the same year his *Infortunios que Alonso Ramírez* . . . was published. Because of this work, Sigüenza, among his many other accomplishments, is considered one of the pioneers of the Mexican novel, although some commentators treat *Infortunios* as a true adventure. The title character, having departed from Acapulco for the Philippines, is captured by English pirates and is forced to circumnavigate the globe. In one scene the English are described as feasting on a human arm.

"perished on the reef of his [Sigüenza's] carelessness," it was due to him (Guzmán) that the *Libra Astronómica* "escaped from this shipwreck" and saw the light of day.

The *Libra Astronómica* contains what is probably the complete text of the *Manifiesto Filosófico* as paragraphs 10 through 27. After these paragraphs, he briefly describes the three books that came out in response to it the same year, and then, in paragraphs 29 and 30, he gives a chapter-by-chapter summary of Kino's book. From paragraph 31 through most of the rest of the book, Sigüenza focuses his attack on Kino's tenth chapter, where the latter had considered arguments for and against the "*fatalidad*" of comets.

Sigüenza gives the nod along the way to some of the other authors we have mentioned, like Enrico Martínez, Gabriel López de Bonilla, and Diego Rodríguez. He relates that Martínez's measurement of the longitude of Mexico using the lunar eclipse on December 20, 1619, was confirmed by later observations of Rodríguez and López de Bonilla.[191] (It is also mentioned that Rodríguez was one of Sigüenza's predecessors in his position at the University.) By Bolton's count, Sigüenza cited "some eighty 'authorities' on comets" in support of his thesis. This work culminates, in paragraphs 388 through 395, with the solutions of spherical triangles with numerous figures and use of the sine and tangent functions. (He also makes use of decimal fractions and logarithms, the first such use we see in the books on our list.[192]) Trabulse's excerpts include this highly trigonometric part of the *Libra Astronómica*. We have already mentioned that paragraphs 320 through 395 here may have been excerpted from the probably unpublished *Belerofonte Matemático* of 1681. In particular, 320 through 328 are marked as a quote from Martín de la Torre's *Manifiesto Cristiano*, now lost.

For a simple example of his use of trigonometry and logarithms, paragraph 391 is instructive. In a right spherical triangle *ABC* with angle *A* opposite side *a*, angle *B* opposite side *b*, and angle *C* the right angle, one of the trigonometric rules is

$$\tan A = \frac{\tan a}{\sin b}.$$

---

191. ". . . segun estos dos Autores, ay 7. hor. 28.' de differencia, [from Mexico to Uraniburg] . . . que difiere de la que deduxe de la observacion de *Henrico Martinez* en . . . 2.' de tiempo, que para tanta distancia, es concordancia estupenda" (paragraph 386).

192. Although see our discussion of Echagary's work of 1682 for a sort of proto-decimal notation.

Sigüenza considers a triangle NEC in the sky with right angle at E. He writes

Seno EN	66°38′28″	9.9628612
Tangente ENC	28°3′40″	9.7267910
Tangente EC	26°4′35″	9.6896522

(We note that Sigüenza actually omits the degree symbol from the numbers that are angles. We have chosen to put it in.) The numbers in the last column are the base 10 logarithms of the respective trig functions, with 10 added.[193] Sigüenza has taken the sine of EN and the tangent of ENC, taken their base 10 logs plus 10, added these together minus 10 to get the last number in the last column, and found the arc whose tangent has a base 10 log plus 10 equal to that number. Thus he has solved for side EC in the triangle. The first number in the last column should end in 3, but that doesn't change the answer.

A more complicated formula employed by Sigüenza is

(1) $$\sin^2 \frac{C}{2} = \frac{\sin \frac{a-b+c}{2} \sin \frac{-a+b+c}{2}}{\sin a \sin b}.$$

This applies to any spherical triangle, regardless of whether there is a right angle present, as long as a, b, and c are the sides and angle C is opposite side c. It can be derived from the Law of Cosines for spherical triangles,

(2) $$\cos c = \cos a \cos b + \sin a \sin b \cos C,$$

but equation (1) was preferable to equation (2) in Sigüenza's time because it does not involve addition after the trig functions have been evaluated, so it is more amenable to solution with logarithms. From the appearance of the first tables of logarithms in 1614 until the invention of handheld calculators in the twentieth century, equation (1) would have been a better choice for most people approaching such a problem.

---

193. Adding the 10 avoids negative numbers, but we must remember from our discussion of trigonometry in the entry for 1606 that the trig functions were defined differently in the seventeenth century from how they are today. Enrico Martínez used sine functions in different applications that ranged up to 60,000 and up to 100,000. Adding 10 to log sin or log tan is equivalent to multiplying our sin or tan by 10^{10}.

Sigüenza applies this in paragraph 389 to a triangle SMC. (S is Scheat, M is Markab, and C is the comet. Scheat and Markab are two of the corners of the Great Square of Pegasus.) He makes the following calculation:

CS	9°32'0''	
MC	9°34'0''	0.7793818
MS	12°52'41''	0.6519291
Suma de los tres lados	31°58'41''	
Semi-suma	15°59'20''	
Difra de la semi-suma, y MS	3°6'39''	8.7345393
Difra de la semi-suma, y MC	6°25'20''	9.0486525
Suma de Logarithmos		19.2145027
Arco de la mitad desta suma	23°52'45''	9.6072513
Su duplo es angulo SMC	47°45'30''	

Knowing the three sides of the triangle, he seeks the angle at M. The first two numbers in the last column are the negatives of the logs base 10 of the sines of MC and MS, but the second number should end in 352 instead of 291. He adds the three sides, or arcs, to get the 31°58'41''. He takes half of this sum in the next line. From this he subtracts MS and MC—the remainders correspond to the $\frac{a-b+c}{2}$ and $\frac{-a+b+c}{2}$ in equation (1), if MS is b and MC is a. Sigüenza takes 10 plus the logs base 10 of these numbers and puts those answers in the third and fourth positions in the last column. The third number should end in 407 instead of 393 and the fourth should end in a 9 instead of a 5. The 19.2145027 is the sum of the four numbers given. By equation (1), it represents $20 + \log_{10} \sin^2 (\angle SMC/2)$, or $20 + 2\log_{10} \sin (\angle SMC/2)$. The number right below that is half as much, or $10 + \log_{10} \sin (\angle SMC/2)$. Sigüenza can now look in his log sine tables to find that $\angle SMC/2 = 23°52'45''$ and so $\angle SMC = 47°45'30''$. The correct answer should have been 47°45'38'', which can be confirmed, with a modern calculator, using either equation (1) or equation (2).

Returning to events in the 1680s, in the absence of a published form of Sigüenza's last book on the subject, the debate continued in 1682 with Evelino's *Especulación Astrológica*. Evelino was a medical doctor, born in Taxco, according to Trabulse (*Apéndices e Índices*). The *Especulación Astrológica* was yet another response to Sigüenza, and it, like the ones before

Kino's, was virtually free of mathematics.[194] Evelino also wrote *Disertación sobre los cometas* in 1683, according to Medina and Trabulse (*Apéndices e Índices*), although we have seen no other reference to it. At the same time the English colonies saw Samuel Willard's *The Fiery Tryal* of 1682 and Increase Mather's *Kometographia* of 1683. All of this was contemporary with an appearance of Halley's Comet in 1682.

Kino was busy with his missionary work and made only two trips back to the capital of the colony, once in 1686 and again in 1695.[195] He is not known to have met with Sigüenza again, but there is a record that he expressed mild astonishment that Sigüenza was upset with him. In the introduction to a later book he wrote that "Don Carlos de Sigüenza is very much offended, charging me in his *Libra Astronómica* that I wrote my *Exposición Astronómica* as an attack on his *Manifiesto Philosóphico*, the fact being that it never entered my mind to pretend to write or to print a single letter in opposition to that *Manifiesto Philosóphico*, nor do I know that I ever read it."[196]

"Sigüenza is a tireless collector of Mexican antiquities; he is the first to draw a complete map of New Spain; he is the first to scrutinize its past in all its aspects," writes Ramón Iglesia.[197] Mexico's past, and in particular the accomplishments of its natives, was an obsession with Sigüenza, to which he devoted much of his life, although little of this work was published. He did not, for most of his life, think the natives to be inferior to Europeans. But that changed with Mexico City's food-shortage riot of June 8, 1692, according to Iglesia. Sigüenza, according to his own account, saved valued records from a burning building during the riot. Iglesia documents how after this point Sigüenza advocated the separation of the native populations from the Spanish.

Sigüenza died in 1700, leaving his impressive collection of books and instruments to the Jesuits. Kino survived until March 15, 1711. When his

---

194. It does start off with the following statement of a central problem under debate: "Question . . . if all the comets are elementary, that is sublunar, or inferior to the moon, or if they are celestial, that is superior, and independent of all matter of the elements . . . [on] its decision depends the judgement of their natural effects if they are elementary, or supernatural, if they are celestial, and formed in that very high ether by only the divine disposition without intervention of the second causes . . ."

195. Irving A. Leonard, *Don Carlos de Sigüenza y Góngora*, Berkeley, 1929: 71.

196. Bolton's translation in *Rim of Christendom*, New York, 1936.

197. Ramón Iglesia, "The Disillusionment of Don Carlos," in *Columbus, Cortés, and Other Essays*, Berkeley, 1969.

tomb was rediscovered, the town in the state of Sonora that is his final resting place was renamed Magdalena de Kino.[198]

As an afterword, the role played by Juana Inés de la Cruz should be noted. Born in 1651, Juana Inés de Asbaje y Ramírez de Santillana came from the small village of Nepantla to Mexico City at the age of eight. Her precocity attracted the attention of the viceregal court, and she was brought to live at the Palace as the favorite of the vicereine.

At the age of fifteen she entered the Order of Discalced Carmelites and became Sor Juana Inés de la Cruz. She left the order after three months because she couldn't adjust to the harshness of their practices. The next year she entered the convent of San Jerónimo, where she was able to line her quarters with books and collect musical and scientific instruments.

Carlos de Sigüenza y Góngora was a frequent visitor and probably brought Eusebio Francisco Kino with him soon after they met. In Leonard's account in *Baroque Times*, Don Carlos and Sor Juana shared an interest in the new approach to knowledge to be found, for instance, in the philosophy of René Descartes. In 1680 Sigüenza praised her highly in his *Theater of Political Virtues*, and she responded with a sonnet praising him.

The following poem by de la Cruz, dedicated to Kino, is quoted in Leonard's *Don Carlos*[199] and in Trabulse's *Ciencia y Religion*.

> Aunque es clara del Cielo la luz pura
> clara la Luna, y claras las estrellas,
> y claras las efímeras centellas,
> que el ayre lleva,[200] y el incendio apura:
>
> Aunque es el Rayo claro, cuya dura
> producción cuesta al viento mil querellas,
> y el relámpago que hizo de sus huellas
> medrosa luz en la tiniebla obscura;
>
> Todo el conocimiento torpe humano
> se estuvo obscuro, sin que las mortales
> plumas pudiessen ser, con vuelo ufano,

---

198. Charles Polzer devotes a whole chapter of *Kino, a Legacy* to the search for, and the verification of the identity of, the Kino skeleton.

199. Irving A. Leonard, *Don Carlos de Sigüenza y Góngora*, Berkeley, 1929: 71, footnote 53.

200. Leonard has *eleva*.

Ícaros de discursos racionales;
hasta que al tuyo, Eusebio soberano,
les dió luz a las luces celestiales.

Leonard speculates that the poem was a response to the gift, from Kino to Juana Inés, of a copy of his *Exposición Astronómica*. Writes Leonard,[201] "Possibly it is lacking in due reverence to that very estimable poetess to presume that her perusal of Kino's little volume was limited to the dedicatory to the viceroy. There is, however, some basis for this supposition when one notes that her sonnet bears the caption 'Applauding the astronomical knowledge of Father Eusebius Francisco Kino of the Company of Jesus who wrote about the Comet which appeared in the year '80 *absolving it of ominousness*'" (italics Leonard's; Trabulse draws the same conclusion). Leonard is referring to the fact that in spite of Kino's theme in the main part of his book, the dedication assures the Viceroy that the comet fails to have dire significance for him. "The comet . . . will be a happy forerunner of your prosperity . . ." writes Kino on [4]v and [5]r.

Could Juana Inés, now hailed as being very much ahead of her time, have been engaging in irony?

In 1693 Sor Juana sold her collection of books and instruments. She died in 1695 during an epidemic, and Sigüenza delivered her funeral oration. Her image is on the 200-peso note in modern-day Mexico.

1691

Diego Pérez de Lazcano.   33

PROPOSICION, / Y MANIFIESTO, / QVE EL CONTADOR DIEGO PEREZ / de Lazcano, que lo es de las cuentas finales, y Visi- / ta de las Reales Caxas de la Villa Imperial / de Potosi, / HAZE / AL REY / NVESTRO SEÑOR: / EXPLICANDO EL MEDIO DE INTRO- / ducir el aumento del valor de la plata, conforme à / la Real Pragmatica de 14. de Octubre del año de / 1686. à la plata en pasta, y barras de todas leyes, pa- / ra que su Real Hazienda perciba los aumentos / quantiosos, y evidentes, que se siguen de conceder- / se a las Indias el aumento del valor de la plata, y la- / bor, y vso de la moneda nueva, conforme a dicha / Pragmatica. En

---

201. Irving A. Leonard, *Don Carlos de Sigüenza y Góngora*, Berkeley, 1929: 71.

que comprueba, que solo in el Rey- / no del Perù tendrà su Magestad (que Dios Guarde) / de aumento al año 400U973 ps. 7. rs. de a ocho de / la nueva moneda, y que vtilizaràn sus vassallos, vn / millon quatrocientos y cincuenta y quatro mil / ciento y ochenta y dos pesos de a ocho, segun el es- / tado presente; y que concedido dicho aumento, / y moneda nueva, seràn mayores estos / aumentos, y vtilidades.

Lima.

Medina, *Lima*, 637. Vargas Ugarte 859.

Original located at: Biblioteca Nacional de Chile.

Microfilm copies are available.

Paging: Total of thirty-six leaves. Eight preliminary pages with 3 to 5 numbered, followed by thirty-two numbered folios.

On 18v, he introduces the tables that will explain the sixteen points he has made up to there.

1693

Nicolás de Olea.   **34**

SUMMA TRI- / PARTITA / SCHOLASTICÆ PHILOSOPHIÆ; / SIVE / CVRSVS PHILOSOPHICVS / TRIENNALIS, / IN LOGICAM, PHYSICAM, ET METAPHY- / sicam Aristotelis; / *IVXTA HODIERNVM SCHOLARVM* / *Soc. IESV morem,breviter,& exactè* / *comprehensus*. / AVCTORE R P. NICOLAO DE OLEA, / Peruano Limensi , ex eadem Societate IESV; / Olim in Cuzcano, & Limano, Collegijs / Theologiæ Primarario Professore / PARS PRIMA. / IN LOGICAM. / IHS. / *Sumptibus Iosephi de Contreras Regio-Limani* / *Typographi. Anno* 1693.

Joseph de Contreras, Lima.

Sabin 57170. Medina, *Lima*, 651. Vargas Ugarte 885.

Original located at: Biblioteca Nacional de Chile.

Paging: Total of 365 leaves. Ten preliminary leaves, pages 1 to 692, five leaves for index, three leaves for errata, one leaf appendix to errata. Among the numbered pages, 25 is misnumbered 23, 44 is misnumbered 41, 216 is misnumbered 116, 219 is misnumbered 119, 311 is misnumbered 31, 323 is misnumbered 32, 445 is misnumbered 443, and 491 is misnumbered 164.

The only known copy has insect damage.

There was a follow-up volume in 1694 on physics (Sabin 57170, Medina, *Lima*, 663, Vargas Ugarte 902). We have examined it and could find nothing mathematical. For the record, the title page reads: SUMMA TRI- / PARTITA / SCHOLASTICÆ PHILOSOPHIÆ; / SIVE / CVRSVS PHILOSOPHICuS / TRIENNALIS, / IN LOGICAM, PHYSICAM, ET METAPHY- / sicam Aristotelis; / *IVXTA HODIERNVM SCHOLARVM / Soc. IESV morem, breviter, & exactè / comprehensus.* / AVCTORE R. P. NICO-LAO DE OLEA, / Peruano Limensi, ex eadem Societate IESV; / Olim in Cuzcano, & Limano, Collegijs / Theologiæ Primarario Professore / PRIMA SECVNDÆ / IN PHYSICAM, / SIVE IN OCTO LIBROS PHYSICORVM. / IHS. / *Sumptibus Iosephi de Contreras Regio Limani Typographi.* / Anno 1693.

1696

Juan Ramón Koenig. 35
CVBVS, ET SPHÆ= / RA GEOMETRICE / DVPLICATA. / LIMÆ, APVD JOSEPHVM DE CONTRERAS, ET / Alvarado extypographia Regia,& Sancti / Officij. Anno 1696.
Joseph de Contreras, Lima.
Medina, *Lima*, 678. Karpinski [11]. Vargas Ugarte 923.

Originals located at: Biblioteca Nacional de Chile, Biblioteca Nazionale Centrale di Roma, Bibliothèque Nacionale de France, Yale University.

Paging: Fourteen preliminary leaves, thirty numbered folios (with 5 unnumbered and 28 misnumbered 38), and a foldout leaf containing Figures I through VII. This leaf of figures is missing in the copy at Yale University. The Chile copy has only figures II to IV and VI to VII.

According to a letter inserted in the front of the copy in Paris, Jean Ramon Conink,[202] Doctor of Theology, was taught mathematics by Jesuits and went to Lima to construct fortifications. The shield of the Universidad de San Marcos on the title page suggests to us that this Dr. Conink is the same as Juan Ramón Koenig,[203] the author of *El Conocimiento de los Tiempos*, who held a chair of mathematics at that institution.

This is, in fact, what we read in an 1807 elegy to Koenig written by Gabriel Morena and included in his *Almanaque Peruano*. According to

---

202. Latinized as Ioannes Ramon Coninkius on verso of second leaf.
203. Also referred to by José Toribio Medina as José Ramón Koening. See the footnote to the 1680 entry for Koenig in Part 2.

Morena, Koenig (born 1623 in Malinas,[204] in present-day Belgium) took the chair of mathematics at the Real Universidad de San Marcos upon the death of Francisco Ruiz Lozano (the first person to hold that position) in September of 1678. At a young age he had learned Hebrew, Latin, and Greek and shown an innate propensity for geometry. Because of his writings on the duplication of the cube, Moreno dubs him the "Euclid of Peru" and mentions that Johann Bernoulli was doing similar work at the same time.

Moreno states that it was, in his time, next to impossible to follow Koenig's argument since the unique copy in Lima lacks the figures. We note that while the copy held at Yale University lacks the figures today, the copy held in Paris still has a plate of figures in the back.

Moreno claims that when Koenig died on July 19, 1709, he had a corrected edition of *Cubus,* which he was planning to have published in Paris. His papers were lost or destroyed after he died.

Let's look at the geometrical problem in more detail. As the title suggests, the doctor claims to have discovered the solution to the problem of doubling the cube and the sphere. A history of that problem is in order.

Two different versions of the origin of the problem can be found in the literature of ancient times—it was either a tomb or an altar that was required to be doubled in volume. Numerous Greek mathematicians tackled the problem. In modern terms, the problem is equivalent to constructing a length that is $\sqrt[3]{2}$ times a given length. This turns out to be possible or impossible, depending on which geometrical tools one is allowed to use.

For example, if one graphs the parabola $y^2 = 2x$ and the hyperbola $xy = 2$, the point of intersection of the two curves is $(\sqrt[3]{2}, \sqrt[3]{4})$. This is related to a method of solution given by the mathematician Menaechmus as reported by Eutocius, although since coordinate geometry did not exist in ancient times, the description of the parabola and hyperbola would have been stated without giving the equations as we just did.

But a crucial point is that parabolas and hyperbolas cannot be drawn with just a compass and straightedge; they require the use of more advanced mechanical drawing devices. Plato is said by Plutarch to have "reproached the disciples of Eudoxus, Archytus, and Menaechmus for resorting to mechanics and instrumental means for resolving the problem..." for "they

---

204. Mechelen in Flemish, or Malines in French.

resorted to a method that was irrational."[205] So the problem of finding a segment of length $\sqrt[3]{2}$ with just a compass and straightedge remained unsolved for millennia.

It fact, it turns out to be possible to construct such a segment using a straightedge with two marks on it and a compass. A simple method for accomplishing this is given on pages 194 and 195 of *The Book of Numbers* by John Conway and Richard Guy. So the classical straightedge refers to an instrument with no marks on it that just allows one to draw a line through two given points.

In 1837, Pierre Wantzel published a proof that the doubling of a solid with just a compass and straightedge was impossible. Likewise, by the same means, the famous problem of trisecting an arbitrary angle with compass and straightedge could be shown to be impossible. This did not stop hordes of amateur mathematicians from continuing to try these problems. See *Mathematical Cranks* by Underwood Dudley for some modern examples.

José Martí. **36**
TABLA GENERAL DE REDVCCIONES DE ORO A / doblones de à dos escudos, formadapor Ioseph Marti,año de 1696.
Blas de Ayessa, Lima.
Medina, *Lima*, 680. Vargas Ugarte 926.
Original located at: Yale University. Medina and Vargas Ugarte both indicate that there is a copy at the Biblioteca Nacional del Perú.
Paging: Two leaves.
The title is a caption title. The text starts immediately below. At the bottom of [1]r is a printer's colophon: *Lima* 8.*de Otubre de* 1696. Imprimase. / EL CONDE. / Don Blas de Ayessa

Some terms used here were introduced in this discussion of the *Sumario Compendioso* of 1556, so this section should be read after that discussion.

Martí's goal in this short tract is to explain a new way of calculating the value of gold. His example on [1]v is 1,373 *pesos* 5 *tomines* of gold at $19\frac{3}{4}$ carats. We have seen earlier that carats are divided into 4 *granos,* and so the $19\frac{3}{4}$ carats are 79 *granos*. He has us now multiply the 79 and the $1,373\frac{5}{8}$ to get $108,516\frac{3}{8}$ or 108,516 *granos* and 3 *octavos*. The parts of this number now should be looked up in the table he provides on [2]r, and the results are to

---

205. www-history.mcs.st-and.ac.uk/HistTopics/Doubling_the_cube.html

be added together. Answers are in *doblones, reales, maravedis,* and fractions of *maravedis* with 34 *maravedis* to a *real* and 32 *reales* to a *doblon.*

100 U		743 d.	21 r.	33 m.	$\frac{775}{2475}$
8 U		59 d.	15 r.	28 m.	$\frac{1250}{2475}$
	500	3 d.	22 r.	33 m.	$\frac{1625}{2475}$
	10		2 r.	12 m.	$\frac{2260}{2475}$
	6		1 r.	14 m.	$\frac{1356}{2475}$
	$\frac{3}{8}$			3 m.	$\frac{339}{9900}$

So the answer is 807 *doblones* 0 *reales* $23\frac{9603}{9900}$ *maravedis.* The last row in this table, giving the value of the 3 *octavos,* is not from the table and so has to be computed or estimated independently.

The size of the denominators in the fractions goes beyond even what Juan de Castañeda was willing to use (see the entry for 1668).

A *peso* of gold at $22\frac{1}{2}$ carats used to be worth 450 *maravedis.* In Martí's table, looking up 90 *granos,* we get 21 *reales* $14\frac{12}{55}$ *maravedis,* or a value of $728\frac{12}{55}$ *maravedis* for a *peso* of gold at $22\frac{1}{2}$. This is the first instance in the books on this list of a change in the price of gold. It could be a result of the devaluation of the currency, which John McCusker discusses in *Money and Exchange.*

Another first is a distinction between *maravedis* of silver and *maravedis* of gold. Those $728\frac{12}{55}$ *maravedis* of silver that were the value of a *peso* of gold at $22\frac{1}{2}$ carats are also equivalent to 589 *maravedis* of gold. The rate of exchange between the two seems to arise from the fact that *doblones* are $32 \times 34 = 1{,}088$ *maravedis* of silver, and yet they are also 2 *escudos,* which are each 440 *maravedis* of gold. So the ratio of their values is 880 to 1,088, or 55 to 68.

There is a mention of a tax of 5% for the king plus the cost of melting and assaying.

Martí uses the *calderón,* U, for thousands and *quento* for millions. In Louis Karpinski's article on our entry for 1649 he wrote, "In numeration the use of Cuento for million is interesting, as the term appears only in Spain."[206] He is, on the face of it, contradicting himself. Anyway, this entry, like the one for 1649, represents a use of *cuento,* or *quento,* in the Americas.

---

206. Louis C. Karpinski, "The Earliest Known American Arithmetic," *Science* 63 (Feb. 19, 1926): 194.

1697

Jacob Taylor. 37

Tenebræ in ... / or, the / ECLIPSES / of the / SUN & MOON / Calculated for Twenty Years / By Jacob Taylor, Amator Studiorum. / New York: Printed and Sold by William Bradford, 1697.

William Bradford, New York.

Evans 39332. This work is mentioned by Wing without a number being assigned; there is only the note "see almanacs."

Originals located at: Library Company (missing title page), New York Public Library (fragments).

Evans Digital Edition shows most of the New York copy. The only eclipse table visible there is for 1711, but the New York copy also has a fragment with the table for 1714.

Paging: Seventy-six pages with 39 and 40 repeated, i.e., 1 to 40, 39 to 74. Pages 1 to 4, 47, 65, and 70 are unnumbered.

There are tables of eclipses for the years 1698 to 1717. There is a table of declinations of the ecliptic on [47], as we saw in our entry for 1660. On page 48 there is a table of ascensional difference, and the method for finding sunrise from ascensional difference, the same method seen in the entry for 1606, is on page 54.

Page 56 (i.e., 58) starts a section called a "Compendium of Mensuration." Here there are methods for solving triangles with applications to other figures. When we discussed Enrico Martínez's *Reportorio* of 1606, we mentioned that trigonometric functions were typically defined in terms of a circle with a large radius until the unit circle became standard. On page 59 is a statement that reflects that practice: "As the Hypothenusal to the Leg, so is the Radius to the Sine of the Angle opposite to that Leg."

This book was the subject of the article "The First North American Mathematical Book and Its Metalcut Illustrations" by Keith Arbour in 1999. The present list shows that title to be a bit of an exaggeration. Arbour claims in particular that "Two of the illustrations are also the earliest mathematical illustrations printed anywhere on the continent."[207] We believe that this record actually goes to Alonso de la Vera Cruz's two entries for 1554. The

---

207. Arbour is familiar with Louis Karpinski's *Bibliography*, and so he considers and rejects the illustrations from Francisco de Belo's work of 1675 as being "technical." Even among the books on Karpinski's list, he could have cited the geometrical illustrations in Diego García de Palacio's *Dialogos Militares* of 1583.

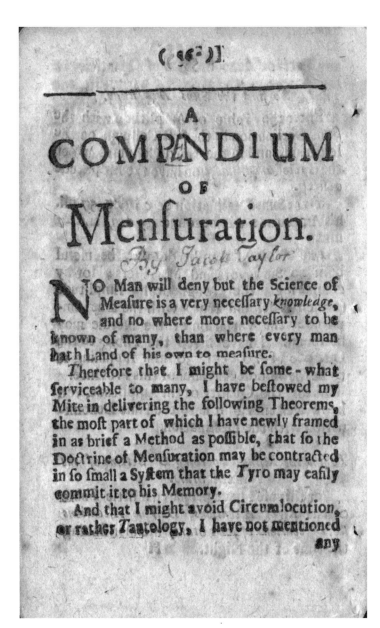

Figure 24. A "Compendium of Mensuration" is the second part of Jacob Taylor's *Tenebrae*, 1697. The Library Company, Philadelphia.

Recognitio Summularum has illustrations of the square of opposition and its variants, and the *Dialectica Resolutio* has marginal illustrations of triangles and parallel lines.

The fragments at the New York Public Library were recovered from the binding of another Bradford imprint along with several other items and identified by Wilberforce Eames. A date of 1698 was assigned to this item at that time. The more complete copy turned up at an auction in 1986, and from this it is evident that the date of publication is most probably 1697.

There is an entry in Part 2 for Jacob Taylor's almanac of 1699, also printed by Bradford in New York. Taylor began writing almanacs for the Press of Reynier Jansen in Philadelphia in 1702, according to Charles Swift Riché Hildeburn. See also the article by Gilbert Cope.[208]

Marion Barber Stowell found a reference to this book in the Daniel Leeds almanac of 1706. Taylor, in his almanac, had attacked the earlier almanacs by Leeds. In reply, writes Stowell, "Leeds suggests that Taylor should 'correct the Boston Almanack too,' for it, too, differs from Taylor's. He contends that Taylor himself has different calculations for the same events in his Book of Eclipses and in his almanac for 1705. . ."[209]

Manuel de Zuaza y Aranguren. 38
REDVCIONES / DE PLATA / COMPUESTO POR EL / THENIENTE DE CAVALLOS / CORAZAS / DON MANUEL / DE ZVAZA / DEDICADO / AL SEÑOR DON LUIS / SANCHEZ DE TAGLE, CAVA- / llero Professo de la Militar Orden, y / Cavalleria de Alcantara, Maestro de / Campo del Tertio Miliciano del Co- / mercio de la Ciudad de Mexico, por / su Magestad, y Comprador de / plata, y oro en ella. / *Con licencia in Mexico en la Imprenta de Juan Joseph / Guillena Carrascoso Año de* 1697.
Juan Joseph Guillena Carrascoso, Mexico.
Medina, *México*, 1686.
Original located at: Biblioteca Nacional de Chile.
Microfilm copies are available.
Paging: Five preliminary leaves followed by forty-three numbered folios with 2 unnumbered, 3 misnumbered 2, 26 misnumbered 9, 33 and 34 unnumbered, and the 3 inverted on 38.

208. Gilbert Cope, "Jacob Taylor, Almanac Maker," *Bulletins of the Chester County Historical Society* (1908): 10–28.
209. Marion Barber Stowell, *Early American Almanacs: The Colonial Weekday Bible*, New York, 1977: 156.

The preliminary material contains an Aprobación by Sigüenza y Góngora.

Some terms used here were introduced in our discussion of the *Sumario Compendioso* of 1556, and so this section should be read after that discussion.

The body of the work contains tables for converting various levels of purity of silver into the purity 2,376 *de ley*, which is exactly 99% pure. The range of purities given for this conversion is 2,210 to 2,375 *de ley*. Answers are given in *marcos*, ounces, *tomines*, and *granos*, with 8 ounces to the *marco*, 8 *tomines* to the ounce, and 12 *granos* to the *tomín*.

As an example of how this works, at the 2,256 level (94%) we see that 9 *marcos* of silver is worth 8 *marcos*, 4 ounces, 2 *tomines*, and 10 *granos* of silver at 2,376. Ninety *marcos* of silver at 2,256 are worth 85 *marcos*, 3 ounces, 5 *tomines*, and no *granos* of silver at 2,376. Totaling these gives 99 *marcos* of 2,256 silver the value of 93 *marcos*, 7 ounces, 7 *tomines*, and 10 *granos* of 2,376 silver, just 2 *granos* short of the right answer of 94 *marcos*.

The last page, 43r, gives what is due to the king if he is taking a *diezmo* or a *quinto*. The rates are 10.9% and 20.8%, respectively. The previous page, 42v, gives *plata de diezmo* and *plata quintada* based on a standard of 2,376. Again, the figures for the first are 10.9% lower than the second.

1700

Francisco de Fagoaga.   **39**
REDUCION / DE ORO / REFORMACION / DE RESCATES DE / PLATA, / Y / DE LAS TABLAS, Y QUENTAS / DE LA PLATA DEL DIEZMO / DISPUESTO / POR EL ALFEREZ FRANCISCO / DE FAGOAGA. / DEDICALO / A MARIA SANTISSI$^{\text{MA.}}$ / S^{RA} N^{RA} DE ARANZAZU. / CON LICENCIA / EN MEXICO, *EN LA IMPRENTA DE* / *Iuan Ioseph Guillena Carrascoso*. Año de 1700.

Juan Joseph Guillena Carrascoso, México.

Medina, *México*, 1764.

Originals located at: Biblioteca Nacional de España, Biblioteca National de México, California State Library, John Carter Brown Library, Library of Congress, University of Texas at Austin (two copies).

Paging: Ninety-nine leaves, unnumbered. There are ten preliminary leaves. Following this, forty-six leaves of tables have signatures on odd leaves, starting with A and running to Z, skipping J, U, and W, and with L missing from the page that should have it. Then on the last forty-three leaves, the alphabet starts over as Aa, Bb, etc., skipping Jj, Uu, Ww, and Zz. The W between Tt and Xx probably was supposed to be Vv. There are three leaves between Oo and Pp, while Pp, Qq, Rr, and Ss are contiguous.

The John Carter Brown copy is incomplete, missing preliminary leaves 2 and 10 and the last leaf.

The title page is printed in red and black. In copy 2 in Austin the red and black parts do not line up as well as in other copies.

An alternate title is given by José Toribio Medina as *Reduccion del Oro y Reforma de las Tablas y Computos para el Rescate de las Platas*.

Fagoaga had later titles that may be regarded as new editions of this entry. There appeared in Seville in 1719 a book by Fagoaga entitled *Nueva Reformación de las Tables, y Quentas del Valor Legítimo de la Plata del Diezmo, y Quintada*. Then in 1729 from Mexico came *Tablas de las cuentas del valor liquido de la plata del diezmo, y del intrinseco, y natural de la que se llama quintada, y de la reduccion de sus leyes a la de 12. dineros. Segun las novissimas ordenanzas de su Magestad, y de los derechos, que de la plata, y oro se le pagan en estos reynos, en conformidad de sus leyes reales, y cedulas*. This edition was reprinted without noticeable change in 1773.

Year	City	Held by
1719	Sevilla	California State Library, University of Minnesota, Yale University
1729	México	Biblioteca Nacional de Chile, Biblioteca Nacional de España, Biblioteca Nacional de México, California State Library, Colegio de México, Columbia University, John Carter Brown Library, University of Texas at Austin, University of Utah
1773	México	Biblioteca Nacional de Chile, Biblioteca Nacional de México, Bibliothèque du Musée de l'Homme, British Library, California State Library, Centro de Estudios de Historia de México Condumex, Consejo Superior de Investigaciones Científicas, John Carter Brown Library, New York Public Library, Southern Methodist University, Stony Brook University, Tulane University, University of California Berkeley, University of Michigan, University of Texas at El Paso, Washington State University, Wellesley College, Wichita State University, Yale University

Microfilm copies of the 1729 and 1773 editions are available. Elías Trabulse reproduces the title page of the 1773 edition in *Historia de la Ciencia en México, Estudios y Textos, Siglo XVI*.

Fagoaga's years are 1679 to 1736 according to various library catalogs. The New York Public Library's catalog gives the name as Francisco Fagoaga Iragorri with those same years, while the University of Texas uses both this and Francisco de Fagoaga. Three of the four books by Fagoaga at the California State Library are catalogued under "Fagoada." According to Jonathan Kandell,[210] the Fagoagas were one of the major mercantile and silver mining families of the eighteenth and nineteenth centuries in Mexico.

According to the statement on the title page of Fagoaga's book of 1719, Fagoaga was from Oyarzún in Guipúzcoa in the Basque Country. He entered the order of Santiago (one of the Spanish military orders) in April of 1736 and died that November. Because of Fagoaga's services to the crown, the king granted his second son, Francisco, the title Marqués del Apartado.

Some terms used here were introduced in the discussion of the *Sumario Compendioso* of 1556, so this section should be read after that discussion.

Altogether there are eleven tables spread over the eighty-nine leaves of the main part of the text. There are two tables, [1] and [3], for the value of gold. In the first, gold at various concentrations is converted to an equivalent amount of 22-carat gold. In the second the conversion is to $22\frac{1}{2}$-carat gold. In both tables the input is in *marcos* (weight) and *quilates* and *granos* (concentration), and output is in *castellanos, tomines,* and *granos*. Here *castellanos* may refer to weight rather than currency—a *castellano* would then be one fiftieth of a *marco* in both tables. A *marco* of gold at 22 carats is given the value of 50 *castellanos* in the first of these tables and a *marco* of gold at $22\frac{1}{2}$ is given the same value of 50 in the second table.

A *tomín* is one eighth of a *castellano* in both tables. A rather odd thing is that in the first table there are 11 *granos* to a *tomín,* and in the second table this ratio is $11\frac{1}{4}$—this is in spite of the fact that Fagoaga says on folio 10 that 12 *granos* make a *tomin*. All of our other texts use 12 or $12\frac{1}{2}$, with the *Sumario Compendioso* using both 12 and $12\frac{1}{2}$.

---

210. Jonathan Kandell, *La Capital*, New York, 1988.

In Fagoaga's tables [1] and [3], an increase of 1 *grano* of purity (one quarter of a carat) ups the value of 1 *marco* of gold by 50 *granos*. What may be happening here is that the *granos* column represents an equivalent weight of pure gold. Fagoaga may be taking *granos* (concentration) times *marcos* times 50 to get the equivalent weight in *granos* of 24-carat gold. Then he uses the 11 *granos* to the *tomín* to convert this number to *castellanos* and *tomines* of 22-carat gold for his first table, and he converts again using $11\frac{1}{4}$ for the second table.

In between these two tables is Table [2], for calculating the *quinto* on gold. It is explained that the values represent the *quinto* plus $1\frac{1}{2}\%$ (presumably for the assayer) plus the royal *señoreaje*. If the effect of the taxes on the net is multiplicative (as we have seen in earlier texts), then the *señoreaje* appears to be about 1.6%, producing a net which is $0.8 \times 0.985 \times 0.984 = 77.5392\%$ of the input, for a total tax of 22.4608%. This is in pretty close agreement with the entry that says the tax on 5,000 *castellanos* is 1,123 *castellanos* 0 *tomines* 3 *granos*. There appear to be 11 *granos* to the *tomín* in this table, as in the one just before it.

Tables [4] and [5] are for "*Rescates de Plata*" and give taxes on silver at rates from one quarter of a *tomín* per *marco* to 1 *peso* per *marco*. "*Un peso menos el marco*" means that 1 *peso de tipuzque* is deducted per *marco* that the silver would weigh if it were at a concentration of 2,376 *de ley*. So, the tax is $\frac{272}{2376}$, or 11.45%. Table [4] converts *reales* to silver and [5] converts back, so it is in [5] that the answers are reduced by the tax.

Tables [6] and [7] give the *diezmo* on silver, and it includes one tenth plus 1% plus a *señoreaje*. Table [10] gives the *quinto* on silver, namely, one fifth plus 1% plus a *señoreaje*. The *senoreajes* for the *quinto* and *diezmo* are practically the same at 1.5474%. So the net after the *quinto* is $0.8 \times 0.99 \times 0.984526 = 77.9745\%$ of the input, for a total tax of 22.0255%, and with the *diezmo* we have $0.9 \times 0.99 \times 0.984526 = 87.7213\%$, for a total tax of 12.2787%. Our guess for how such a number arises for the *señoreaje* is that the *señoreaje* represents 1 *real* for each *marco* that the silver would weigh if it were at 2,220 *de ley*, and that this charge is applied to the full weight, not what is left after the assayer's 1%. Thus,

$$\left(1 - 0.01 - \frac{34}{2,220}\right) \times 0.8 = 77.9748\%.$$

Tables [8] and [9] deal with mercury.

The last table, [11], is interesting for its use of a sort of floating point notation. It gives values of *plata del diezmo* in *pesos de tipuzque* for different weights and concentrations. The entry 1—712735 for the concentration 2,210 tells us that one *marco* is 7.12735 *pesos*, 10 *marcos* is 71.2735 *pesos*, 100 *marcos* is 712.735 *pesos*, etc. The decimal fractions would be converted by the user into *tomines* and *granos* as needed. These numbers include the same taxes as in tables [6] and [7], i.e.:

$$\frac{2{,}210}{272} \times 0.9 \times 0.99 \times 0.984526 = 7.12735.$$

For ounces, a hyphen is used for a decimal point. So 7 ounces of 2,210 is valued at 6-23, or 6.23 *pesos*.

Decimal notation is used on folio 10r in an explanation of the how to use Table [11]. The example is 1,247 *marcos* 2 *onças* at a concentration of 2,214. We show Fagoaga's first three columns here, and we have added the fourth column to show what one actually sees when looking in Table [11].

1000 Ms		7140.25	714025
200		1428.05	142805
40		285.61	285610
7		49.98	499817
	2 Onç	1.78	1-78
		8905.67	

Then the 67 is interpreted as *granos* at $12\frac{1}{2}$ to the *tomin*, and so the answer is 8,905 *pesos de tipuzque* 5 *tomines* $4\frac{1}{2}$ *granos*.

PART TWO

Almanacs, Ephemerides, and Lunarios

> No book or publication has ever been the subject of more ridicule
> and contempt than the almanack, yet no book has been more
> universally read, or more highly valued, or more serviceable
> to its day and generation.
>
> —Samuel Briggs, 1887

The years given in the headings here correspond to the calendar year discussed by the almanac, not the year of first printing. Since almanacs were generally available at the beginning of the year, whether that was March or January,[1] one can assume that they were almost always printed toward the end of the previous year. Charles Evans puts some almanacs in the year of the calendar and some in the previous year. To avoid that sort of inconsistency, and to make it easy to compare almanacs for the same year, we have adopted the present policy.

Almanacs of the era would nearly always have a calendar for the year, and following the number for each day there would be room for other information, typically including the signs that the sun and moon were in, possibly the sign that each of the other planets was in as well. Solstices and equinoxes would be indicated. Somewhere there would be a list of all the eclipses, of both the sun and moon, for the year. Even the shortest almanacs of eight leaves would use six leaves for the calendar and then have the verso of the title page and the whole last leaf for other material. This could include essays, poems, more astronomical tables, or advertisements. The high and low tides and the rising and setting of the sun were other events that might be tabulated or discussed.

---

1. The year still began in March for many regions using the Julian calendar, but where the Gregorian calendar had been adopted, the year began in January.

Two of the entries in Part 1 contain ephemerides, those for 1587 and 1606. They are in Part 1 because they contain other material not usually associated with almanacs.

Several sources refer to a *Kalendario Perpetuo* by Alejo García of Mexico. He was a Dominican, according to Nicolás Antonio.[2] A date of publication is not known, but García apparently died about 1579. José Toribio Medina lists this as number 199 in the section of *La Imprenta in México* for works of uncertain date. His title there is *Kalendarium perpetuum. A Fratre Alexus Garcia. Mexici,* . . .

Another speculative entry of Medina's is a series of *Lunarios* attributed by one early source to "Licenciado Brambila" (Medina, *México*: 180). There were supposed to be several issues in the late sixteenth century, and Medina guesses that the author may have been one Antonio Brambila de Arriaga.

According to José Miguel Quintana, an edict of 1647 required that almanacs be presented to the Inquisition to check that they dealt only with those subjects that were allowed: navigation, agriculture, medicine, and predictions of eclipses, rains, dry spells, etc.[3] Because of this edict, records exist of almanacs printed from 1649 on. Quintana reproduces the entire edict in Appendix 1 in his *La Astrología en la Nueva España*. The next eighty-seven appendices, numbered 2 to 88, correspond to the almanacs themselves and so that is the numbering that we use in the formal citations in this text—"Quintana 88," for example, refers to Appendix 88 in Quintana.

Clearly there must have been almanacs in Mexico before 1649, or the edict would not have been necessary. However, other than Enrico Martínez's for 1604, and a couple of authors just mentioned, there are no records of them. We might guess with some confidence at some of the authors, but titles and years would be pure speculation.

Quintana writes that the Inquisition records now at the Archivo General de la Nación show that the Mexican authors who applied for permission to write almanacs in the seventeenth century after the edict were: Felipe de Castro, Gabriel López de Bonilla (author of one of the comet books mentioned in Appendix B), Martín de Córdoba (suspected by Quintana to be a pseudonym of Diego Rodríguez, also an author of one of the comet books

2. Nicolás Antonio, *Biblioteca Hispana Nueva*, Madrid, 1788: 11.
3. A large number of the titles after the edict refer to *pronóstico de temporales*, prediction of storms.

mentioned in Appendix B), Juan Ruiz (son of Enrico Martínez and another author of one of the comet books mentioned in Appendix B), Nicolás de Matta, Carlos de Sigüenza y Góngora (author of the entry for 1690, in Part 1; sometimes writing his almanacs under a pseudonym, Juan de Torquemada), Juan de Saucedo, José de Escobar Salmerón y Castro (yet another author of one of the comet books mentioned in Appendix B), Antonio Sebastián de Aguilar Cantú, Juan de Avilés Ramírez, José de Campos (initially under the name Michael Henrico Romano), and Marco Antonio de Gamboa y Riaño.

It is strange that both Quintana in his *La Astrología* and Carmen Corona in her *Lunarios* omit the name of Feliciana Ruiz from their lists of almanac authors. Quintana has an appendix for her almanac of 1676 and calls her the granddaughter of Juan Ruiz (but she calls Juan Ruiz her *padre*). Quintana may be implying that she is simply printing a posthumous almanac of Juan Ruiz, but there is no reason to reject her as the author of the almanac, and, therefore, as the first woman to author a mathematical work printed in the New World.

Mexican almanac titles, their authors, and their printers listed here are from the Inquisition records when possible. In one particular volume, 670, of Inquisition records, there are two sets of leaf numbers. The numbers in blue pencil are not consecutive. Someone has renumbered the leaves consecutively in regular pencil. Throughout this section the regular pencil numbers are used, whereas Quintana's list uses the blue pencil numbers. We owe a debt to Quintana in that we have not rechecked all the Inquisition records but only the ones that he points to in his appendices.

For early English-language works, the Evans Digital Edition is an online resource available by subscription. Its ultimate aim is to put online the complete text of every extant entry in Evans's bibliography and its supplements. (Evans numbers that are more than three digits are taken from the update of Evans by Clifford Shipton and James Mooney.) In most cases the online version is identical to what can be found on the Readex microprint cards, which are indexed by the Evans number. For a few of the earlier entries a Readex copy exists but its image does not appear online. Seventy-one of the entries in this section for the English colonies are available in this online form, and six of those are reproduced twice owing to variant titles, or, in the case of Daniel Leeds's 1700 almanac, because of mistaken identity.

Another online resource is Early English Books Online. Sixteen of our English-language almanacs are available there.

1604

Enrico Martínez. **A1**
Lunario y Regimiento de Salud.
Enrico Martínez, México.
González de Cossío, 121.

This may or may not be the first almanac printed in the New World, depending on the truth about the Alejo García and Brambila de Arriaga references mentioned in the introduction to Part 2. Irving Leonard says this series ran until 1620, but neither Francisco González de Cossío nor Juan Pascoe lists it after 1604. Leonard's remark on page 76 of *Don Carlos de Sigüenza* may refer to the lunar tables for the years 1606 to 1620 in Martínez's *Reportorio de los Tiempos* (Part 1, 1606).

In his *Reportorio* of 1606, at the very end of Chapter VII of Treatise 4 on the age of the moon, Martínez refers to the *Lunario y Regimiento de salud* that he had printed and which he would print in the future "con el favor de Dios." Readers may look to the *Reportorio*, with its lunario starting on folio 56, for some idea of what Martínez's almanacs would have been like.

See the entry for the *Reportorio*, under 1606 in Part 1, for more information about this author and printer.

1639

William Pierce. **A2**
An Almanack for the year of our Lord 1639. Calculated for New England. by Mr. William Pierce, mariner.
Stephen Daye, Cambridge.
Sabin 59555, 62743, and note to 25791. Evans 2. Nichols [1]. Drake 2815.

This almanac was the second printed work in the English colonies of America. (The first was *The Oath of a Free-man*.) Isaiah Thomas, in his *The History of Printing in America*, gives some information about the printer, Daye, mostly taken from court records. Daye died in 1668. The title given is the form used by Evans, who indicates that this was probably a broadside, or sheet almanac.

The press in Cambridge was controlled by Harvard College. This press, under various printers, had no competition in the almanac field until 1676. Some write that Stephen Daye was the printer in name only and that his son Matthew did the actual printing.

Samuel Danforth's almanac of 1648 gives a death date of 1641 for a "Mr. William Peirce, [sic] an expert mariner . . ."

1640

Anon.   A3
An Almanack.
Stephen Daye, Cambridge.
Evans 3. Nichols [2]. Drake 2816.

Since almanacs were such a lucrative venture it is easy to believe that an almanac was produced each year once a press was set up. George Parker Winship[4] argues for the existence of almanacs for some but not all of the years 1640 through 1645 based on the inventory of paper used by Daye.

1641

Anon.   A4
An Almanack.
Stephen Daye, Cambridge.
Evans 5. Nichols [3]. Drake 2817.

1642

Anon.   A5
An Almanack.
Stephen Daye, Cambridge.
Evans 8. Nichols [4]. Drake 2818.

1643

Anon.   A6
An Almanack.
Stephen Daye, Cambridge.

---

4. George Parker Winship, *The Cambridge Press*, Philadelphia, 1945.

Evans 11. Nichols [5]. Drake 2819.

For this year there was printed an *Almanack for Rhode Island and Providence Plantations*. It was printed, however, in London, England.

1644

Anon.   A7
An Almanack.
Stephen Daye, Cambridge.
Evans 14. Nichols [6]. Drake 2820.

1645

Anon.   A8
An Almanack.
Stephen Daye, Cambridge.
Evans 16. Nichols [7]. Drake 2821.

1646

Samuel Danforth (1626–1674).   A9
[MDCXLVI. / AN / ALMANACK / FOR THE YEAR OF OUR / LORD / 1646 / Calculated for the Longitude of 315 / degr. and Elevation of the Pole Ar- / ctick 42 degr. & 30 min. & may ge- / nerally serve for the most part of / New-England. / By Samuel Danforth of Harvard Colledge / Philomathemat.]

Stephen Daye, Cambridge.

Sabin 47209, 62743. Evans 18. Nichols [8]. Drake 2822. Wing A1573.

Original located at: Huntington Library (third through seventh leaves).

Readex microprint and photostat copies are available. Early English Books Online shows the Huntington copy.

Paging: Eight leaves.

Since the title page is missing in the only copy, the title here is based on the titles for 1647 to 1649. This copy includes manuscript pages signed, "Rev. Samuel Sewall."

Runs March to February.

There are two Samuel Danforths who authored almanacs included in this list, so the range of years is indicated for each one. In fact, both John and Samuel, the first two of the elder Samuel Danforth's children to survive to adulthood, were authors of almanacs.

The Samuel Danforth who authored this entry was born in Suffolk County, England, and came to Massachusetts with his father in 1634. He graduated from Harvard in 1643.

The early almanacs in Massachusetts tended to be authored by Harvard graduates, and since in nearly every case the almanac was produced during the three years of study leading to a master's degree, it was likely to be, in each case, their first publication. For instance, Danforth got his master's on July 28, 1646, three years after his first degree. He remained at Harvard after his second degree and produced more almanacs until he left in 1650.

By Charles Nichols's count, forty-one of the forty-four Massachusetts almanacs produced before 1687 were the creation of twenty-six Harvard graduates, ten of whom were tutors at Harvard at the time of publication.[5] Nichols himself lists fifty-five almanacs for Massachusetts before 1687, forty-three of which can be attributed to a specific author. William Pierce and John Sherman were the only two of those authors who were not Harvard graduates.

Samuel Danforth was later the author of *An Astronomical Description of the Late Comet or Blazing Star as it appeared in New-England in the 9th, 10th, 11th, and in the beginning of the 12th Moneth, 1664. Together with a brief Theological Application thereof* (Cambridge, 1665). It was reprinted in London in 1666. Samuel Danforth's son Samuel wrote an essay on comets for his almanac of 1686.

At the bottoms of the calendar pages are short notes discoursing on the year, the month, the week, the day. There is here an explanation of the Julian and Gregorian calendars with precise values given for the tropical year (365 days, 5 hours, 48 minutes) and the sidereal year (365 days, 6 hours, 9 minutes, 39 seconds). The first value is only one minute low and the second value is about half a minute too high.

Within these notes on calendars there is a mention of a synonym for leap-year, ". . . so that every 4th yeare one day is added to February and counted before the 6th of the Kalends of March, wherefore it is called Bissextile or Leap-year." Apparently in the Roman calendar the extra day added to February was not the 29th but was six days before the calends (first

---

5. Charles Nichols, "Notes on the Almanacs of Massachusetts," *Proceedings of the American Antiquarian Society*, new series 22 (1912): 6.

day) of March. This sextile was doubled, giving rise to the term "bissextile" (Spanish *bisiesto*). Nachum Dershowitz and Edward Reingold give the complete Latin phrase for this day as *ante diem bis sextum Kalendas Martias*.

After each new moon the beginning of the Hebrew month is indicated. Most of the names of the months agree with the names that Daniel Leeds started using in 1694 (see our discussion for that entry).

These early Cambridge almanacs are called the "Philomath" almanacs by Robb Sagendorph and Marion Barber Stowell because of the frequent use of the word Philomath or its variations as part of the author's signature. Stowell presents a contrast between the Philomath almanacs of Cambridge, which tended to be both Puritan and scientific, and the later "farmer's" almanacs of John Foster, William Bradford, Leeds, and John Tulley, which tended to be both lighter in tone and more practical.

1647

Samuel Danforth (1626–1674).   **A10**

MDCXLVII. / AN / ALMANACK / FOR THE YEAR OF OUR / LORD / 1647 / Calculated for the Longitude of 315 / degr. and Elevation of the Pole Ar- / ctick. 42 degr. & 30 min. & may ge- / nerally serve for the most part of / New-England. / By *Samuel Danforth of Harvard Colledge* / Philomathemat. / CAMBRIDGE / Printed by *Matthew Day*. / Are to be solde by *Hez. Usher at Boston*. / 1647

Matthew Day, Cambridge.

Sabin 1873, 62743. Evans 21. Nichols [9]. Drake 2823. Wing A1574.

Original located at: Huntington Library.

Readex microprint and photostat copies are available. The complete text of the Huntington copy is available online at Evans Digital Edition. There is a reproduction of the title page in Joseph Blumenthal's *The Printed Book in America*.

Paging: Eight leaves.

Runs March to February.

The copy is interleaved with notes about the weather.

According to Samuel Abbott Green[6] and others, the printer Matthew Day is the son[7] of Stephen Daye but he has dropped the final "e" of the

---

6. Samuel Abbott Green, *John Foster*, Boston, 1909: 23.
7. Joseph Sabin says brother.

name. Matthew Day died in 1649. Shearer writes that this is the "first appearance of a printer's name in an American imprint" and that this is the only appearance for Matthew Day.[8]

The years for the bookseller, Hezekiah Usher, are 1615 to 1676. According to Richard Clement, Usher arrived in Cambridge in 1638 and moved to Boston in 1642. His bookstore in Cambridge in 1638 was the first bookstore in the English colonies in America.[9]

The longitude given in the title must be intended as degrees east of London. Today, longitudes in the Western Hemisphere are given as west from the Prime Meridian. The correct longitude of Boston is about 71° (the 71° meridian appears to go right through Logan Airport). Going the long way around should be 360 − 71 = 289°, so the 315° is too high. Yet the 315° figure is repeated in almanac titles from Cambridge and Boston through 1691. In 1694 we begin to see the figure 69°20' "to the westward of London," which is a little low but much closer than 315° east. Before the eighteenth century, fixing the longitude of a place was much more difficult than finding its latitude (see Dava Sobel's *Longitude*).

1648

Samuel Danforth (1626–1674). A11

MDCXLVIII. / AN / ALMANACK / FOR THE YEAR OF OUR / LORD / 1648 / Calculated for the Longitude of 315 / degr. and Elevation of the Pole Ar- / ctick. 42 degr. & 30 min. & may ge- / nerally serve for the most part of / New-England. / By Samuel Danforth of Harvard Colledge / Philomathemat. / Printed at CAMBRIDGE. / 1648.

Matthew Day, Cambridge.

Sabin 18473, 62743. Evans 23. Nichols [10]. Drake 2824. Wing A1575.

Original located at: Huntington Library.

Readex microprint and photostat copies are available.

Paging: Eight leaves.

Runs March to February.

A table of high tides in Boston uses the age of the moon.

The chronology for 1641 has "Mr. William Peirce, [sic] an expert mariner and Samuel Wakeman slain at Providence, newly taken by $\overset{e}{y}$ Spainyard."

---

8. Mark S. Shearer, "Stephen Day (or Daye)," in *Boston Printers*, ed. B. Franklin V, Boston, 1980.

9. Richard Clement, *The Book in America*, Golden, CO, 1996.

This may be the Pierce who authored the 1639 almanac. For 1646 we see, "Mr. Eliot began to preach to y̆ Indians in their own language." This is the author of the entry for 1672 in Part 1.

Drake claims in his introduction (p. [vii]) to be reading from this almanac when he quotes "The twelfth month called February hath xxix dayes." This is an error in that Danforth's almanac for 1648 in fact says that "The twelfth month called February hath xxviij dayes." See our discussion of leap year in the English colonies under our entry for 1650.

1649

Felipe de Castro. A12
Lunario y Repertorio de salud para el año venidero de 1649.
México.
Quintana 2.

The records that pertain to this title at the Archivo General de la Nación in Mexico City are *Inquisición* 670, leaves 249, 250v, and 251. Leaf 249 is a list of things that should be deleted, signed by Francisco Calderón.

This is the first of the 87 seventeenth-century Mexican almanac titles listed by Quintana. Of Felipe de Castro, Quintana says that he was an Augustinian and was the astrology teacher of Melchor Pérez de Soto.

Samuel Danforth (1626–1674). A13
MDCXLIX. / AN / ALMANACK / FOR THE YEAR OF / OUR LORD / 1649 / Calculated for the Longitude of 315 / degr. and Elevation of the Pole Ar- / ctick. 42 degr. & 30 min. & may ge- / nerally serve for the most part of / *New-England.* / By Samuel Danforth of Harvard Colledge / Philomathemat: / Printed at CAMBRIDGE. / 1649.
Samuel Green, Sr., Cambridge.
Sabin 18473, 62743. Evans 27. Nichols [11]. Drake 2825. Wing A1576
Original located at: New York Public Library.

Readex microprint and photostat copies are available. The complete text of the New York copy is available online at Evans Digital Edition.

Paging: Eight leaves.

Runs March to February.

The printer Samuel Green took over the press that had been run by Stephen Daye. Isaiah Thomas believes that he was assisted by Daye at first.

On the other hand, it has been proposed that since Matthew Day didn't die until May 10, 1649, he may have printed this work.

Richard Clement writes that Green "supplemented his income as printer by selling books and supplies to Harvard students and by cutting their hair."[10]

Gabriel López de Bonilla.   A14
El Diario y Discurso Astronomico para El Año que viene de 1649.
México.
Quintana 3.

The record that pertains to this title at the Archivo General de la Nación is *Inquisición* 670, leaf 250r.

Quintana writes that López de Bonilla was from the town of Alcolea de Tajo in Spain and was a professor of astronomy and mathematics. He also wrote *Discurso, y Relación Cometographia del repentino aborto de los Astros, que sucedió del Cometa que apareció por Diziembre de 1653*, which was printed in México in 1654. It is listed in our Appendix B.

López de Bonilla became related by marriage to one of the major authors on our list, Carlos de Sigüenza y Góngora, and thus was mentioned in the latter's will. There is some confusion in the literature, however, as to exactly how he did it. Elías Trabulse[11] and Irving Leonard[12] write that Gabriel López de Sigüenza was Carlos de Sigüenza's nephew and executor of his will. Leonard adds that the mother of Gabriel López de Sigüenza was Inés, the sister of Carlos de Sigüenza.[13] This would all suggest that Gabriel López de Bonilla, likely the father of Gabriel López de Sigüenza, married the sister of Sigüenza y Góngora, and that is exactly what Francisco Pérez Salazar says happened in his biography of Sigüenza for *Obras*.

But Quintana insists on page 49 of *La Astrología* that Gabriel López de Bonilla married the niece of Sigüenza y Góngora, quoting a clause in Sigüenza's will that refers to López as the husband of "María Guadalupe, daughter of Doña Inés de Sigüenza, my deceased sister." He then contradicts himself by agreeing with the other sources that Gabriel López de Bonilla [Sigüenza?] was Carlos de Sigüenza's nephew and heir.

---

10. Richard Clement, *The Book in America*, Golden, CO, 1996.
11. In his prologue to the 2001 facsimile edition of Sigüenza's *Libra Astronómica*.
12. In his "Sigüenza y Góngora and the Chaplaincy" and in *Don Carlos*, p. 13.
13. *Don Carlos*, p. 8.

1650

Urian Oakes. **A15**

MDCL. / AN / ALMANACK / FOR THE YEAR OF / OUR LORD / 1650 / Being the third after Leap year / and from the Creation 5582. / Calculated for the Longitude of 315 / degr. and Elevation of the Pole Ar- / ctick 42 degr. & 30 min. & may ge- / nerally serve for the most part of / *New-England*. / *Parvum parva decent: sed inest sua / gratia parvis.* / Printed at CAMBRIDGE. / 1650.

Samuel Green, Sr., Cambridge.

Sabin 47210, 62743. Evans 32. Nichols [12]. Drake 2826. Wing A2002.

Original located at: Huntington Library (seven leaves).

Readex microprint and photostat copies are available. Early English Books Online shows the Huntington copy.

Paging: Eight leaves.

Runs March to February.

There is a handwritten signature, "By President Oakes" on the only copy. Oakes's years are 1631 to 1681. Born in England, he graduated from Harvard in 1649 and was the fourth president of the College.

This is the first of the almanacs from the English colonies to mention the leap year in the title. In the regions of the world where the beginning of the year was celebrated in January, the last leap day prior to 1650 fell in February of 1648. In the English colonies, with the year starting in March, the very same February was still February 1647, and so Oakes can say that 1650 is the third after leap year instead of second after leap year. See John Foster's discussion of this point in the entry under 1679.

1651

Anon. **A16**

AN / ALMANACK.
Samuel Green, Sr., Cambridge.
Nichols [13]. Drake 2827.

Francisco Ruiz Lozano **A17**

Other Author: Diego Rodríguez?

Reportorio anual para el reino de Mexico. Por el Capitan Francisco Ruiz Lozano.

Juan Ruiz, México.

Medina, *México*, 740.

This title and the similar entries in succeeding years were for the use of navigators in the waters off the Pacific coast of South America. This series may not have been covered by the edict of 1647 (see the introduction to Part 2). No copies are known, so we cannot be sure, but if it had no astrological predictions, then that might have made it exempt from the edict. After Ruiz Lozano moved to Lima in 1653 he continued the series there for six more issues, starting in 1654.

This entry and the similar one in the next year are attributed to Diego Rodriguez by Gumersindo Placer López. "Este libro y el siguiente aparecen bajo el nombre de un su discípulo, pero que en realidad son del mercedario, según atestigua el P. Hardá."[14] For more about Rodríguez, see the entry here for Martín Córdoba in 1655.

Besides these *Reportorios*, Ruiz Lozano also published in Lima the comet book *Tratado de Cometas, Observationes, y iuicio del que se vio en esta ciudad de los reyes, y generalmente en todo el mundo, por los fines del año de 1664. Y principios deste de 1665*, which is listed in Appendix B.

He also wrote in 1662 that he "had arranged to give to the press" *Derrotero general de esta Mar del Sur, desde el estrecho de Maire, sito en la parte más austral del polo, hasta el cabo mendocino, que es lo más septentrional de la América, con las tablas de las declinaciones del sol y estrellas de primera magnitud, nuevamente corregidas, y que será de grande utilidad para la navegación de estas costas*. José Toribio Medina quotes the document where Ruiz Lozano says this but fails to list this work as having been printed in Lima in 1662. If it was printed, then it might be something we could have listed in Part 1.

The document cluster "Lima 170" at the Archivo General de las Indias contains four petitions from Ruiz Lozano to the viceroy of Peru. The earliest, dated 1662, appears to be asking for the position of Cosmógrafo Mayor of the Kingdom of Peru. Ruiz Lozano repeatedly stresses his knowledge of "las Mathematicas, especialmente la Arizmetica arte mayor del Algebra y los seis primeros libros de la Geometria de Euclides, y explicacion de la Esphera elemental, y celeste." In the later petitions he uses the title Cosmógrafo Mayor as well as Capitán. Ruiz Lozano was the first to hold the chair of Mathematics at the University of San Marcos. He died in 1678.

This author also comes up in Part 1 in the discussion of Juan Ramón Koenig's 1696 work and here, under the 1655 entry for Martín Córdoba.

14. Gumersindo Placer López, *Bibliografía Mercedaria*, Madrid, 1968.

1652

Anon.   A18
AN / ALMANACK.
Samuel Green, Sr., Cambridge.
Nichols [14]. Drake 2828.

Francisco Ruiz Lozano.   A19
Other author: Diego Rodríguez?
Reportorio anual para el reino de Mexico. Por el Capitan Francisco Ruiz Lozano.
Juan Ruiz, México.
Medina, *México*, 762.

1653

Anon.   A20
AN / ALMANACK.
Samuel Green, Sr., Cambridge.
Nichols [15]. Drake 2829.

Juan Ruiz.   A21
Reportorio.
Juan Ruiz, México.
In his *Discurso hecho sobre la significación de dos impressiones meteorológicas que se vieron el año passado de 1652. La primera de un arco que se terminava de oriente a occidente a 18 de noviembre y la segunda del cometa visto por todo el orbe terrestre desde 17 de diziembre del mesmo año de 1652* of 1653, which we have listed in Appendix B, Ruiz makes numerous references to this *Reportorio* of 1653. We are guessing that when Francisco Ruiz Lozano left for Lima, his printer, Juan Ruiz, took up the task of continuing the title in Mexico for at least one more year.

Elías Trabulse (*Apéndices e Índices*), Francisco González de Cossío (appendix to the reprint of the *Reportorio de los Tiempos*), and Quintana (*La Astrología en la Nueva España*) all say that Juan Ruiz was the son of Enrico Martínez, and they probably all draw this from Francisco de la Maza's 1943 biography of Martínez. Quintana, for example, says that Ruiz is the son of Enrico Martínez and Juana Leonor and that this is stated in Ruiz's will of 1670.

Note that Ruiz was the printer for the entries in Part 1 for 1615 and 1623. We assume that he was the printer for all of his own almanacs in this section. He probably learned the printing trade from Martínez. Louis Karpinski, in his review of Florian Cajori's *The Early Mathematical Sciences*, credits Ruiz as the printer of the first map printed in the New World.[15] This was a 1658 map of part of the West Indies.

Karpinski also mentions Juan Ruiz's 1653 *Discurso* as the first work in the Americas on the Aurora Borealis; this must be referring to the *arco* of the title. We wonder about this interpretation when we read Ruiz's claim that the arc appeared at 4:30 in the afternoon. We might guess that Ruiz is describing a rainbow, except that he says it moves from west to east. Trabulse (*Historia de la Ciencia en México, Estudios y Textos, Siglo XVII*) gives an excerpt from this *Discurso*.

Trabulse writes that Ruiz received a medical doctorate in 1610. Ruiz calls himself "Professor de Mathematicas" on the title page of his *Discurso*. Quintana states on page 50 of *La Astrología* that Ruiz died June 17, 1675. Francisco Pérez Salazar has published the list of holdings of Ruiz's library.[16]

1654

Anon.   A22
AN / ALMANACK.
Samuel Green, Sr., Cambridge.
Nichols [16]. Drake 2830

Francisco Ruiz Lozano.   A23
Reportorio anual para el reino del Perú. Por el Capitan Francisco Ruiz Lozano.
Lima.
Medina, *Lima*, 361. Vargas Ugarte 486.

José Toribio Medina lists a copy at the Archivo General de Indias, Sevilla, with number 70-5-26. This number leads to what is, in fact, a heap

---

15. Karpinski was an expert on early maps of the Americas, so this is probably right, but it seems odd that Ruiz's father, Enrico Martínez, was both a printer and a cartographer, and yet he never printed a map. There are thirty-three Martínez maps at the Archivo General de las Indias, all hand drawn.

16. Francisco Pérez Salazar, "Dos Familias," *Mémoires Société Scientifique "Antonio Alzate"* 43 (1925): 447–511.

of manuscripts, now numbered Lima, 170, four of which are by Ruiz Lozano. See our entry here for 1651 for Ruiz Lozano for more about these manuscripts.

Ruben Vargas Ugarte indicates that there were six such *Reportorios* by Ruiz Lozano printed in Lima but does not give numbers to the others.

1655

Anon.   A24
AN / ALMANACK.
Samuel Green, Sr., Cambridge.
Nichols [17]. Drake 2831.

Martín de Córdoba.   A25
Pronostico para 1655.
Paula de Benavides, Widow of Bernardo Calderón, México.
Quintana 4.
The record that pertains to this title at the Archivo General de la Nación is *Inquisición* 670, leaf 279.

Quintana suspects that the name Martín de Córdoba (he uses Córdova) is an assumed name for the mercedarian Diego Rodríguez (1596–1668), because Jaén de la Plaza states that Rodríguez wrote prognosticaciones under the name "El Cordobés." He also points out that the publication dates of the five Córdoba almanacs were all within Rodríguez's known lifetime. Also suspicious is that all of the permission slips were submitted by the printer in a time when other such requests to the Inquisition came from the author. He adds that one of the readers for the almanac of this year expressed suspicion about the name.

These five Córdoba almanacs are also attributed to Rodríguez by Gumersindo Placer López, who cites P. Hardá as his authority. We noted in an earlier entry that Placer López adds the *Reportorios* of Ruiz Lozano in 1651 and 1652 to the works of Rodríguez as well.[17]

Elías Trabulse has a mention of Córdoba (*Apéndices e Índices*, p. 58).

In addition to his alleged authorship of almanacs, the only other known printed work by Rodríguez is his comet book, *Discurso Etheorológico del Nuevo Cometa, Visto en aqueste Hemisferio Mexicano; y generalmente en todo el*

---

17. Gumersindo Placer López, *Bibliografía Mercedaria*, Madrid, 1968.

mundo. Este Año de 1652, which we have listed in Appendix B. (We have examined the only known copy of the *Discurso Etheorológico* at the Biblioteca Nacional de México and found no mathematics.) Medina gives the title as *Discurso ethereologico sobre el cometa aparecido en México en 1652* and gives the printer as Hipolito de Ribera (it should be Widow of Bernardo Calderón). Trabulse reproduces the title page and gives an excerpt in his *Historia de la Ciencia en México, Estudios y Textos, Siglo XVII*. He also shows copies of manuscript pages that he says are from the *Discurso*. These pages deal with parallax and triangles, including a mention of the sine function, but those topics do not appear in the *Discurso*. Juan Ruiz mentions Diego Rodríguez and his *Discurso Eteorológico* in his own *Discurso* of 1653, but he gives the year of Rodríguez's work as 1653 instead of 1652.

Rodríguez is known to have written many manuscripts that might have made our list in Part 1 had they been published. Trabulse lists the following such manuscripts in *Historia de la Ciencia en México, Estudios y Textos, Siglo XVI*: *Tractatus Proemialium Mathematices y de Geometría, Tratado de las Equaciones, Fabrica y Uso de la Tabla Algebraica Discursiva, De los Logaritmos y Aritmética, Tratado del Moda de Fabricar Reloxes, Modo de Calcular Qualquier Eclipse de Sol y Luna*, and *Doctrina General Repartida por Capítulos de los Eclipses de Sol y Luna*. Alternatively, Placer López gives the following titles for manuscripts: *Geometría Speculativa, De Aritmética, Tratado de los Ecuaciones, Arte de Fabricar Relojes*, and *Tractatus Proaemialium Mathematicarum*.

José Toribio Medina quotes José Mariano Beristáin de Souza as saying that Rodríguez gave some manuscripts, including one entitled *Logaritmos*, to Francisco Ruiz Lozano, who was leaving Mexico for Lima. Ruiz Lozano was to publish these in Lima, but this apparently never occurred. However, Placer López says *Logaritmi Astronomi* was printed as two volumes in Mexico in 1665. Florian Cajori says that *Logaritmos* was printed in Mexico about 1668.[18] In his review of Cajori's book, Louis Karpinski reiterates that "according to Medina... Logaritmos... was never printed either in Mexico or elsewhere."[19] Trabulse (*Siglo XVII*) gives an excerpt entitled "De los logaritmos y aritmética," from a manuscript at the Biblioteca Nacional de México.

In his *La Ciencia Perdida*, Trabulse says that Rodríguez wrestled with the concept of imaginary numbers. To solve certain third-degree equations it is

---

18. Florian Cajori, *The Early Mathematical Sciences*, Boston, 1928.
19. Louis Karpinski, "Review of Florian Cajori's *The Early Mathematical Sciences in North and South America*," *Isis* XII, I/37 (1929): 163–165.

necessary to introduce imaginary numbers. The status of these numbers would not be fully settled until much later.

To solve the third-degree equation $x^3 - ax - b = 0$, one can use Cardano's Formula,

$$x = \sqrt[3]{\frac{b}{2} + \sqrt{\frac{b^2}{4} - \frac{a^3}{27}}} + \sqrt[3]{\frac{b}{2} - \sqrt{\frac{b^2}{4} - \frac{a^3}{27}}}$$

published in Gerolamo Cardano's *Ars Magna* in 1545. This is fine for an equation such as $x^3 + 3x - 2 = 0$, for which the formula tells us that

$$x = \sqrt[3]{1 + \sqrt{2}} + \sqrt[3]{1 - \sqrt{2}}.$$

But in the case of a simple equation such as $x^3 - 3x = 0$, whose solutions are obviously $-\sqrt{3}$, 0, and $\sqrt{3}$, the formula gives

$$x = \sqrt[3]{\sqrt{-1}} + \sqrt[3]{-\sqrt{-1}}.$$

What is to be made of $\sqrt{-1}$? In modern notation, letting i stand for $\sqrt{-1}$, the solution becomes

$$x = \sqrt[3]{i} + \sqrt[3]{-i}.$$

With the benefit of current methods for manipulating these sorts of numbers, we can say that every number of the form $a + bi$ has three cube roots, unless a and b are both 0. The cube roots of i are $-i$, $\frac{1}{2}(\sqrt{3} + i)$, and $\frac{1}{2}(-\sqrt{3} + i)$, and similarly the cube roots of $-i$ are i, $\frac{1}{2}(\sqrt{3} - i)$, and $\frac{1}{2}(-\sqrt{3} - i)$. If we add pairs of these in just the right way we get the correct solutions:

$$\frac{1}{2}(\sqrt{3} + i) + \frac{1}{2}(\sqrt{3} - i) = \sqrt{3},$$

$$\frac{1}{2}(-\sqrt{3} + i) + \frac{1}{2}(-\sqrt{3} - i) = -\sqrt{3}$$

and

$$-i + i = 0.$$

This illustrates how the application of Cardano's Formula is not always straightforward. Using it to find solutions to cubic equations sometimes makes necessary use of so-called imaginary numbers (of the form *bi*) and even so-called complex numbers (of the form *a* + *bi*). This even happens when the desired answers have no imaginary term but are only so-called real numbers. Thus Cardano's Formula created a strong reason for mathematicians to begin thinking about these imaginary numbers.

Francisco Ruiz Lozano. **A26**
Reportorio anual para el reino del Perú. Por el Capitan Francisco Ruiz Lozano.
Lima.
Medina, *Lima*, 370.

1656

Gabriel López de Bonilla. **A27**
Lunario y discurso Astronomico para El Año que Viene de 1656.
México.
Quintana 5.
The records that pertain to this title at the Archivo General de la Nación are *Inquisición* 670, leaves 96 to 97. The petition to the Inquisition for permission to publish this almanac is actually held at the Newberry Library. The items at the Archivo General de la Nación are the Inquisition's response.

Francisco Ruiz Lozano. **A28**
Reportorio anual para el reino del Perú. Por el Capitan Francisco Ruiz Lozano.
Lima.
Medina, *Lima*, 373.

Thomas Shepard. **A29**
MDCLVI. / AN / ALMANACK / FOR THE YEAR OF / OUR LORD / 1656. / Being first after Leap year, and / from the Creation 5588. / Whose Vulgar Notes are , / Golden number 4. Cycle of the Sun 13. / Romane [sic] Indict: 9. The Epact 14. / Dominicall letter E. / Calculated for the Longitude of 315 / gr: and 42 gr: 30 min: of N. Lat: / and may Generally serve for / the most part of / New England. / By T. S. Philomathemat: / *CAMBRIDG* / Printed by *Samuel Green* 1656.

Samuel Green, Sr., Cambridge.
Sabin 62743. Evans 43. Nichols [18]. Drake 2832. Wing A2381aA.
Original located at: American Antiquarian Society.
Readex microprint and photostat copies are available. The complete text of the AAS copy is available online at Evans Digital Edition.
Paging: Eight leaves.
Runs March to February.
The copy is signed by Rev. Flynt, 1756.
This entry was attributed to Shepard by Charles Nichols.[20] Shepard was born April 5, 1635, in London. He graduated from Harvard in 1653 and died of smallpox December 22, 1677. His father, Thomas Shepard (1605–1649), was a founder of Harvard College.

The definition for the golden number used by these almanac writers is the same as that used by Diego García de Palacio in his *Instrucción Nautica* (see the discussion for 1587 in Part 1). The relevant calculation here is that $1656 + 1 = 87 \times 19 + 4$.

However, the definition of epact is different. As we mentioned in Part 1 within the entry for 1587, the relevant equation for the English colonial almanac writer, using the Julian calendar instead of the Gregorian, is

$$\text{Epact} \equiv 11 \times \text{Golden Number} \pmod{30}$$

instead of

$$\text{Epact} \equiv 11 \times \text{Golden Number} - 10 \pmod{30}$$

This difference of 10 is exactly the difference between the Julian and Gregorian calendars in the seventeenth century. If the rule set out in the *Instrucción Nautica* gives the age of the moon on January first on the Gregorian calendar, then since the moon is ten days older on the first of January in the Julian calendar, ten days later, both rules are giving the age of the moon, equally well or poorly, on January first for their respective calendars.

The Roman Indiction is a fifteen-year cycle originating with the Roman tax calendar, according to Jerónimo de Chávez.[21] The formula is

$$\text{Roman Indiction} \equiv \text{Year} + 3 \pmod{15}.$$

20. Charles Nichols, "Notes on the Almanacs of Massachusetts," *Proceedings of the American Antiquarian Society* new series 22 (1912): 13.
21. Jerónimo de Chávez, *Chronographia o Reportorio*, Sevilla, 1572.

The cycle of the sun and the dominical letter are based on a twenty-eight year cycle arising from the seven-day week and the four-year leap cycle. We have

$$\text{Cycle of the Sun} \equiv \text{Year} + 9 \pmod{28}.$$

The dominical letter varies with what day of the week the year starts on, and so it also follows a twenty-eight year cycle. The days of the year are lined up with the letters a through g, with January 1 next to a and so on. The letter that falls by Sunday is the dominical letter. For leap years it doubles up, e.g., DC in the title of Israel Chauncy's almanac for 1663 (see later entry). This is because the cycle of letters skips February 29, and so there is one dominical letter for January and February and another one for the rest of the year. In the tables given by Chávez, a year with Solar Cycle 13 should be a leap year, and so its dominical letter is always FE and the previous year's would be G. For Shepard the leap year was 1655, so that year would get GF, leaving just the E for 1656. (See the discussion of leap year in the entry for 1650.) Note that even though Shepard's calendar starts with March, the association of letters with days still matches the letter a with January 1. (However, the Julian and Gregorian dominical letters still disagree due to the ten-day difference in their dates—see the mention of this in the entry for 1665.)

1657

Samuel Bradstreet.   A30

  · AN / ALMANACK / FOR THE YEAR OF / OUR LORD / 1657. / Being Second after Leap-year. / The Vulgar Notes whereof are , / *Golden number* 5. / *Cycle of the Sun* 14. / *The Epact* 25. / *Dominicall letter* D. / Calculated for the Longitude of 315 / gr: and 42 gr: 30 *min:* of N. Lat: / and may Generally serve for / the most part of / *New-England*. / By *S. B.* Philomathemat: / CAMBRIDG. / Printed by *Samuel Green* 1657.

Samuel Green, Sr., Cambridge.

Sabin 62743. Evans 44. Nichols [19]. Drake 2833. Wing A1374B.

Original located at: American Antiquarian Society (leaves 1, 3, 4, 7).

Readex microprint and photostat copies are available. Four leaves of the text (from the American Antiquarian Society copy) are available online at Evans Digital Edition.

Paging: Eight leaves.

Runs March to February.

The American Antiquarian Society copy is signed by Rev. Flint.

This almanac was attributed to Bradstreet by Charles Nichols.[22] John Langdon Sibley has a respectable amount to say about him in his *Graduates of Harvard University*, for the class of 1653[23]; Sibley does not, however, list any writings for Bradstreet. It could be that Nichols's attribution stems from the fact that there is no other "S. B." in the Harvard classes of 1653 to 1657.

Samuel Bradstreet was the son of Anne Bradstreet (the poet) and Simon Bradstreet (governor of Massachusetts). Sibley does not give a birth year for Samuel, but a Web site for Anne Bradstreet (www.rootsweb.com/~nwa/bradstreet.html) asserts that Samuel was born in Cambridge in 1633/4. He was the oldest of eight children.

Harvard had two commencements in 1653, on August 9 and 10. The first group was given two years to complete a second degree, and the second group was required to spend three years. Increase Mather was in the first group and Samuel Bradstreet was in the second. Thus he might have been working on the almanac for 1657 while he was wrapping up his second degree.

He was in England from 1657 to 1661, probably studying medicine. After 1670 he went to Jamaica, and he died there in 1682.

Joseph Dudley, author of an almanac for 1668, was the son of Governor Thomas Dudley, therefore the brother of Anne Dudley Bradstreet, and therefore an uncle to Samuel Bradstreet.

Francisco Ruiz Lozano. **A31**
Reportorio anual para el reino del Perú. Por el Capitan Francisco Ruiz Lozano.
Lima.
Medina, *Lima*, 380.

1658

Anon. **A32**
AN / ALMANACK.
Samuel Green, Sr., Cambridge.
Nichols [20]. Drake 2834.

---

22. Charles Nichols, "Notes on the Almanacs of Massachusetts," *Proceedings of the American Antiquarian Society* new series 22 (1912): 13–14.

23. John Langdon Sibley, *Biographical Sketches of Graduates of Harvard University, in Cambridge, Massachusetts*, Vol. I, Cambridge, 1885: 360–361.

Francisco Ruiz Lozano. **A33**
Reportorio anual para el reino del Perú. Por el Capitan Francisco Ruiz Lozano.
Lima.
Medina, *Lima*, 387.

1659

Zecharia Brigden. **A34**
AN / ALMANACK / OF THE COELESTIAL MOTIONS / FOR THIS PRESENT YEAR / of the Christian Æra / 1659. / Being (in our account) Bissextile or Leap- / year, and from the Creation (according to / truest Computation) 5608. / Whose Vulgar Notes are / Golden number 7. Cycle of the Sun 16. / The Epact 17. Dominical Letter B. / Fitted to 315 degrees of Longitude, the (supposed) / Meridian of the MASSACHUSETS BAY, where the / Pole Artique is Elevated 42 gr: and 30 *min*: and / may without any sensible errour be applyed to / any part of New-England. / By *Zech: Brigden* Astrophil: / *Et dixit Deus, sunto Luminaria in Expanso Coeli, ad / distinguendum inter Diem & Noctem: & sint in signa / & Tempora, in Dies & Annos.* / CAMBRIDG / Printed by *Samuel Green* 1659.

Samuel Green, Sr., Cambridge.

Sabin 62743. Evans 54. Nichols [21]. Drake 2835. Wing A1379.

Original located at: Library of Congress.

Readex microprint and photostat copies are available. The complete text of the LC copy is available online at Evans Digital Edition.

The copy is signed by Rev. Flint.

Paging: Eight leaves.

Runs March to February.

Since this is a leap year, the dominical letter should be BA.

Brigden's years are 1639 to 1662. He graduated from Harvard in 1657.

This work contains a description of a solar system with the sun in the center and calls it the "philolaick" system, after Philolaus.

Juan Ruiz. **A35**
El Pronostico del Año venidero de 1659.
Juan Ruiz, México.
Quintana 6.

The records that pertain to this title at the Archivo General de la Nación are *Inquisición* 670, leaves 380 to 381.

This is the first Ruiz almanac listed by Quintana, but we have listed an earlier almanac for him. See the Ruiz entry for 1653 for more about Ruiz.

Francisco Ruiz Lozano. **A36**
Reportorio anual para el reino del Perú. Por el Capitan Francisco Ruiz Lozano.
Lima.
Medina, *Lima*, 398.

1660

Samuel Cheever. **A37**
MDCLX. / *AN* / ALMANACK / FOR THE YEAR OF / OUR LORD / 1660. / *Being first after Leap year, and from* / *the Creation* 5609. / Whose Vulgar Notes are / *Golden number* 8. *Cycle of the Sun* 17. / *Roman Indict.* 13. *The Epact* 28. / *Dominicall letter G.* / Calculated for the Longitude of 315 / gr: and 42 gr: 30 *min:* of N: Lat: / and may Generally serve for / the most part of / *New-England.* / By *S. C.* Philomathemat: / *CAMBRIDG* / Printed by *Samuel Green* 1660.
Samuel Green, Sr., Cambridge.
Sabin 62743. Evans 57. Nichols [22]. Drake 2836. Wing A1399.
Original located at: Library of Congress.
Readex microprint and photostat copies are available. The complete text of the Library of Congress copy is available online at Evans Digital Edition.
The copy is signed by Rev. Flint.
Paging: Eight leaves.
Runs March to February.
Cheever's years are 1639 to 1724. He graduated from Harvard in 1659.
There is an essay, "A brief discourse concerning the various Periods of time," that includes a discussion of the Gregorian calendar.

Juan Ruiz. **A38**
El Pronostico del Año venidero de 1660.
Juan Ruiz, México.
Quintana 7.
The records that pertain to this title at the Archivo General de la Nación are *Inquisición* 670, leaves 185 to 186.

Ruiz uses the phrase "en continuación de años passados" on his request for permission to print. So he must have done at least one other almanac before 1659. Does he mean 1653?

1661

Samuel Cheever. A39

MDCLXI. / AN / ALMANACK / FOR THE YEAR OF / OUR LORD / 1661. / Being second after Leap year, and from / the Creation 5610. / Whose Vulgar Notes are / Golden number 9. Cycle of the Sun 18. / Roman Indict: 14. The Epact 9 / Dominicall letter F. / Rectifyed to the Longitude of 315 / gr: and 42 gr: 30 min. of N: Lat: / and may Generally serve for / the most part of / New-England. / By S. C. Philomathemat: / CAMBRIDG / Printed by S. G. and M. I. [sic] 1661.

Samuel Green, Sr., and Marmaduke Johnson, Cambridge.
Sabin 62743. Evans 66. Nichols [23]. Drake 2837. Wing A1400.
Original located at: American Antiquarian Society.
Readex microprint and photostat copies are available. The complete text of the American Antiquarian Society copy is available online at Evans Digital Edition.
Paging: Eight leaves.
Runs March to February.
Contains "A brief Discourse of the Rise and Progress of Astronomy," which argues for Nicolaus Copernicus.
The printer, Marmaduke Johnson, was brought over from England to assist Green with the printing of a Bible in Algonquian. According to Richard Clement,[24] Johnson was fined for "alluring the daughter of Samuel Green" in 1662. Johnson went back to England but returned permanently to Massachusetts in 1665 with the intention of operating his own press.

Juan Ruiz. A40
El Reportorio del Año venidero de 1661.
Juan Ruiz, México.
Quintana 8.
The record that pertains to this title at the Archivo General de la Nación is Inquisición 670, leaf 181.

24. Richard Clement, The Book in America, Golden, CO, 1996.

1662

Nathaniel Chauncy. **A41**

AN / ALMANACK / FOR THE YEAR OF / OUR LORD / 1662. / *Being second after Bissextile or Leap year,* / *and since the Creation* 5611. / Wherein is Contained the Longitude, Latitude, / and Aspects of the Planets, together with the / time of Suns rising and setting, for / every day in the Year. / Whose Vulgar Notes are / *Golden number* 10. *Cycle of the Sun* 19. / *Roman Indict:* 15. *The Epact* 20. / *Dominicall letter C.* / Rectifyed for Longitude of 315 gr. / and 42 gr: of *New-England* Latit. / *Fœlix qui potuit rerum cognoscere causas.* / By *Nathaniel Chauncy.* / *CAMBRIDG* / Printed by *Samuel Green* M.DC.LXII.

Samuel Green, Sr., Cambridge.

Sabin 62743. Evans 69. Nichols [24]. Drake 2838. Wing A1398.

Originals located at: American Antiquarian Society (first leaf supplied in facsimile), Library of Congress (leaves [1]–[7]).

Readex microprint and photostat copies are available. The composite text of the American Antiquarian Society copy is available online at Evans Digital Edition.

Paging: Eight leaves.

Runs March to February.

Nathaniel Chauncy's years are c. 1639 to 1685. He graduated from Harvard in 1661.

Note "second after Bissextile" instead of third (see the discussion of leap year in the entry for 1650). This could be a printer's mistake. Also, the dominical letter should have been E.

An essay, "The Primum Mobile," describes the heliocentric solar system.

Martín de Córdoba. **A42**

El Pronostico y Lunario de temporales para 1662.
Paula de Benavides, Widow of Bernardo Calderón, México.
Quintana 9.

The record that pertains to this title at the Archivo General de la Nación is *Inquisición* 670, leaf 197.

Gabriel López de Bonilla. **A43**

Diario y Discurso Astronomico para El Año de 1662.
México.
Quintana 10.

The records that pertain to this title at the Archivo General de la Nación are *Inquisición* 670, leaves 170 to 171.

1663

Israel Chauncy. **A44**

MDCLXIII. / AN / ALMANACK / OF / The Cœlestial Motions for the Year of the / CHRISTIAN ÆRA / 1663. / Being(in our Account) Bissextile,or Leap-year, / and from the Creation 5612. / Whose Vulgar Notes are / Golden Number 11. Cycle of the Sun 20 / Domin. Letters D.C. Roman Indict. 1. / Epact 1. Num.of Direct.29. / Fitted to New-England Longitude 315 gr. / and Latitude 42 d. 30 m. / By Israel Chauncy φιλομαθής. / Fælices animæ quibus hæc cognoscere primum, / Inque domos superas scandere cura fuit. / Ovid. Lib 1. Fast. / CAMBRIDGE: / Printed by S. Green and M. Johnson. 1663.

Samuel Green, Sr. and Marmaduke Johnson, Cambridge.

Sabin 62743. Evans 76. Nichols [25]. Drake 2839. Wing A1396.

Originals located at: American Antiquarian Society, Harvard University, Huntington Library, Library of Congress.

Readex microprint and photostat copies are available. The complete text of the American Antiquarian Society copy is available online at Evans Digital Edition. Early English Books Online shows the Harvard copy.

Paging: Eight leaves.

Runs March to February.

Israel Chauncy's years are 1644 to 1703. He graduated from Harvard in 1661.

There is a handwritten signature of Peter Easton in the American Antiquarian Society copy.

The rising and setting of the sun are given for six days out of every month.

Chauncy includes "The Theory of Planetary Orbs," which discusses the view that the planets move on solid spheres and lists Tycho Brahe's arguments against this view.

This is the first mention here of the "number of direction" on a title page. According to the article on the calendar in the eleventh edition of the *Encyclopedia Britannica,* the number of direction is the number of days from March 21 to Easter. This may have been a way to get around the Puritan ban on mentioning any holy days other than Sunday.

Since March 21 is always a C day in the calendar (see our explanation of dominical letter in Shepard's entry for 1656) and Easter is always a Sunday then

3 + Number of Direction ≡ Dominical Letter (mod 7)

where 1 corresponds to A, 2 to B, and so on (using the first of the two dominical letters in the case of leap years). This allows for a quick check on whether the number of direction can be right. While this works for most of the entries that give the number of direction, it fails for Chauncy's entry in the very next year.

Based on information in the aforementioned *Encyclopedia Britannica* article, we can construct the following table relating epacts to the possible range of values for the number of direction. (See our discussion in Part 1, 1587, for more about the epact.) Recall that Easter should fall on the first Sunday following the first ecclesiastical full moon on or after March 21. So the first number in each range shown in the following table is one more than the number of days from March 21 to the next ecclesiastical full moon.

Epact	Number of Direction	Epact	Number of Direction
1	23–29	16	8–14
2	22–28	17	7–13
3	21–27	18	6–12
4	20–26	19	5–11
5	19–25	20	4–10
6	18–24	21	3–9
7	17–23	22	2–8
8	16–22	23	1–7
9	15–21	24	29–35
10	14–20	25	29–35
11	13–19	26	28–34
12	12–18	27	27–33
13	11–17	28	26–32
14	10–16	29	25–31
15	9–15	30	24–30

These ranges fit the information for many of our almanacs here but fail in the consecutive cases of Samuel Brackenbury's of 1667 and Joseph Dudley's of 1668. In principle, if the number of direction is reconciled with both the epact and the dominical letter using the table and the modular rule shown here, then that should determine the number of direction uniquely.

This table represents the rules that came into effect with the Gregorian calendar. To be consistent with the pattern seen in the last six rows, the epact 24 should give a range of numbers of direction of 30 to 36. This would have meant that Easter could fall on April 26 in some years; under the Julian calendar the latest Easter could fall was April 25. (In the Julian calendar only the epacts 1, 3, 4, 6, 7, 9, 11, 12, 14, 15, 17, 18, 20, 22, 23, 25, 26, 28, and 29 are possible.) It was deemed at the institution of the Gregorian calendar that the traditional range of dates for Easter should be preserved, and so two special rules were put into effect. Under the first, epacts of 24 were to be treated the same as 25. It was also deemed to be esthetically pleasing if no epact occurred twice in a nineteen-year Metonic cycle. So the second rule was that whenever a 24 was treated as a 25 in the first eight years of the cycle, the 25 that always occurs eleven years later was changed to a 26.

The first time that the first rule was implemented was 1590, when a 24 occurred in the fourteenth year of the cycle. The first time this affected the date of Easter was 19 years later, in 1609, when April 26 was a Sunday, but Easter was April 19. The first time the second rule applied was 1916, when the epact 25 must be changed to a 26 in the seventeenth year of the cycle. The first time, and only time so far, that this rule had an effect on the date of Easter was 1954, when April 25 was a Sunday but Easter was April 18.

The Web site aa.usno.navy.mil/data/docs/MoonPhase.php notes that in 1954 the actual full moon occurred on April 18 at 5 A.M. GMT (5:49 A.M. according to the NASA Web site by Fred Espenak). So in certain time zones, including all of Europe, Asia, Africa, Australia, and South America, a full moon occurred on Easter, contrary to the intent of the Council of Nicaea.

The next year that this rule causes Easter to fall on April 18 is 2049, and the full moon will occur at 1:04 A.M. GMT on that date according to Fred Espenak.

Martín de Córdoba. **A45**
El Pronostico de Temporales para el ano que viene de sesenta y tres.
Paula Benavides, Viuda de Bernardo Calderón, México.
Quintana 11.

The records that pertain to this title at the Archivo General de la Nación are *Inquisición* 670, leaves 182 to 183.

The printer now spells Córdoba's name Córdova on the request for permission to print.

Gabriel López de Bonilla. **A46**
Diario para 1663.
México.
Quintana 12.
The record that pertains to this title at the Archivo General de la Nación is *Inquisición* 670, leaf 160.

Juan Ruiz. **A47**
El Lunario y Regimiento de salud para el año que Viene de mil y seiscientos y sesenta y tres.
Juan Ruiz, México.
Quintana 13.
The record that pertains to this title at the Archivo General de la Nación is *Inquisición* 670, leaf 184.

1664

Israel Chauncy. **A48**
MDCLXIV. / AN / ALMANACK / OF / The Cœlestial Motions for the Year of the / CHRISTIAN ÆRA / 1664. / Being in our Account first from Leap-year, / and from the Creation 5613. / Whose Vulgar Notes are / Golden Number 12. / Cycle of the Sun 21. / Dominical Letter B. / {} / Roman Indiction 2. / Epact 12. / Number of Direction 16. / Fitted to New-England Longitude 315 gr / and Latitude 42 gr. 30 m. / By ISRAEL CHAUNCY. / Pronaq; cum spectent animalia cætera cuncta / Os homini sublime dedit Cœlumq; tueri, / Jussit, & erectos ad sydera tollere vultus. Ovid. / Isai. 40. Levate in excessum oculos vestros & videte quis / creavit hac? qui educit in numero militiam eorum, / & omnio suis nominibue vocat. / CAMBRIDGE / Printed by S. Green and M. Johnson. 1664.

Samuel Green, Sr., and Marmaduke Johnson, Cambridge.
Sabin 62743. Evans 87. Nichols [26]. Drake 2840. Wing A1397.
Original located at: American Antiquarian Society (eighth leaf is two-thirds gone).

Readex microprint and photostat copies are available. The first seven leaves of the original American Antiquarian Society copy are available online at Evans Digital Edition.
Paging: Eight leaves.
Runs March to February.

1665

Martín de Córdoba.   A49
El Pronostico para el año Venidero de sesenta y cinco.
Paula de Benavides, Widow of Bernardo Calderón, México.
Quintana 14.
The record that pertains to this title at the Archivo General de la Nación is *Inquisición* 495, leaf 37.
Again, the spelling is Córdova.

Gabriel López de Bonilla.   A50
Diario / y / Discursos Astronomicos / segun la Reuolucion y Ecclip [sic] / ses desde Año que Viene de 1665 / D.D. / AL y LL.^{Mo} y Ex^{Mo} señor Doctor Don / Diego Ossorio, de Escouar y llamas, / Obispo de La Puebla de Los Angeles, / de Consejo de Su Mag̃. Virrey, / Gouernador y Capitan General desta / Nueua España, y Presidente dela / R! Chancilleria de Mexico — / Por Gabriel Lopez de Bonilla As / tronomo y Mathematico Vezino / desta ciudad, y Natural de la Vi^a / de Alcolea de Tajo Arçobispado de / Toledo, En Los Reynos de Castilla
México.
Quintana 15.
The records that pertain to this title at the Archivo General de la Nación are *Inquisición* 670, leaves 57 to 72 and 143 to 145. The text of the almanac in manuscript form is on leaves 58 to 67.

López applied twice for permission to print this almanac. The first application is on leaf 143 and the second, in response to criticism by the Inquisitors, on 57.

Quintana reproduces the entire text in his *La Astrología*.

The words Morales y Po / líticos, originally between "Astronomicos" and "segun" in the title, have been crossed out.

This is the earliest Latin American almanac that is available in manuscript form. For comparison with the English almanacs, the "Vulgar Notes" are reproduced here; these would have been several pages into this almanac, rather than on the title page.

Epacta 13, *Letra Dominical* D, *Septuagésima* February 1, *Ceniza* [Ash Wednesday] February 18, *Pascua de Resurrección* April 5, *Letanías* May 11, *La Ascension del Señor* May 14, *Pascua de Espíritu Santo* [Pentecost] May 24, *La Santísima Trinidad* May 31, *Corpus Christi* June 4, *Adviento* November 29.

There is clearly more emphasis on holy days here than in the Puritan almanacs, from which mention of them is banned. Note that both the epact and the dominical letter are different from those in Nowell's title below because of the ten-day difference between the Julian and Gregorian calendars. Until the year 1700, when the gap increases to 11, the Gregorian epact should always be the Julian epact minus 10 (mod 30) and the Gregorian dominical letter should always be three letters beyond the Julian dominical letter (mod 7). (Epact is explained more fully in Part 1, in the 1587 entry, and dominical letter is described in this section in the entry for 1656.)

Alexander Nowell. **A51**

MDCLXV. / *AN* / ALMANACK / OF / Cœlestial Motions for the Year of the / *CHRISTIAN EPOCHA* / 1665. / *Being in our Account second from Leap-* / *year, and from the Creation* 5614. / Whose Vulgar Notes are / *Golden Number* 13. / *Roman Indiction* 3. / *Dominical Letter* A / {} / *Cycle of the Sun* 22. / *Epact* 23. / *Number of Direction.* 5. / Fitted for Longitude 315. *gr.* and 42. / *gr.* 30 *m.* of North Latitude. / By *Alex. Nowell* φιλόμασος.[25] / Gen. 1. 17, *Et collocavit ea Deus in expanso* / *Coe i, ad afferendum lucem super Terram* : / Ver. 18. *Ad presidendum diei & nocti, & ad* / *distinctionem saciendum inter lucem & tenebras* : / *viditque Deus id esse bonum.* / *CAMBRIDGE* Printed by *Samuel Green* 1665

Samuel Green, Sr., Cambridge.

Sabin 56207, 62743. Evans 104. Nichols [27]. Drake 2841. Wing A1990.

Originals located at: American Antiquarian Society, Library of Congress.

Readex microprint and photostat copies are available. The complete text of the American Antiquarian Society (AAS) copy is available online at Evans Digital Edition. Early English Books Online supposedly shows the Library of Congress copy but the image has the Peter Easton signature seen in the AAS copy.

---

25. The alpha appears rotated 90° counter-clockwise. Joseph Sabin gives the word as φιλόμουσος.

Paging: Eight leaves.

There is a handwritten signature of Peter Easton in the AAS copy. "Peter Easton His Book cast 3 penc."

Runs March to February.

Nowell's years are 1645 to 1672. He graduated from Harvard in 1664.

There is a table of the rising, southing, and setting for twenty-two stars.

Nowell says, "Comets ... proceed from naturall causes, but they oft preceed preternatural effects." He also argues that planets do not produce their own light.

Juan Ruiz. A52

El Lunario, Regimiento de Salud, y Pronostico de los Temporales del Año venidero de 1665.

Juan Ruiz, México.

Quintana 16.

The records that pertain to this title at the Archivo General de la Nación are *Inquisición* 670, leaves 146 to 147.

There is an alternate title in the Inquisition records, which follows the word "Temporales" with "segun La Revolucion, y Disposición del año de 1665."

1666

Martín de Córdoba. A53

El Lunario, y Pronostico de temporales para el año de 1666.

Paula de Benavides, Widow of Bernardo Calderón, México.

Quintana 18.

The records that pertain to this title at the Archivo General de la Nación are *Inquisición* 670, leaves 119 to 120.

Benavides uses the spelling Córdova again in her application, but the inquisitors vary between Córdoba and Córdova in their responses.

Josiah Flint. A54

1666. / AN / ALMANACK / OR, / *Astronomical Calculations* / Of the most remarkable Celestial Revo- / lutions,&c. visible in our Horizon. / Together with the *Scripture* and *Jewish* / Names (wherein though we agree not with / their *Terms*, yet we follow their *Order*) / for the ensuing Year 1666. / *Being in our Account third from Leap-year, / and from the Creation 5615.* / Whose

Vulgar Notes are / *Golden Number* 14. / *Epact* 4. / *Cycle of the Sun* 23. / {} / *Roman Indict.* 4. / *Numb. of Direct.* 25. / *Dominical Letter* G. / Calculated to *N. E. Longit.* 315 *degr.* / and *Latit.* 42 *degr.* 30 *min.* / By *JOSIAH FLINT* φιλομαϑης. / Job 26. 13. *By his Spirit hath he garnished the* / *Heavens.* / Psal.19.1. *The Heavens declare the glory of God,* / *and the Firmament sheweth his handy-work.* / *CAMBRIDGE:* / Printed *Anno Dom.* 1666.

Samuel Green, Sr., Cambridge.

Sabin 62743. Evans 107. Nichols [28]. Drake 2842. Wing A1660.

Original located at: American Antiquarian Society (missing leaf [5]), Harvard University.

Readex microprint and photostat copies are available. A composite copy is available online at Evans Digital Edition.

Paging: Eight leaves.

Runs [March] to [February].

Flint's years are 1645 to 1680. He graduated from Harvard in 1664.

There is a handwritten signature of Peter Easton on the American Antiquarian Society copy.

It is a shame that the author did not use Roman numerals for the year in this title, since MDCLXVI uses each letter of the Roman numeral system just once and in decreasing order. There is, however, a poem entitled "A short Discourse about 66" on [8]v.

Flint's almanac has a unique feature in that the months are numbered, not named. This may be what is referred to in the part of the title which reads, "we agree not with their terms, yet we follow their order."

This entry broke from the trend of the philomath almanacs, which had been promoting Copernican theory for some years.

Gabriel López de Bonilla. A55

El Diario y Discursos Astronomicos para El Año que Viene de 1666.
México.

Quintana 17.

The records that pertain to this title at the Archivo General de la Nación are *Inquisición* 670, leaves 142 and 382 to 384.

Again, López de Bonilla had to file his application twice.

Juan Ruiz. A56

Lunario, Regimiento de Salud, y temporales del Año de 1666.
Juan Ruiz, México.

Quintana 19.

The record that pertains to this title at the Archivo General de la Nación is *Inquisición* 670, leaf 198.

1667

Samuel Brackenbury.   A57

1667. / AN / ALMANACK / FOR / The Year of our LORD / 1667. / *Being in our account* Bissextile, *or* Leap- / year : *and from the Creation* 5616, / Whose Vulgar Notes are, / Golden Number 15.[26] / Cycle of the Sun 24.[27] / Dominical Letters FE / {} / Roman Indiction 5 / Epact 15 / Numb : Direction 17 / Fitted for the *Longitude* of 315 *gr.* / and 42 *gr.* 30 *m.* of North Lat : / and may serve without sensible / errour for most part of *N-England*. / By *Samuel Brakenbury* Philomath. / Job 38. 31, *Canst thou bind the sweet Influences* / *of* Pleia des, *or loose the bands of* Orion ? / Ver.3 2. *Canst thou bring forth* Mazzaroth *in his* / *season*[28] *or canst thou guide* Arcturus *with his sons*&c. / *CAMBRIDGE* / Printed by *Samuel Green* 1667.

Samuel Green, Sr., Cambridge.

Sabin 62743. Evans 113. Nichols [29]. Drake 2843. Wing A1375.

Originals located at: American Antiquarian Society, Boston Public Library, Massachusetts Historical Society.

Readex microprint and photostat copies are available. The complete text of the American Antiquarian Society copy is available online at Evans Digital Edition.

Paging: Eight leaves.

Runs March to February.

Brackenbury's years are 1646 to 1678. (A note on the title page of the Massachusetts Historical Society copy says, "The worthy Author Dyed Jan. 16. 1677.") He graduated from Harvard in 1664. Many catalogs omit the c in his name, as he does on his title page, but John Langdon Sibley gives the name with the c.

There is a "Table of the Suns Altitude for every hour of the day" on [8]v, and "The Use of the Almanack" on [8]r.

---

26. No period in Boston Public Library copy.
27. No period in Boston Public Library copy.
28. Comma and no space in Boston Public Library and Massachusetts Historical Society copies.

The Massachusetts Historical Society copy comes with a metal plate negative of the title page mounted on a wooden block. This could make it possible to stamp an image of the title page onto new paper.

Gabriel López de Bonilla.   A58

Diario / y / Discursos Astronomicos fecho / por Gabriel Lopez de Bonilla / Vez. desta Ciudad de Mexico de / La Nueva España. No solo por / la Rebolucion del Año de / 1667 / Sino por los dos Ecclipses [sic] de Luna / que secedexan En el. y de los / accidentes que estan porpassax / de las Causas sucedidas En los / Años antecedentes

México.

Quintana 20.

The records that pertain to this title at the Archivo General de la Nación are *Inquisición* 670, leaves 148 to 154 and 254 to 268. The text of the almanac in manuscript form is on leaves 254 to 267. Most of 266 and 267r are scratched out.

Quintana reproduces the entire text in his *La Astrología*.

Juan Ruiz   A59

El Pronostico del Año venidero de 1667.

Juan Ruiz, México.

Quintana 21.

The records that pertain to this title at the Archivo General de la Nación are *Inquisición* 670, leaves 155 to 158.

1668

Joseph Dudley.   A60

MDCLXVIII. / AN / ALMANACK / OF / The *Cœlestial Motions* for Year of / the *Christian Epocha.* / 1668. / Being in our account first from / *Leap-year,* and from the Creation / 5617. / Whose Vulgar Notes are / Golden Number 16 / Roman Indiction 06 / Dominical Letter D. / }{ / Cycle of the Sun 25 / Epact 26 / Number Direction 01 / Calculated for Longitude 315 *gr.* and Latitude / 42 *gr.* 30 *m.* North.[29] / By *Joseph Dudley* Astrophil. / *Job* 38. 33. *Knowest thou the Ordinances of Heaven? / Canst thou set the dominion thereof in the Earth.* / Per varios usus artem experientia fecit, / Exemplo monstrante viam: Manlius. / Cambridge: Printed by *Samuel Green* / 1668.

29. No period in Boston Public Library copy.

Samuel Green, Sr., Cambridge.
Sabin 62743. Evans 121. Nichols [30]. Drake 2844. Wing A1646.
Originals located at: American Antiquarian Society, Boston Public Library, Library of Congress.
Readex microprint and photostat copies are available. The complete text of the American Antiquarian Society copy is available online at Evans Digital Edition.
Paging: Eight leaves.
Runs March to February.
This issue contains an essay entitled, "The Beginning of the Year."
Dudley's years are 1647 to 1720. He was the son of Thomas Dudley, governor of Massachusetts, who was 70 when Joseph was born. The author of this almanac was therefore a younger brother to Anne Bradstreet, the poet, whose husband was Governor Simon Bradstreet, and so he was an uncle to their son, almanac writer Samuel Bradstreet, whose almanac for 1657 has already been discussed. Joseph was himself made governor of Massachusetts, New Hampshire, and Rhode Island in 1701.

Dudley graduated from Harvard in 1665. He took a second degree in 1668. According to John Langdon Sibley, he was briefly imprisoned in 1689 and 1690, went to England in 1693 (his third trip), and served as governor of the Isle of Wight for about eight years before returning to New England for the governorship there. Sibley does not mention his writing the almanac.[30]

Gabriel López de Bonilla.   A61
Diario y discursos astronomicos para el año 1668.
México.
Quintana 22.
The records that pertain to this title at the Archivo General de la Nación are *Inquisición* 670, leaves 222 to 224.

1669

Joseph Browne.   A62
1669. / AN / ALMANACK / OF / *Cœlestial Motions* / For the Year of the Christian Æra, / 1669. / *Being (in our Account) second after* Leap- / year, *and*

---

30. John Langdon Sibley, *Biographical Sketches of the Graduates of Harvard University*, vol. II, Cambridge, 1885: 166–188.

from the Creation / 5618. / The Vulgar Notes whereof are / *Golden Number* 17. / *Roman Indiction* 07. / *Dominicall Letter* C. / {} / *Cycle of the Sun* 26. / *Epact* 07. / *Numb.*[31] of Direction 21. / Calculated for the Longitude of 315. gr. / and 42 gr. 30 m. North Latitude. / By *J. B. Philomathemat.* / *Astra regunt Mundum, sed regit Astra Deus.* / CAMBRIDGE: / Printed by S. G. and M. J. 1669.

Samuel Green, Sr., and Marmaduke Johnson, Cambridge.

Sabin 62743. Evans 135. Nichols [31]. Drake 2845. Wing A1380.

Originals located at: American Antiquarian Society, Boston Public Library, Yale University.

Readex microprint and photostat copies are available. The complete text of the American Antiquarian Society copy is available online at Evans Digital Edition.

Paging: Eight leaves.

Runs March to February.

Browne's years are c. 1646 to 1678. He graduated from Harvard in 1666.

There is a table of sunrise and sunset times for six days out of each month.

The chronology mentions a great comet of 1664.

Yale's copy is bound with manuscript pages containing a glossary of the Pequot language, attributed to Rev. James Noyes of Stonington.

Juan Ruiz. **A63**

El Lunario, Regimiento de Salud, y Pronostico de temporales del Año venidero de 1669.

Juan Ruiz, México.

Quintana 23.

The records that pertain to this title at the Archivo General de la Nación are *Inquisición* 670, leaves 179 to 180.

1670

Nicolás de Matta. **A64**

Lunario y pronostico de temporales para el a.° que Viene de seiscientos y settenta.

México.

---

31. Period very faint in Boston Public Library copy.

Quintana 24.
The records that pertain to this title at the Archivo General de la Nación are *Inquisición* 670, leaves 277 to 278.
Matta calls himself a parish priest.

John Richardson. **A65**
1670. / AN / ALMANACK / OF / *Cœlestiall Motions* / For the Year of the Christian *Æra*, / 1670. / *Being (in our Account) third after* Leap- / year , *and from the Creation* / 5619. / The Vulgar Notes whereof are / *Golden Number* 18. / *Roman Indiction* 08. / *Dominicall Letter* B. / {} / *Cycle of the Sun* 27. / *Epact* 18. / *Numb. of Direct.*13. / Calculated for the Longitude of 315 gr. / and 42 gr. 30 m. North Latitude. / By J. R. φιλομαθής. / *Invigilate viri; tacito nam temporagressu* / *Diffugiunt, nulloq; sono convertitur annus.* Juven[32] / *CAMBRIDGE:* / Printed by S. G. and M. J. 1670.
Samuel Green, Sr., and Marmaduke Johnson, Cambridge.
Sabin 62743. Evans 154. Nichols [32]. Drake 2846. Wing A2243.
Originals located at: American Antiquarian Society, Boston Public Library, Massachusetts Historical Society (leaves [1] and [8] supplied in facsimile).
Readex microprint and photostat copies are available. The complete text of the American Antiquarian Society copy is available online at Evans Digital Edition.
Paging: 8 leaves.
Runs March to February.
Richardson's years are 1647 to 1696. He graduated from Harvard in 1666.
Sunrise and sunset are given for six days out of every month.
Marion Barber Stowell cites this entry as the earliest attempt in the English colonies to inject "humorous satire" into the almanac format.[33] On 8[v] is a spoof in the form of "A *Perpetuall Calender,* [sic] fitted for the Meridian of BABYLON, where the *Pope* is Elevated 42 *Degr.*" This sort of anti-Catholic matter appears again in the almanacs of Benjamin Harris in the 1690s.

Juan Ruiz. **A66**
El Pronostico del Año venidero de 1670.

---

32. Appears to be I ven in American Antiquarian Society copy.
33. Marion Barber Stowell, *Early American Almanacs*, New York, 1977: 46.

Juan Ruiz, México.
Quintana 25.
The records that pertain to this title at the Archivo General de la Nación are *Inquisición* 670, leaves 114 to 115.

1671

Daniel Russell. A67
1671. / AN / ALMANACK / OF / *Cœlestiall Motions* / For the Year of the *Christian Æra,* / 1671. / *Being* (*in our Account*) Leap-year , / *and from the Creation* / 5620. / The Vulgar Notes whereof are / *Golden Number* 19. / *Roman Indiction* 09. / *Dominic. Letters* A.G / {} / *Cycle of the Sun* 28. / *Epact* 29. / *Numb. of Direct.*33. / Calculated for the Longitude of 315 gr. / and 42 gr. 30 m. North Latitude. / By D.R. Philomathemat. / *—Qui detrahit astris,* / *Rastris, non astris, crede mihi aptus erit.* / CAMBRIDGE: / Printed by S. G. and M. J. 1671.

Samuel Green, Sr., and Marmaduke Johnson, Cambridge.

Sabin 62743. Evans 164. Nichols [33]. Drake 2847. Wing A2306.

Originals located at: American Antiquarian Society (first leaf supplied in facsimile), Boston Public Library, Library of Congress, Trinity College.

Readex microprint and photostat copies are available. The composite text of the American Antiquarian Society copy is available online at Evans Digital Edition.

Paging: Eight leaves.

Runs March to February.

There is a table of tides (for Boston and Charlestown) and a table of the Moon's southing from the Moon's age, both on [1]v.

Russell's years are 1642 to 1679. He graduated from Harvard in 1669.

Carlos de Sigüenza y Góngora. A68
El Lunario y Pronostico de temporales para el Año de 1671.
Paula de Benavides, Widow of Bernardo Calderón, México.
Quintana 26.
The records that pertain to this title at the Archivo General de la Nación are *Inquisición* 670, leaves 187 and 188.

According to Carmen Corona and Francisco Pérez Salazar, Sigüenza began publishing almanacs in 1671 and continued annually until a posthumous one in 1701. (See the discussion for 1692.) Elías Trabulse says thirty-one almanacs were published from 1671 until Sigüenza's death.³⁴ These statements give us the confidence to fill in the gaps in the Inquisition record so that Sigüenza will have an unbroken chain of almanacs for 1671 to 1701.

1672

Jeremiah Shepard.   A69

AN / EPHEMERIS / OF THE / Cœlestial Motions for Year of the / *CHRISTIAN EPOCHA* / 1672. / *Whereto are numbred* [sic] *from* / The {³⁵ Creation of the World. 5621 / General Deluge. 3965. / Constitution of the *Julian* Year. 1716. / Passion and Death of Christ. 1639. / Correction of the Calender by / Pope *Gregory*. 90 / Planting of the *Massachusets* Colony. 44 / Bissextile or Leap year. 01 / *The Vulgar Notes are* / Golden Number. 01 / Cycle of the Sun. 01 / Dominical Letter. F / }{ / Roman Indiction 10 / Epact. 11 / Numb. of Direct. 17 / Calculated for the Longitude of *315. gr.* / and Elevation of the Pole *Artick 42 gr.* / and *30 m.* and may generally serve for / the most part of *New-England.* / By *Jeremiah Shepard.* φιλομαθής. / —*Si quid novisti rectiusistis,* / *Candidus imperti*³⁶ *si non his utere mecum.* / CAMBRIDGE: / Printed by *Samuel Green.* 1672.

Samuel Green, Sr., Cambridge.

Sabin 62743. Evans 172. Nichols [34]. Drake 2848. Wing A2381.

Originals located at: American Antiquarian Society, Trinity College.

Readex microprint and photostat copies are available. The complete text of the American Antiquarian Society copy is available online at Evans Digital Edition.

Paging: Eight leaves.

Runs March to February.

Shepard's years are 1648 to 1720. He graduated from Harvard in 1669.

The altitudes of the following five stars are given for each month: Lucid plei, Aldebaran, Regel, Hydr. hear, and Spic. virg. Known as "the Virgin's Spike" in almanacs from the period and later, Spica is indeed a star in the constellation of Virgo. Hydra is a constellation, and "Hydra's Heart" is

---

34. Elías Trabulse, prologue to the facsimile edition of Sigüenza's *Libra Astronómica*, México, 2001: xi.
35. The brace spans the next eight lines.
36. Possibly a faint comma in the American Antiquarian Society copy.

what the table is referring to. This is probably the star Alphard, the brightest star in Hydra and right in the middle of the constellation. It is also called "Cor. Hydr." in the calendar of the same almanac. The six brightest stars in the Pleiades are Alcyone, Atlas, Electra, Maia, Merope, and Taygeta. They are close enough together that "Lucid plei" may refer to all of them. "Regel" looks like Rigel but it could also be Regulus, the "Lion's heart," a very important star to almanac makers. Rigel, Aldebaran, and Spica are the fifth, ninth, and tenth brightest stars seen from mid-northern latitudes.[37]

Shepard explains that to find the time of the night, measure the altitude of one of these stars. Then look in the table on [8]v to get the hours and minutes from southing. Look in the calendar to get the southing of the star for every five days and interpolate for the current date. Adding or subtracting the time from southing gives the time of the night.

It is in years like this one, divisible by 19 (1672 = 19 × 88), that the epact jumps by twelve days instead of eleven. This happens whenever the golden number rolls from 19 back to 1. Then under the Julian calendar the epact will go from 29 to 11 in consecutive years, an increment of twelve days when the arithmetic is done modulo 30. In the Spanish almanacs, using the Gregorian calendar, the epact will be less by 10, and so the transition will be from an epact of 19 to an epact of 1.

Carlos de Sigüenza y Góngora. **A70**
El Lunario y pronostico de temporales del año que biene de 1672.
México.
Quintana 27.
The record that pertains to this title at the Archivo General de la Nación is *Inquisición* 670, leaf 215.

1673

Nehemiah Hobart. **A71**
1673. / AN / ALMANACK / OF / Cœlestial Motions for the Year of the / *CHRISTIAN ÆRA*. / 1673. / *Being second after Leap year, and from* / *the Creation,* / 5622. / *The Vulgar Notes are* / Prime 02 / Roman Indiction 11 / Dominical.Letter. E / }{ / Cycle of the Sun. 02 / Epact. 22 / Numb.of Direct.[38] 09 / Calculated for the Longitude of *315.gr.* / & North Latitude 42.

37. Steven L. Beyer, *The Star Guide*, Boston, 1986: 373.
38. No period in Harvard copy.

gr. 30. *m.* / By *N. H.* / Gen.1.16. *Fecit Deus duo Luminaria magna* ; *Luminare* / *maius ad præfecturam diei, & Luminare minus ad præ-* / *fecturam noctis* ; *fecit etiam stellas* / Astra inclinant, non necessitant. / CAMBRIDGE: / Printed by *Samuel Green. 1673.*

Samuel Green, Sr., Cambridge.

Sabin 62743. Evans 175. Nichols [35]. Drake 2849. Wing A1823.

Originals located at: American Antiquarian Society, Boston Public Library, Harvard University, Trinity College.

Readex microprint and photostat copies are available. The complete text of the American Antiquarian Society copy is available online at Evans Digital Edition.

Paging: Eight leaves.

Runs March to February.

Hobart's years are 1648 to 1712. He graduated from Harvard in 1669.

Note that the "Prime" is the same as the Golden Number.

Times of sunrise and sunset are given for six days out of every month.

Hobart gives the "half shining" for eleven stars on [1]v. Adding the half shining to the southing gives the setting, subtracting the half shining from the southing gives the rising. There is also an "Explanation of the Almanack" on [8]v.

The chronology includes the death of almanac writer Alexander Nowell in 1672.

Juan Ruiz. A72

Pronostico del año Venidero de 1673.

Juan Ruiz, México.

Quintana 28.

The records that pertain to this title at the Archivo General de la Nación are *Inquisición* 670, leaves 272, 275, and 276.

Quintana reproduces letters from the inquisitors when he finds them in the records of the Archivo General de la Nación. For this almanac, the letter from the inquisitor Antonio Núñez censuring Ruiz for making overly grand claims for the influence of the stars on human fate seems especially harsh.

Juan de Saucedo. A73

El Juico astronomico de el año proximo venidero de 1673.

México.

Quintana 29.

The records that pertain to this title at the Archivo General de la Nación are *Inquisición* 670, leaves 273 to 274.

The author is a professor of mathematics according to Paula de Benavides on *Inquisición* 670, leaf 253 (for the 1677 almanac). We relate in Part 1, 1690, how he and José de Escobar Salmerón y Castro competed against Carlos de Sigüenza y Góngora for the Chair of Mathematics and Astrology at the University.

Carlos de Sigüenza y Góngora. A74
Lunario del año que viene de 73.
México.
Quintana 30.
The records that pertain to this title at the Archivo General de la Nación are *Inquisición* 670, leaves 269 to 270.

1674

Juan Ruiz. A75
Lunario de el Año venidero de 1674.
Juan Ruiz, México.
Quintana 31.
The records that pertain to this title at the Archivo General de la Nación are *Inquisición* 670, leaves 378 to 379.

Juan de Saucedo. A76
El Pronostico para el año que biene de setenta y quatro.
Paula de Benavides, México.
Quintana 32.
The records that pertain to this title at the Archivo General de la Nación are *Inquisición* 670, leaves 167 to 169.

John Sherman. A77
1674. / AN / ALMANACK / OF / Cœlestial Motions *viz.* of the Sun and / Planets , with some of their Principal / Aspects, for the Year of the / CHRISTIAN ÆRA / 1674 / *Being (in our account) third after Leap* / Year, and *from the Creation 5623.* / *The Vulgar Notes whereof are* / Cycle of the Moon 03 / Roman Indiction 12 / Dominic. Letter D / }{ / Cycle of $\overset{e}{y}$ Sun 03 / Epact. 03 / Numb. Direction[39] 29 / Calculated for the Longitude of *315.gr.* /

---

[39]. n inverted in the Boston Public Library copy.

and 42. gr. 30.m. North Latitude / Gen. 1. 14. *And God said, Let there be Lights in the / Firmament of the Heaven*[40] *to divide the day from the night / And let them be for signes and for seasons , and for / dayes and for years. / Scientia non habet inimicum nisi ignorantem.* / Compiled by J. S. / CAMBRIDGE: / Printed by *Samuel Green. 1674.*

Samuel Green, Sr., Cambridge.

Sabin 940, 62743. Evans 196. Nichols [36]. Drake 2850. Wing A2382.

Originals located at: Boston Public Library, Massachusetts Historical Society (two copies; copy 2 is leaves [1] and [8] only), Trinity College.

Readex microprint and photostat copies are available. The complete text of a "Huntington" copy is available online at Evans Digital Edition.

Paging: Eight leaves.

Runs March to February.

Leaf [7] is bound in after leaf [1] in the Boston Public Library copy, making it appear that the calendar starts with January.

Among the Cambridge almanacs now extant, this one is the first whose supposed author was not a recent graduate of Harvard. For that reason, George Parker Winship assigns the authorship of this item to Jeremiah Shepard, the author of the 1672 entry.[41] The title is certainly more similar, though, to Sherman's almanacs of 1676 and 1677 than it is to Jeremiah Shepard's of 1672.

Sherman's years are 1613 to 1685. He wrote "To the Reader" for Increase Mather's *Kometographia* of 1683.

Note that the "cycle of the moon" is the same as the golden number.

There is "A Postscript to the Preceeding Kalendar" with two pages of explanation. It discusses Johannes Kepler and the elliptical orbits of the planets. Here Sherman says that because of the eccentricity of the Earth's orbit, the time from the vernal equinox to the autumnal equinox is 186 days, 11 hours, 51 minutes, and 30 seconds, whereas the time from the autumnal equinox to the vernal equinox is 178 days, 17 hours, 57 minutes, and 34 seconds. The sum of these, 365 days, 5 hours, 49 minutes, and 4 seconds, Sherman calls a tropical year. This is within half a second of the modern value of the Vernal Equinox Tropical Year, as opposed to the Mean Tropical Year.

Sherman's value of 186 d., 11 hr., 51 min., and 30 s. for the combined length of spring and summer is very close to our calculation for the length

---

40. Comma in copy 1 at Massachusetts Historical Society but not in copy 2.
41. George Parker Winship, *The Cambridge Press*, Philadelphia, 1945: 80.

MATHEMATICAL WORKS PRINTED IN THE AMERICAS, 1554-1700

of those seasons when the winter solstice coincides with the perihelion.[42] The accuracy of Sherman's numbers raises questions about why Juan de Figueroa in 1660 should give a value for the length of spring and summer that is significantly lower (see our entry in Part 1) and why Thomas Brattle should give a significantly higher value in his almanac of 1678 (see entry for that year).

A chart on [8v] gives the tides and southing of the moon against the age of the moon.

Carlos de Sigüenza y Góngora.   A78
El Lunario que pasa el año venidero de 1674.
México.
Quintana 33.
The records that pertain to this title at the Archivo General de la Nación are *Inquisición* 670, leaves 165 to 166.

1675

John Foster.   A79
1675. / AN / ALMANACK / OF / Cœlestial motions for the Year of the / *CHRISTIAN ÆRA* / 1675. / *Being (in our Account) Leap-Year, / and from the*

---

42. The time from vernal equinox to autumnal equinox is shrinking slowly owing to the fact that the perihelion (the time when the earth is closest to the sun) falls later each year. Currently, the perihelion is close to January 3. In the seventeenth century, the perihelion would have been much closer to the winter solstice. The assumption that the perihelion occurs at the winter solstice produces the maximum discrepancy in duration between spring and summer versus autumn and winter. Using Kepler's laws we can calculate that the length of spring and summer, divided by the length of a year, would then be

$$\frac{1}{2} + \frac{\sin^{-1}e + e\sqrt{1-e^2}}{\pi},$$

where e is the eccentricity of the Earth's orbit (currently $e = 0.0167$) and the inverse sine should be evaluated in radians. Multiplying this by the Vernal Equinox year of 365.2424 days gives the combined length of spring and summer as 186 days, 12 hours, 6 minutes with a margin of error of about 15 minutes. The current figure for the average time from the vernal equinox to the autumnal in a year is about 186 days, 9 hours, and 45 minutes, although the U.S. Naval Observatory figures show that this interval can vary by as much as 15 minutes from this average.

Creation 5624. / *The Vulgar Notes whereof are* / Golden Number 4 / Cycle of the Sun 4 / Dominic.Letter C / } / Epact 14 / Roman Indict. 13 / Numb. Directiō 14 / Calculated for the Longitude of 315 *gr* / and 42 *gr*. 30 *m*. North Latitude. / *By J. Foster.* / *Os homini sublime dedit Cœlumque tueri* / *iussit ,& erectos ad Sydera tollere Vulsus.* Ovid. / *CAMBRIDGE.* / Printed by *Samuel Green.* 1675.

Samuel Green, Sr., Cambridge.

Sabin 62743. Evans 198. Nichols [37]. Drake 2851. Wing A1706.

Originals located at: American Antiquarian Society, Boston Public Library (missing last leaf, first is $\frac{3}{4}$), Trinity College.

Readex microprint and photostat copies are available. The complete text of the American Antiquarian Society copy is available online at Evans Digital Edition.

Paging: Eight leaves.

Runs March to February.

Because this is a leap year, the dominical letter should be CB. However, Foster is consistent in that he again gives a single letter in 1679.

Green shows a copy of the title page where "Longitude" is spelled "Lonitude" (as it is in the American Antiquarian Society and Trinity copies) and notes that the copy at the Boston Public Library has it spelled correctly.[43]

Samuel Sewall signed the Trinity copy.

Foster's years are 1648 to 1681. He graduated from Harvard in 1667.

There is an illustration on [1]v to explain eclipses.

The Copernican System is discussed and objections answered. On [8]r he writes, "The *Ptolemaik Hypothesis* having for many *Centuryes* of years been the Basis of *Astronomical Calculations*, is now in this latter age of the World by *Astronomers* wholly rejected. . ."

John Sherman. A80

[1675. / AN / ALMANACK / OF / Cœlestial motions.]

Samuel Green, Sr., Cambridge.

Drake 2852.

Paging: Eight leaves.

This is Drake's first listing for Massachusetts that is not in the article by Charles Nichols.

---

43. Samuel Abbott Green, *John Foster*, Boston, 1909.

Carlos de Sigüenza y Góngora. **A81**
El Lunario y pronostico de Temporales para el año Venidero de 1675.
México.
Quintana 34.

The records that pertain to this title at the Archivo General de la Nación are *Inquisición* 670, leaves 163 to 164, 199, and 210. Page 199v continues on 163r.

Francisco Pérez Salazar says that there is a copy at the Archivo General de la Nación, but we did not find it. Quintana includes three letters from the inquisitors concerning this issue.

Sigüenza mentions this almanac in paragraph 318 of the *Libra Astronómica* as one of the places where he had already spoken against astrology.

1676

John Foster. **A82**
1676. / AN / Almanack / OF / Cœlestial Motions for the Year of the / Christian *Epocha* / 1676. / *Being first after* Leap-year *and from the* / Creation 5625. / *The Vulgar Notes are* / Golden Number 5 / Cycle of the Sun 5 / Dominic.Letter A / }{ / Epact 25 / Roman Indict. 14 / Numb of Direct. 5 / Calculated for the Longitude of 315.gr. / and 42.gr. 30.m. North Latitude. / By J. F. / *Sunt ex terra homines, non ut habitatores sed specta-* / *tores rerum superarum & Cœlestium.* / BOSTON, / Printed by *John Foster.*
John Foster, Boston.
Sabin 62743. Evans 212. Nichols [38]. Drake 2853. Wing A1707.
Original located at: Trinity College.
Readex microprint and photostat copies are available. The complete text of the Trinity copy is available online at Evans Digital Edition.
Paging: Sixteen leaves.
Runs March to February.

John Foster, whose 1675 almanac was printed by Samuel Green, began with this issue to print his almanacs on his own press in Boston, in competition with Samuel Green in Cambridge. Marmaduke Johnson had been printing in Cambridge, both on his own and in partnership with Samuel Green, for some years. He applied for and received a license to print in Boston in 1674. He died, however, before anything appeared in Boston with his imprint. Foster bought Johnson's press in December 1674, probably just too late to have printed his 1675 almanac himself.

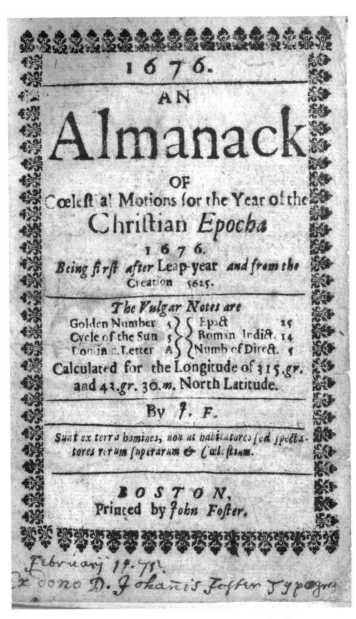

Figure 25. Title page of John Foster's almanac of 1676, the first to appear from his own press. Watkinson Library, Trinity College, Hartford, CT.

There is a handwritten "Februarii 11. '75." on the title page. Samuel Abbot Green attributes this to Samuel Sewall.[44] Sewall made notes about the weather for every day from September to February.
"The Use of the Almanack" can be found on [2]r.

Feliciana Ruiz. **A83**
El lunario, Regimiento de Salud, y Pronostico de temporales, del Año venidero de 1676.
Feliciana Ruiz, Heir of Juan Ruiz, México.
Quintana 35.
The record that pertains to this title at the Archivo General de la Nación is *Inquisición* 670, leaf 363.

According to Quintana, Feliciana Ruiz was the granddaughter of Juan Ruiz and therefore the great granddaughter of Enrico Martínez,[45] but on her application for permission to publish she refers to Ruiz as her padre. José Toribio Medina gives the first name Feliciano and guesses that this is Juan Ruiz's son. Francisco Pérez Salazar notes this error in his article on two families of Mexican printers and shows a copy of Feliciana Ruiz's signature.[46]

Quintana further notes that she was the widow of José de Butragueño and that she died in 1677, two years after her "grandfather." After mentioning again the almanacs of Juan Ruiz, he writes, "Su nieta pidió autorización para publicar uno suyo correspondiente al año de 1676."[47] The word "suyo" is ambiguous, in that we don't know without reading between the lines whether he is saying that the almanac of 1676 was his or hers.

Quintana does not seem to credit Feliciana with being an author, thinking, perhaps, that the Ruiz almanac of 1676 was a posthumous work of Juan. We do not agree. Juan Ruiz died June 17, 1675,[48] so if he had already made the almanac for 1676 he was working way ahead of deadline. In Feliciana's request to the Inquisition, she doesn't ask to print her father's almanac. She asks to print "in continuation of" her father's years of almanacs. She specifically says that she "has made" the almanac. This

---

44. Samuel Abbott Green, *John Foster*, Boston, 1909.
45. José Miguel Quintana, *La Astrología*, México, 1969: 49–50.
46. "Francisco Pérez Salazar Dos Familias," *Mémoires Société Scientifique "Antonio Alzate"* 43 (1925): 447–511.
47. José Miguel Quintana, *La Astrología*, México, 1969: 50.
48. Ibid.: 50.

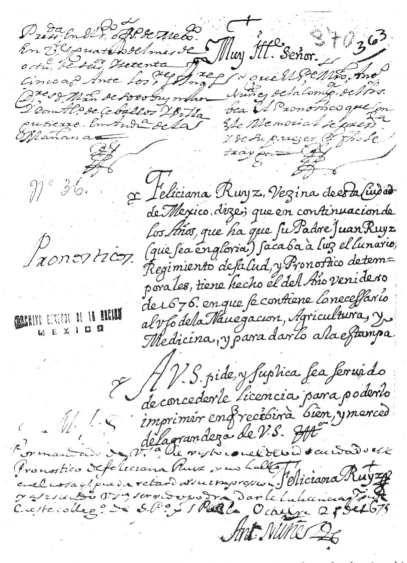

Figure 26. Feliciana Ruiz, possibly the first woman to author a book printed in the New World, asks permission from the Inquisition to print the almanac that she has created. Archivo General de la Nación, México, D. F.

makes her the first woman to author a mathematical work, and probably any kind of scientific or technical work, printed in the New World.

We might even tentatively call Feliciana Ruiz the first woman author of a book printed in the Americas. A quick look at some of the many Web sites devoted to Juana Inés de la Cruz reveals that her first two Mexican publications, a letter (published without her knowledge) and a play, were printed in 1690 and that her books of poetry were published only in Spain at first, starting in 1689. However, Juana Inés did have some sonnets appear in other people's books as early as 1668.[49] There is a reference to a *Villancicos* [Carols] by Juana Inés printed in México in 1677. A *Perfecta Religiosa* (Puebla, 1662) is supposed to be the story of a nun in the Philippines, but Medina (*La Imprenta en la Puebla*, # 59) gives the author as Bartolomé de Letona. In British America, the much-studied genre of captivity narratives started with the publication of Mary Rowlandson's story in 1682. The poems of Anne Bradstreet were first published in England in 1650, and her first book published in America was done posthumously in 1678.

John Sherman.   **A84**

1676. / AN / ALMANACK / OF / Cœlestial Motions of the Sun and Planets, / with some of their principal Aspects. / *For the Year of the CHRISTIAN ÆRA* / 1676 / *Being in our Account the first after* Bis- / sextile *or* Leap-year *and from the* / Creation, 5625. / The Vulgar Notes of which are / Cycle of ☾ or Golden numb. 5 / Roman Indiction 14 / Dominical Letter A / For January & February G / }{ / Cycle of ☉ 5 / Epact 25 / Number of Dire- / ction. 5 / Calculated for Longitude 315. *gr.* and 42. *gr.* 30. *min* / of North Latitude / By J. S. / The Heavens declare the Glory of God, and the Firma- / ment sheweth his handy work. Day unto day ut- / tereth speech, and night unto night sheweth know- / ledge, *Psal.* 19.1,2. / *Cambridge* Printed by S. *Green.* 1676.

Samuel Green, Sr., Cambridge.

Sabin 62743. Evans 223. Nichols [39]. Drake 2854. Wing A2383.

Originals located at: American Antiquarian Society (two copies; copy 2 is $3\frac{1}{2}$ leaves), Trinity College.

Readex microprint and photostat copies are available. The complete text of the American Antiquarian Society copy is available online at Evans Digital Edition. Early English Books Online shows a "Newberry Library" copy.

Paging: Eight leaves.

---

49. Dorothy Schons, *Bibliografía de Sor Juana Inés de la Cruz*, México, 1927.

The Trinity copy has handwritten notes in Latin throughout the calendar and has Gregorian dates written in next to the printed Julian dates.
Runs March to February.
Sherman starts splitting the dominical letter, even though it is not a leap year. He is saying that January and February (the last two months on his calendar) get next year's dominical letter. This is a small step toward starting the new year with January.

Carlos de Sigüenza y Góngora. **A85**
Lunario para el año venidero de 1676.
México.
Quintana 36.
The record that pertains to this title at the Archivo General de la Nación is *Inquisición* 670, leaf 376.
The letter from inquisitor Antonio Núñez mentions Juan de Torquemada. This may be when Sigüenza began using that name on his almanacs.

1677

John Foster. **A86**
1677. / AN / ALMANACK / OF / Cœlestial Motions.
John Foster, Boston.
Evans 229. Nichols [40]. Drake 2855.

Juan de Saucedo. **A87**
El pronostico para el año que biene de setenta y siete.
Paula de Benavides, México.
Quintana 37.
The record that pertains to this title at the Archivo General de la Nación is *Inquisición* 670, leaf 253.

John Sherman. **A88**
1677. / AN / ALMANACK / OF / Cœlestial Motions of the Sun and Planets, / with some of their principal Aspects / *For the Year of the CHRISTIAN ÆRA* / 1677. / *Being in our Account the second after* / Leap-year *and from the* Creation, / 5626. / The Vulgar Notes of which are / Cycle of ☾ or Golden numb. 6 / Roman Indiction 15 / Dominical Letter G / For January & February F }{ / Cycle of ☉ 6 / Epact 6 / Number of Dire- / ction. 25 /

Calculated for Longitude 315. gr. and 42. gr. 30. min / of North Latitude / By J. S. / Psal.8.3,4. When I consider thy Heavens[50] the work of / thy Fingers, the moon and the stars which thou hast / ordained: what is man that thou art mindful of him? / and the Son[51] of man that thou visitest him? / *Cambridge* Printed by S. *Green*. 1677.

Samuel Green, Sr., Cambridge.

Sabin 62743. Evans 241. Nichols [41]. Drake 2856. Wing A2384.

Originals located at: American Antiquarian Society, Trinity College.

Readex microprint and photostat copies are available. The complete text of the American Antiquarian Society copy is available online at Evans Digital Edition. Early English Books Online shows a "Huntington" copy.

Paging: Eight leaves.

Runs March to February.

Carlos de Sigüenza y Góngora. **A89**
El Lunario para el año de 1677.
México.
Quintana 38.

The records that pertain to this title at the Archivo General de la Nación are *Inquisición* 670, leaves 243 to 244.

1678

Thomas Brattle. **A90**

*Sic in se sua per vestigia volvitur annus.*Virg / 1678: / AN / ALMANACK / OF / Cœlestial Motions of the Sun and Planets, / with their principal Aspects, / For the Year of the CHRISTIAN ÆRA / 1678. / Being in our Account the third after / Leap-year and from the Creation, / 5627. / The Vulgar Notes of which are / Cycle of ☾ or Golden numb. 7 / Roman Indiction 1 / Dominical Letter F / For January & February E / }{ / Cycle of ☉ 7 / Epact 17 / Number of Dire- / ction. 10 / Calculated for the Longitude of 315. gr and 42. gr. 30. / min of North Latitude / By T. B. / *Quid potest esse tam apertum, tamque perspicuum, cum cœlum / suspeximus, cœle stiaque contemplati sumus, quam aliquoa esse / numen præstantissimae mentis quo haec reguntur.* Tul in Lib. / 2. *de natura deorum* / Cambr : Printed by S. *Green* & S. *Green*: / 1678.

50. Very faint semicolon in American Antiquarian Society copy.
51. The n is inverted in both copies.

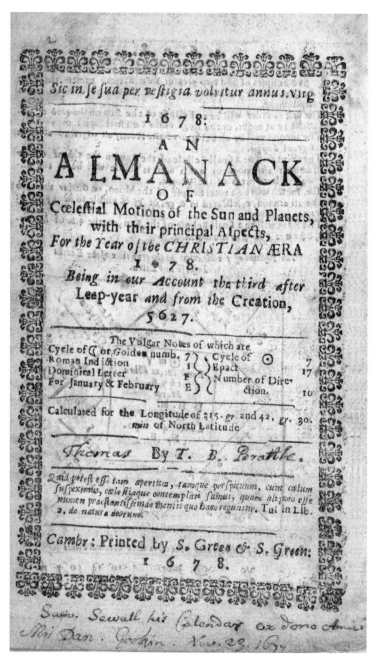

Figure 27. Title page of Thomas Brattle's almanac for 1678. Watkinson Library, Trinity College, Hartford, CT.

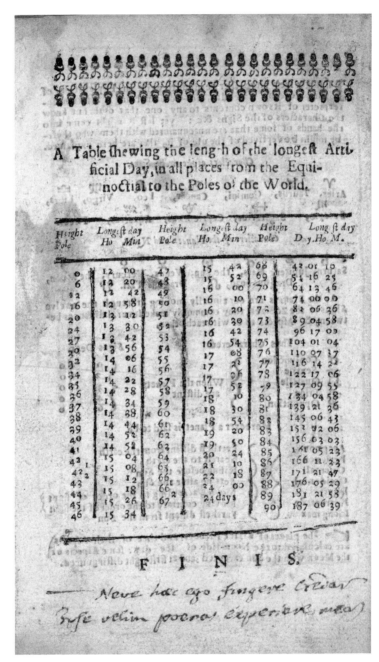

Figure 28. Brattle's table of the length of the longest period of daylight of the year at various latitudes. Watkinson Library, Trinity College, Hartford, CT.

Samuel Green, Sr. and Samuel Green, Jr., Cambridge.
Sabin 62743. Evans 245. Nichols [42]. Drake 2859. Wing A1376.
Original located at: Trinity College.
Readex microprint and photostat copies are available. The complete text of the only copy is available online at Evans Digital Edition.
Paging: Eight leaves.
Runs March to February. Continues Sherman's practice of splitting the dominical letter.

Handwritten "Thomas Brattle" on the only copy: there is also the signature of Samuel Sewall dated Nov. 23, 1677.

Samuel Green, Jr., had been apprenticed to his father, but could not find work as a printer on his own. He ran a general store in Hartford for a while. In 1677 he married and moved to New London to run another store, probably soon after this printing.

It is generally accepted that Thomas Brattle (June 20, 1658–May 18, 1713) was the person alluded to in Newton's *Principia* who made such thorough observations of the comet of 1680. This is the same comet that sparked the debate in print between Eusebio Francisco Kino and Carlos de Sigüenza y Góngora, which we have discussed elsewhere.

Brattle graduated from Harvard in 1676 and was treasurer of the College from 1693 until he died. He spoke out against the Salem witch trials in a letter of 1692. His brother William is also an author of an almanac on our list.

There is an explanation of the almanac on [8]r.

The last page has a table giving the length of the longest day of the year by latitude. The heading is "A Table shewing the length of the longest Artificial Day, in all places from the Equinoctial to the Poles of the World." Entries include: $0°$, 12 hours; $42\frac{1}{2}°$ (the latitude of Boston), 15 hours and 8 minutes; $66\frac{1}{2}°$ (the Arctic Circle), 24 hours; and $90°$, 187 days, 6 hours, and 39 minutes.

That last time span seems a little long. The daylight period at the North Pole should last from the Vernal Equinox to the Autumnal Equinox (not counting twilight—we are only talking about the time that the center of the sun is above the horizon, since that is how one gets the answers of 12 hours for the equator and 24 hours at the Arctic Circle). Therefore we again are led to consider a question that arose in Part 1 in our discussion of Juan de Figueroa's work of 1660 and in this section for the Sherman almanac in 1674.

John Sherman gave a value for the combined length of spring and summer in 1674 that we suspect is fairly close to correct. Figueroa gave such a value in 1660 that is too low, and here Brattle gives a value, if that is what he is doing, that is considerably high. This led us to try to analyze Brattle's entire table to make sense of his numbers.

The U.S. Naval Observatory Web site also gives a table for the longest day (see http://aa.usno.navy.mil/faq/docs/longest_day.php). They explain that their figures for sunrise and sunset include an adjustment corresponding to angles of 16' for the radius of the sun and 34' owing to refraction by the atmosphere. As we said earlier, Brattle is not adjusting for this, based on his answers for the equator and the Arctic Circle. Here is a comparison of the U.S. Naval Observatory values with Brattle's.

Latitude	Longest Day (U.S. NO)	Longest Day (Brattle)
25° N	13h 42m	
27° N		13h 42m
35° N	14h 31m	14h 22m
40° N	15h 02m	14h 52m
45° N	15h 38m	15h 26m
50° N	16h 23m	16h 10m
60° N	18h 53m	18h 30m

We suspect that Brattle is not using the same rules for his calculation at the North Pole that he used from the equator to the Arctic Circle. Here is a way that the problem could be broken down. From the equator to the Arctic Circle, one could assume that the earth is rotating but ignore its revolution about the sun. The maximum error this assumption produces is about four minutes, i.e., the difference between the solar and sidereal days. From the Arctic Circle to the North Pole, one could consider the earth to be revolving around the sun but ignore the earth's rotation, simply asking for what part of the year would the sun be higher than a celestial circle of latitude tangent to the horizon, i.e., for what part of the year would the sun's declination be greater than 90° minus the latitude. The maximum error for this assumption would be 48 hours, since, if the timing is right, the sun could be up for nearly one full day before and after the time when its declination guarantees that it should stay above the horizon.

The first of these scenarios leads to the formula for hours in the longest day of

$$24 - \frac{2}{15}\cos^{-1}\frac{\tan L}{\tan 66.5°},$$

where L stands for latitude and the trigonometric functions are evaluated in degree mode. This formula yields numbers that are off by less than a minute from the corresponding times in Brattle's column in the table shown here. (The assumptions here are the same as those underlying the solution, by Enrico Martínez in the entry for 1606 in Part 1, to the question of how long the sun will be up on any given date and at a given latitude. Therefore our formula here, for the longest day at latitudes up to the Arctic Circle, is a special case of the solution there.)

Our earlier description for how to treat the latitudes above the Arctic Circle does not really explain Brattle's numbers. For example, at a latitude of 69°, the sun's declination will be high enough, from 4.0° of Gemini to 26.0° of Cancer, to keep it above the horizon. Employing interpolation in the sun's positions given in Brattle's calendar, we find that this should last about 54 days and 14 hours. This fails to shed much light on Brattle's answer of 54 days, 16 hours, and 25 minutes. It could be that Brattle is allowing for effects that he did not consider for lower latitudes, like the width of the sun or refraction, or it could be that he is allowing for some rotation of the earth while the sun has a declination below that given earlier, but we have not yet found a combination of effects that explains the numbers he gives.

John Danforth. **A91**
AN / ALMANACK.
Samuel Green, Sr., Cambridge.
Drake 2857. This work is not listed by Charles Nichols, nor is it in the list of publications given for John Danforth in John Langdon Sibley's *Biographical Sketches*.

John Danforth's years are 1660 to 1730. He is the oldest of the children of Samuel Danforth (1626–1674) to survive to adulthood and is the older brother of Samuel Danforth (1666–1727). He graduated from Harvard in 1677.

José de Escobar Salmerón y Castro. **A92**
El Lunario y regimiento de Salud Con el prognostico delos temporales para el año que Viene de 1678.

México.

Quintana 39.

The records that pertain to this title at the Archivo General de la Nación are *Inquisición* 670, leaves 246 to 247. The name used here by the author is Joseph Salmerón de Castro y Escouar. We favor, for consistency's sake, the arrangement of names used on his comet book in 1681, although we've used José for his first name instead of the Joseph used there.

Escobar writes that he is on the Faculty of Philosophy, Theology, and Medicine at the Royal University of Mexico. Elías Trabulse says Escobar was born in New Spain and he held a Chair of Anatomy and Surgery from 1678 to his death in 1684. He wrote the *Discurso Cometológico*, which we list in Appendix B, in 1681. See our entry for 1690 in Part 1 for more about the content of this book.

Quintana writes on page 51 of *La Astrología* that Escobar had a son, José Escobar y Morales. He was a doctor of law and medicine and Chair of Mathematics at the University, and he published almanacs from 1728 to 1736. He died in a smallpox epidemic, one which Quintana jokingly says that he couldn't foretell.

John Foster. A93

1678. / AN / ALMANACK / OF / Cœlestial motions for the year of the / Christian *Epocha* / 1678 / Being (in our account) third after / Leap-year, and from the Creation / 5627. / The Vulgar Notes are / Golden numb. 7. / Cycle of the Sun 7 / Dominic.Let. F. / }{ / Epact 17. / Rom Indict. 1. / Num.direct. 10. / Calculated for the longitude of 315 gr. / and 42 gr. 30 min. north latitude. / J. F. / Printed by *J.Foster*, for *John Usher* of / *Boston*. 1678.

John Foster, Boston.

Sabin 62743. Evans 247. Nichols [43]. Drake 2858. Wing A1708.

Originals located at: American Antiquarian Society, Trinity College.

Readex microprint and photostat copies are available. The complete text of the American Antiquarian Society copy is available online at Evans Digital Edition.

Paging: Sixteen leaves.

Runs March to February.

The Trinity copy is signed by Samuel Sewall. Sewall writes in the calendar after December 15, "Returned to my own Bed after my Sickness of ye Sm. Pox."

"Directions for the Use of the Kalendar" on [14]v.

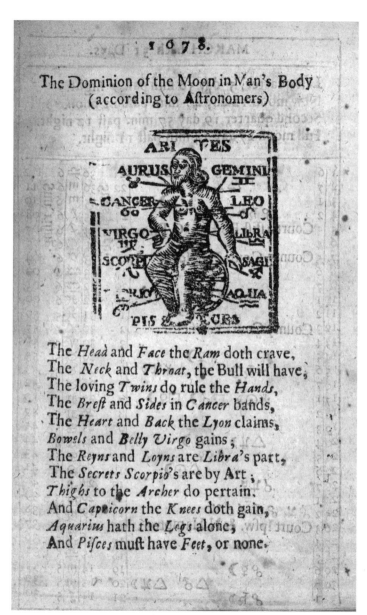

Figure 29. The "Man of Signs" from Foster's almanac of 1678. Aries is associated with the head and Pisces with the feet, with the other signs in order in between. The poem is offered as an aid to memory. Almanac users could check the moon tables to see what sign the moon was in and thus know which part of the body the moon was currently influencing. Watkinson Library, Trinity College, Hartford, CT.

There is a woodcut on [2]r illustrating the "Dominion of the Moon in Man's Body" that was later used in John Tulley's almanacs starting in 1693 and Samuel Clough's almanacs starting in 1702. Parts of the body are associated with the signs of the zodiac, with Aries at the head and Pisces at the feet. A poem in rhymed couplets is given, perhaps as a mnemonic device for remembering the diagram. This is the first use of the "Man of Signs" in British North America, according to Charles Nichols.[52]

Carlos de Sigüenza y Góngora.   A94
El pronostico para el año venidero de 1678.
México.
Quintana 40.
The records that pertain to this title at the Archivo General de la Nación are *Inquisición* 670, leaves 11 to 17. The calendar is 12 to 17.
Quintana reproduces the daily calendar from this text.

1679

John Danforth.   A95
AN / ALMANACK / OR / REGISTER OF / *Cœlestial Configurations &c:* / For the year of our Lord GOD / 1679. / *And of the World* / 5628: / Being *(*in our account*)* Leap year, And / from the beginning of the Reign of our / Soveraign Lord *CHARLES* II, by the / Grace of God, King of great Brittain, France / and Ireland, &c: the 31$^{st.}$ year. / *The Vulgar notes of which are* / Cycle of ☉ : & Cycle of ☽, or Golden Number 8 / The Epact, 28 / Number of Direction, 30 / Dominical Letters, E D C. / Calculated for the Longit. of 315 degr. and 42 degr. / 30 min. North Lat. in New England. / By *J. D.Philomath.* / Psal. 148 1, 3 — *Praise ye the Lord from the Heaven's* / *praise him in the heights.* / *Praise ye him*[53] *Sun & Moon* ; *praise him all ye Stars of light* / Cambridge printed by *Samuel Green* 1679.
Samuel Green, Sr., Cambridge.
Evans 265. Nichols [44]. Drake 2861. Wing A1572.
Originals located at: American Antiquarian Society, Boston Public Library (missing first and last leaves), Massachusetts Historical Society (missing last leaf), Trinity College, Yale University (missing last leaf).

52. Charles Nichols, "Notes on the Almanacs of Massachusetts," *Proceedings of the American Antiquarian Society* new series 22 (1912): 15–134.
53. Comma in the Yale copy.

Readex microprint and photostat copies are available. The complete text of the American Antiquarian Society copy is available online at Evans Digital Edition.

Paging: Eight leaves.

Runs March to February.

The three dominical letters in the title may be explained by combining the practice of splitting the dominical letter, seen in the last three Cambridge almanacs, with the fact that this is a leap year. Thus the dominical letter is E for January and February of 1678/9, D for March through December of 1679, and C for January and February of 1679/80.

A table correlates the number of direction (which runs from 1 to 35) with the election day in Boston. The number 1 corresponds to April 29, and 35 is June 2. Therefore election day is always 38 days after Easter. (See our discussion of number of direction in an entry in 1663.) Another table gives the length of twilight for various days of the year. Both are on [8]r.

The table of recent events on [8]v includes several highlights of King Philip's War. Also listed are the deaths of Danforth's father, Samuel, in 1674, and the deaths of two other almanac writers, Thomas Shepard and Joseph Browne, in 1677 and 1678. The appearance of a comet is listed for February 1676.

Samuel Sewall signed the Trinity copy on [2]r and dated it March 13, 1678/9. The Massachusetts Historical Society copy is interleaved with manuscript pages (some in code) attributed to Thomas Shepard.

José de Escobar Salmerón y Castro. **A96**
Lunario y regimiento de Salud para el año que viene de 79.
México.
Quintana 41.

The record that pertains to this title at the Archivo General de la Nación is *Inquisición* 670, leaf 377. The name used here by the author is Salmerón de Castro y Escobar.

Now Escobar says he is Doctor of the Court and Chair of Surgery and Anatomy at the Royal University.

John Foster. **A97**
MDCLXXIX. / AN / ALMANACK / OF / Cœlestial Motions for the Year of the / *Christian Æpocha,* / 1679. / Being in our Account *Bissextile,* or

Leap-year : / And from the Creation, / 5628. / The Vulgar Notes. / Golden Numb. 8. / Cycle of the Sun. 8. / Dominic.Let. E. / }{ / Epact 28. / Rom. Indict. 2. / Num. Direct. 30. / *Calculated for the Meridian of* Boston *in* New- / England, *where the* Artick Pole *is elevated* / 42 Degrees & 30 Minutes. / *J. F.* / BOSTON: / Printed by *J. Foster*, and sold by *Henry Phillips* / in the west end of the *Exchange*. 1679.

John Foster, Boston.

Sabin 62743. Evans 268. Nichols [45]. Drake 2860. Wing A1709.

Originals located at: Boston Public Library (missing first and last leaves), Trinity College, Yale University (leaves [1]–[14], title $\frac{2}{3}$).

Readex microprint and photostat copies are available. The complete text of the Trinity copy is available online at Evans Digital Edition. There is a listing for the last leaf only, priced at $50, in the sale of the Phelps collection.[54]

Paging: Sixteen leaves.

The Yale copy is a different state. The last two lines of the title page are replaced by "Printed by *John Foster* . 1679."

There is a handwritten signature of Samuel Sewall on the Trinity copy and an entry in the calendar for the birth of Hannah Sewall.

Runs March to February.

"DIRECTIONS for the use of the following ALMANACK" is on [2]r. The sine and tangent functions are mentioned on [15]r.

On [16]r Foster says, "The Reason why this is called Leap-Year with us, and not in other parts of the World, is, because our Almanacks (*I* know not for what Reason) do differ two Moneths in their beginning from others, so taking in *January* and *February* of the next year, 1680, and consequently the Day which gives it the Name of Leap-year."

The calendar notes historical events, many from the recent King Philip's War: June 24, 1675, "Several persons murdered at *Swanzy* [Swansea], which was the first English blood that was spilt by the Indians in an hostile way." August 12, 1676: "*Philip* Sachem of *Mount-Hope*, who first began the war with the English, was slain by Capt. *Church* of *Plymouth*."

Carlos de Sigüenza y Góngora. **A98**

Lunario y pronostico de temporales para el año venidero de 1679.
México.

---

54. David L. O'Neal, *Early American Almanacs, The Phelps Collection 1679–1900*, Petersborough, NH, 1978?

Quintana 42.
The records that pertain to this title at the Archivo General de la Nación are *Inquisición* 670, leaves 347 to 348.

1680

José de Escobar Salmerón y Castro.   A99
El lunario y regimiento de Salud para el año que viene de 1680.
México.
Quintana 43.
The record that pertains to this title at the Archivo General de la Nación is *Inquisición* 670, leaf 280. The name used here by the author is Salmerón de Castro y Escobar.

John Foster.   A100
MDCLXXX. / AN / ALMANACK / OF / Cœlestial Motions for the Year of the / *Christian Æpocha,* / 1680. / Being in our Account first after *Leap-year*: / And from the Creation, / 5629. / The Vulgar Notes. / Golden Numb. 9. / Cycle of the Sun. 9. / Dominic.Let. C. / }{ / Epact 9. / Rom. Indict. 3. / Num. Direct. 21. / *Calculated for the Meridian of* Boston *in* New- / England, *where the* Artick Pole *is elevated* / 42 Degrees & 30 Minutes. / Printed for, and sold by *Henry Phillips* in the / west end of the *Exchange* in Boston.1680.
John Foster, Boston.
Sabin 940, 62743. Evans 283, 284. Nichols [46]. Drake 2862, 2863. Wing A1710, A1711.
Originals located at: American Antiquarian Society, Trinity College (Evans 284, Drake 2862, Wing A1711). Another impression, with title page identical except that the last two lines above are replaced by the last line "Printed for *John Usher* of *Boston.* 1680." is represented by an original at Massachusetts Historical Society (Evans 283; Drake 2863, Wing A1710).
Readex microprint and photostat copies are available. The complete text of the American Antiquarian Society copy is available online at Evans Digital Edition. Early English Books Online says the Huntington Library has a copy of Evans 284 but doesn't show an image. They show a "Newberry Library" copy for Evans 284. The Readex card says that Evans 283 is at the Boston Public Library.

Paging: Eight leaves.
Runs March to February.
"To the Reader" is an explanation of the calendar on [1]v. Spring tides are tabulated on [8]v.
The Trinity copy has handwritten notes about the weather. "Hath been a very Severe Winter for Snow." on [7]v. Also "Thanks-Giving" for Thursday, Nov. 25.

Juan Ramón Koenig.[55]   **A101**
El Conocimiento de los Tiempos.
Lima.
A survey of early Peruvian almanacs by Federico Schwab gives the following list of authors for the almanac that started as *El Conocimiento de los Tiempos*: Juan Ramón Koenig, ¿1680–1708?; Pedro de Peralta Barnuevo y Rocha, ¿1708?, 1721–1743; José de Mosquera y Villaroel, 1744–1749; Juan Rer, 1750–1756; Cosme Bueno, 1757–1798; Gabriel Moreno, 1799–1809; José Gregorio Paredes, 1810; Francisco Romero, 1811–1813; José Gregorio Paredes, 1814–1825; Eduardo Carrasco, 1826; Nicolás de Piérola, 1827–1828; José Gregorio Paredes, 1829–1839; Eduardo Carrasco, 1840–1857; Pedro M. Cabello, 1858–1874.[56] Clearly the series founded by Koenig had quite a long run. Note that we have already listed six Lima *Reportorios* by Francisco Ruiz Lozano that were not listed by Schwab.

Schwab also speculates that the original title may have been *Efemérides de Lima* and that it was changed later by Peralta, but he's not sure.

The record in José Toribio Medina's books is spottier. His first listing for a *Conocimiento* is 1721, *El Conocimiento de los Tiempos. Prognostico, y*

---

55. This is the name given by Gabriel Moreno in the elegy included in his almanac of 1807. The name also appears this way in Federico Schwab's 1948 survey of Peruvian almanacs. In contrast, we read on page 459 of José Toribio Medina's *Imprenta in Lima*, Tomo II, where he is discussing a 1750 entry, Juan Rer, *Conocimiento de los Tiempos*, that the *Conocimiento* was founded by a José Ramón Koening in 1680 and that this Koening was one of the people who had held the chair of mathematics at the Universidad de San Marcos. Florian Cajori gives the name as José Ramón Koenig and gives Medina as his source. For what it's worth, on page 359 of Medina's Tomo III he gives the name as Juan Ramón Koenig when he notes the presence of the aforementioned elegy in Moreno's *Almanaque Peruano*.

56. Federico Schwab, "Los Almanaques Peruanos ¿1680?–1874," *Boletín Bibliográfico* XIX/1–2 (1948).

Lunario Semestre . . . by Pedro de Peralta (listed as Pedro de Peralta, Barnuevo y Rocha in his index of authors). The next year the title becomes *El Conocimiento de los Tiempos. Ephemeride del Año* . . . and remains more or less the same through 1798. Peralta is listed as author for 1721 to 1723, 1725, 1727 to 1729, and 1732 to 1743. José de Mosquera y Villaroel is the author for 1744 through 1749.[57] Rer[58] takes over for 1750 to 1756, followed by Bueno (1757–1798). Then Gabriel Moreno is listed for 1799 to 1809 under the title *Almanaque Peruano* . . . (Moreno has an elegy to Bueno in his first number). Gregorio Paredes takes over in 1810 and 1811. For the years 1814 to 1821 he becomes José Gregorio Paredes (but Medina doesn't distinguish between them in his index of authors). For 1822 through 1824 the title changes to *Calendario y Guia de Forasteros*. This takes us to the end of Medina's Tomo IV.

Irving Leonard compiled a book of works by Peralta (1663–1743) and says in the notes that Peralta produced almanacs from 1719 on and perhaps before. He refers to Koenig as "Juan Raimundo Koenig" and says that Peralta succeeded Koenig in the Chair of Mathematics at San Marcos when the latter died in 1709.[59]

The Biblioteca Nacional del Perú has copies listed in their catalog back to the 1733 issue by Pedro de Peralta Barnuevo y Rocha. We have examined a copy held by the Wellcome Library of *El Conocimiento de los Tiempos; Efemeride del Año de 1780* by Cosme Bueno; this is the earliest issue listed by WorldCat.

According to Gabriel Moreno in his *Almanaque Peruano* in 1807, Juan Ramón Koenig (1623–1709) held the professorship of mathematics[60] at the Real Universidad de San Marcos in Lima, the second person to hold that position, after the Mexican scholar Francisco Ruiz Lozano.[61] He started publishing the *Conocimiento de los Tiempos* in 1680 and continued until 1708. It was, he says, based on a periodical with a similar title published by

---

57. Medina does mention a Luis Godin as Rer's predecessor when he says that Godin's journey to Spain in 1749 meant that Rer was made the new holder of the seat of mathematics. Mosquera y Villaroel on his title page calls himself *substituto de la catedra de prima de mathemáticas*.
58. Rher in Vargas Ugarte.
59. Irving A. Leonard, "Introduction," in Pedro de Peralta Barnuevo, *Obras Dramáticas con un Apéndice de Poemas Inéditos*, Santiago de Chile, 1937: 9.
60. *La Catedra de Prima de Matemáticas*
61. See the second entry for 1651 in Part 2 for more about Ruiz Lozano.

the Academy of Sciences of Paris from 1669 to 1792.[62] Moreno gives the title *Conocimiento de los Tiempos y Efeméride Anual* as having been used by Koenig's successor Peralta.

See the first entry for 1696 in Part 1 for more information about this author.

Carlos de Sigüenza y Góngora.   A102
El Lunario y pronostico de temporales para el año proximo venidero de 1680.
México.
Quintana 44.

The records that pertain to this title at the Archivo General de la Nación are *Inquisición* 670, leaves 117 to 118.

Francisco Pérez Salazar says there is a copy at the Archivo General de la Nación, but we did not find it.

1681

John Foster.   A103
MDCLXXXI. / AN / ALMANACK / OF / Cœlestial Motions for the Year of the / *Christian Epocha,* / 1681. / Being in our Account second after *Leap-year* : / And *fr* om the Creation, / 5630. / The Vulgar Notes. / Golden Numb. 10 / Cycle of the Sun. 10 / Dominic.Let. B. / }{ / Epact 20. / Rom. Indict. 4. / Num. Direct. 13. / *Calculated for the Meridian of* Boston *in* New- / England, *where the* Artick Pole *is elevated* / 42 Degrees & 30 Minutes. / By *John Foster,* Astrophil. / Eph.5.16 *Redeem the time,because the dayes are evil.* / BOSTON; Printed by *J.F.* 1681.

John Foster, Boston.

Sabin 62743. Evans 300, 301. Nichols [47]. Drake 2864, 2865. Wing A1712, A1713.

Originals located at: American Antiquarian Society (leaves 1, 3–10), Massachusetts Historical Society (Evans 300; Drake 2864; Wing A1712). Another impression, with title page identical except that the last line is replaced by "BOSTON; Printed by *J. F.* for *Samuel Phillips* / in the west end of the *Exchange.* 1681.," is represented by an original at Massachusetts Historical Society (Evans 301; Drake 2865; Wing A1713).

Readex microprint and photostat copies are available. The Evans Digital Edition image of Evans 300 is supposed to be the American Antiquarian

---

62. The Bureau des Longitudes has been publishing *La Connaissance des Temps* since 1795. Their Web site gives 1679 for the founding of the series.

Society copy but it matches the Massachusetts Historical Society (MHS) copy. Evans Digital Edition claims to show a copy of Evans 301 but it is actually the title page of the MHS copy of 301 and [2]v–[12]v of the MHS copy of Evans 300.
Paging: Twelve leaves.
Runs March to February.
There are "Directions for the Use of the following EPHEMERIS" on [2]v. Also we see "Of Comets, Their Motion, Distance & Magnitude" on [9]r–[10]r and "Observations of a Comet seen this last Winter 1680" on [10]v–[11]v. (This must be the comet we discussed in Part 1 in the entry for 1690.) Foster says that comets move in straight lines until they come under the influence of the sun, which then forces them to follow curved paths. On [12]r are listed the spring tides and a correction to the latitude of Boston: the standard 42°30' should be 42°24', he says. On [12]v there is a woodcut entitled "The Copernican System."

There are signatures of Samuel Sewall on the titles pages of the American Antiquarian Society copy and the MHS copy of Evans 300; the latter has Sewall's notes throughout: "Rain Plenty," "Soaking," "Serene pleas. Wether," for example. The MHS copy of Evans 301 is interleaved with manuscript pages, and Samuel Sewall signed one of these as well.

Foster died this year. His headstone, as reported by Samuel Abbott Green, reads, "The Ingenious Mathematician & Printer Mr John Foster Aged 33 Years Dyed Septr 9th 1681."[63] The date of Foster's death was also inserted by Sewall in two places into what is now the MHS copy of Evans 300.

Juan Ramón Koenig. **A104**
El Conocimiento de los Tiempos.
Lima.

Carlos de Sigüenza y Góngora. **A105**
Lunario para 1681.
México.
Quintana 45.
Quintana made this listing based on a reference by Sigüenza to a lunario for 1681. There is no record for it at the Archivo General de la Nación.
Irving Leonard says that Mariano Veytia reported owning a copy in his *Historia Antigua de México*.[64]

---

63. Samuel Abbott Green, *John Foster*, Boston, 1909.
64. Irving J. Leonard, *Ensayo Bibliográfico*, México: 13.

Elías Trabulse used this year as an example of Sigüenza's using abbreviated histories at the head of his almanacs.[65]

1682

Antonio Sebastián de Aguilar Cantú.   A106
Pronostico de los Temporales de el Año de Mil, Seiscientos y ochenta y dos.
México.
Quintana 46.
The records that pertain to this title at the Archivo General de la Nación are *Inquisición* 670, leaf 312.

William Brattle.   A107
Unius labor multorum laborem allevat / AN / EPHEMERIS / OF / Cœlestial Motions , Aspects , / Eclipses , &c. For the Year of the / Christian Æra 1682. / Being from / The { Creation of the World. — 5631. / Floud [sic] of Noah. — 3975 / Constitution of the Julian Year 1726 / Suffering of Christ. — 1649 / Correction of Calend. by P.Greg. 100 / Laying the foundation of Harv.Col 40 / Leap year (in our account) — 03 / The Vulgar Notes are / Cycle of Sun & Moon 11 Epact — 1 / Dominical Letter — A Number of Direct. 26 / Calculated for $\overset{e}{y}$ Meridian of Boston in N-England / where the North Pole is Elevated 42 degr. / 30 min. Longitude 315 Degr. / By W. Brattle Philomath. / Genesis 1. 14. And God said, Let there be Lights in / the Firmament of the Heavens, to divide the Day / from the night , and let them be for Signes & / for Seasons , for Dayes & for Years. / CAMBRIDGE / Printed by Samuel Green 1682.
Samuel Green, Sr., Cambridge.
Other printer: Samuel Sewall.
Sabin 62743. Evans 314. Nichols [48]. Drake 2866. Wing A1378.
Originals located at: American Antiquarian Society, Library of Congress, Massachusetts Historical Society, Yale University.
Readex microprint and photostat copies are available. The complete text of the American Antiquarian Society copy is available online at Evans Digital Edition.
Paging: Twelve leaves. Pages 16 through 23 are numbered 2 through 9. Runs March to February.

---

65. Prologue to Carlos de Sigüenza y Gongora, *Libra Astronómica y Filosófica*, México, 2001: xii.

A handwritten note on the title page of the Massachusetts Historical Society (MHS) copy reads, "ex dono Authoris Ja⁻ 30. 1681."

There is a considerable explanation of calendars, pre-Julian, Julian, and Gregorian. On [9]v, Brattle proposes skipping the next thirteen leap years to set the calendar right. Otherwise, he says, if the Julian calendar is kept for 10,000 more years, "the vernal Equinox would be on the latter end of December . . ."

On [8r] is a spoof of a typical almanac formula. "AN Explanation of the Preceding Ephemeris, fitted to the Meridian of their Pates whose Poles are least Elevated, Longitude little or none." What follows is a discussion of what is rational and irrational. Brattle proposes to show, for example, not just that there will be an eclipse but also, "what the Reason of an Eclipse of the Sun or Moon is . . ."

An advertisement on the last page announces a book of verses dedicated to almanac writer, and printer, John Foster.

Ola Elizabeth Winslow writes that this came from Sewall's press in Boston,[66] but, according to the handwritten note attributed to Sewall on the last page of the MHS copy, only the last half sheet was printed in Boston. Sewall acquired the press there on October 12, 1681, after the death of John Foster. Samuel Abbott Green writes that "last half sheet" means the last four leaves.[67]

Sewall made many notes on the calendar about weather and about deaths and burials. Most interesting are the notes of a "Blazing Starr" for August 17 to 23. This is almost certainly Halley's comet.

William Brattle (1662–1717) was the brother of Thomas Brattle, author of a previous almanac on our list. Their father was also named Thomas. William graduated from Harvard in 1680.

William Brattle also wrote several works on logic, used as texts at Harvard. These were not printed right away, as the custom instead was that students transcribed manuscript copies into their notebooks. His *A Compendium of Logick, According to the modern Philosophy, extracted from Le-grand; & others their Systems*, according to Rick Kennedy, "made Cartesian logic dominant at Harvard" from "1687 until at least 1743, probably until 1767."[68] (The MHS has two manuscript copies of this book. One has

---

66. Ola Elizabeth Winslow, *Samuel Sewall of Boston*, New York, 1964.

67. Samuel Abbott Green, "Early American Imprints," *Proceedings of the Massachusetts Historical Society* second series IX (1894–1895): 410–540.

68. Charles Morton and William Brattle, *Aristotelian and Cartesian Logic at Harvard, Charles Morton's* A LOGICK SYSTEM *and William Brattle's* COMPENDIUM OF LOGICK, ed. Rick Kennedy, Boston, 1995.

"*Philosophers*" instead of "*Philosophy.*") His *Compendium Logicæ secundum Principia D: Renati Des-Cartes propsitum in usum Pupillorum* was a Latin catechism done about 1682 in response to Harvard president Increase Mather's encouraging a return to Latin. The first printed version was the *Compendium Logicæ secundum Principia D. Renati Cartesii Plerumque Efformatum et Catechistice Propositum* in 1735.

José de Escobar Salmerón y Castro.   A108
El Lunario y pronostico de temporales del año que viene.
México.
Quintana 47.
The record that pertains to this title at the Archivo General de la Nación is *Inquisición* 670, leaf 245. The name used here by the author is Joseph de Escobar Salmerón y Castro, in contrast to the one used in 1680.

Juan Ramón Koenig.   A109
El Conocimiento de los Tiempos.
Lima.

Carlos de Sigüenza y Góngora.   A110
El Lunario y pronostico de temporales para el año venidero de 1682.
México.
Quintana 48.
The record that pertains to this title at the Archivo General de la Nación is *Inquisición* 670, leaf 98.

1683

Anon.   A111
CAMBRIDGE Ephemeris. / AN / ALMANACK.
Samuel Green, Sr., Cambridge.
Drake 2868. Not listed by Nichols.

José de Escobar Salmerón y Castro.   A112
El Lunario y pronostico del año que viene de 83.
México.
Quintana 49.

The records that pertain to this title at the Archivo General de la Nación are *Inquisición* 670, leaves 8 to 10. The name used here by the author is Joseph Salmerón de Castro y Escobar.

Juan Ramón Koenig.   A113
El Conocimiento de los Tiempos.
Lima.

Cotton Mather.   A114
M.DC. LXXXIII. / *The BOSTON EPHEMERIS.* / AN / ALMANACK / FOR / The (*Dionysian*) Year of the Christian / ÆRA. M^{69} DC. LXXX III. / And of the Worlds Creation 5632. / *Anno Oppidi inchoati* 53. / Of which the Vulgar Notes are. / Cycle of the Sun 12. / Dominic.Let. G.F. / Golden Numb. 12. / Epact 12. / Num. of Direct. 18. / Serving the Meridian of *Boston* in *New-Engl.* / Latitude, 42.*gr.*^{70} 30.*min.* / Longitude 315.*gr.* / Eph. 5. 16. *Redeeming the Time.* / Damna fleo rerum,sed plus fleo Damna Dierum; / Quisq; potest rebus succurrere ; nemo Diebus. / BOSTON IN NEW-ENGLAND / Printed by *S. G.* for *S. S.* 1683.

Samuel Green, Jr., Boston.

Sabin 46239, 62743. Evans 351. Nichols [49]. Drake 2867. Wing A1938.

Originals located at: American Antiquarian Society, Boston Public Library, Harvard University, Library of Congress, Massachusetts Historical Society, University of Virginia (missing last leaf).

Readex microprint and photostat copies are available. The complete text of the American Antiquarian Society copy is available online at Evans Digital Edition.

Paging: Twelve leaves.

Runs March to February.

The Boston Public Library, Harvard, Massachusetts Historical Society, and Virginia copies have a variant title page where the second to last line is "*BOSTON IN NEWENGLAND,*".

The reference in the title to Dionysian year refers to Dionysius Exiguus, who suggested, in 526, the current numbering of years.

This work contains a description of the 1682 comet (Halley's) and an announcement of the publication of *Kometographia* by Cotton Mather's

---

69. Faint period in Massachusetts Historical Society copy.

70. The period after the 42 is very faint in the American Antiquarian Society and Virginia copies.

father, Increase Mather, printed by Samuel Green, Sr., in this same year. Mather says that he observed the comet on August 13, 1682, in the east and then on August 23 in the west, and he continued seeing it into mid-September.

Mather gives an etymology of the word "almanac" from "almonaght," meaning observations of all the moons.[71] He also allows that some people have said that "almanac" is an Arabic word. A peek at a modern dictionary gives a descent from the Arabic word *al-manākh*, which Marion Barber Stowell translates as "the count." Stowell quotes some other fanciful etymologies for the word "almanac" that have been offered.[72] Recently, a Wikipedia article insisted that the word was Greek in origin, but now Wikipedia takes the position that the ultimate ancestor of the word *al-manākh* is unknown. The son of Increase Mather, Cotton Mather's years are 1663 to 1728. His birth was recorded as February 12, 1662, on the Julian calendar (February 22, 1663, on the Gregorian calendar).[73]

Cotton Mather graduated from Harvard in the class of 1678. John Langdon Sibley wrote in 1885 that Mather had been, at that time, the third youngest Harvard graduate ever.[74] He received a second degree in 1681. He knew French, Spanish, and Iroquois, according to Sibley. (This must have been in addition to the classical languages that were part of the curriculum at Harvard.)

This was his second publication. The first was a poem, published in 1682, dedicated to the memory of fellow almanac writer Urian Oakes, who was still president of Harvard when Mather graduated and whose son, Urian, graduated with him. Sibley lists 456 works for Mather, some of which are manuscripts and letters.

The appearance of Samuel Sewall (S.S.) on the title page is a result of the fact that Sewall took over John Foster's business in Boston after Foster's death. He hired Samuel Green, Jr., to do the actual printing for him. This Samuel Green had learned printing from his father (see our first

---

71. Robb Sagendorph (*America and her Almanacs*, Dublin, NH, 1970) quotes George Putnam to the effect that this *almonaght* is a Saxon word.

72. Marion Barber Stowell, *Early American Almanacs*, New York, 1977.

73. Sixty-nine years later George Washington would be born on Feb. 11, 1731, Julian (Feb. 22, 1732, Gregorian) and his birthday would come to be celebrated on Feb. 22. The British Empire switched to the Gregorian calendar in September of 1752.

74. John Langdon Sibley, *Biographical Sketches*, vol. III, 1678–1689, Cambridge, 1885.

entry for 1678) but had been running a store in New London before coming to work for Sewall, since there had not been enough business for more printers.

The American Antiquarian Society copy has the handwritten note, "By Cotton Mather." A handwritten note on the Harvard copy (used for the photostat) says "By Mr. Cotton Mather." Sibley says that this last was Judge Sewall's copy.[75] Many of the early Massachusetts almanacs that still exist were once in his collection, according to notations he made on them.

Sewall's years are 1652 to 1730. Sewall was later one of the judges at the Salem witch trials, which means that his life again intersected with Cotton Mather's. When, in 1692, numerous people in Salem Village (now Danvers, Massachusetts) were put on trial (and eventually executed) for witchcraft, Cotton Mather wrote a letter of advice to the court describing what sort of evidence they should consider.[76] In 1697, Sewall, through an act of public confession, apologized for his role in the trials.

In 1700, Sewall was the author of *The Selling of Joseph*, which has been called the first anti-slavery tract printed in New England. He kept a diary from 1673 to 1729, and all but eight years of it are extant. See the biography by Ola Elizabeth Winslow for more about him.

Carlos de Sigüenza y Góngora.   A115
El Lunario y Pronostico de temporales para el año vendidero de 1683.
México.
Quintana 50.
The records that pertain to this title at the Archivo General de la Nación are *Inquisición* 670, leaves 192 to 195.

1684

José de Escobar Salmerón y Castro.   A116
El pronostico o Lunario de los temporales, y guarda de Salud con sus medicinas segun indican los Astros el año que viene de 84.
México.
Quintana 51.

---

75. Ibid.: 43.
76. Chadwick Hansen, *Witchcraft at Salem*, New York, 1969: 96–98.

The records that pertain to this title at the Archivo General de la Nación are *Inquisición* 670, leaves 374 to 375. The names used here by the author are Joseph Salmerón de Castro y Escobar and Joseph de Escobar Salmerón y Castro. He also changes his title from Bachiller to Doctor this year.

Benjamin Gillam.   A117

1684. / The *BOSTON EPHEMERIS.* / AN / ALMANACK / FOR / The Year M DC. LXXX IV. / And of the Worlds Creation 5633. / *Oppidi Inchoati,* 55. / *Being the first after* Leap-year. / Of which the Vulgar Notes are. / Cycle of the Sun 13. / Dominic.Let. E. / Epact 23. Prime 13. / Calculated for the Meridian of *Boston* in *New-Engl.* / where the North Pole is elevated 42 *gr.* 30 *m.* / Longitude 315 *gr.* / By *Benjamin Gillam* Philonauticus. / *BOSTON IN NEWENGLAND,* / Printed by *Samuel Green* for *Samuel* / *Philips,* and are to be Sold at his Shop at / the West end of the *Town-House.*[77] 1684.

Samuel Green, Jr., Boston.

Sabin 62743. Evans 359. Nichols [50]. Drake 2869. Wing A1792.

Originals located at: American Antiquarian Society, Massachusetts Historical Society, Yale University.

Readex microprint and photostat copies are available. The complete text of the MHS copy is available online at Evans Digital Edition.

Paging: Eight leaves.

Runs March to February.

Samuel Sewall was released from management of the press in Boston on September 12, 1684. There is a handwritten "Samuel Sewall" on the MHS copy.

There is, on [8]r, a "Table of Expence" where daily payments of 1 penny to 20 shillings are taken to the week, month, and year.

Juan Ramón Koenig.   A118
El Conocimiento de los Tiempos.
Lima.

Noadiah Russell.   A119

MDC LXXX IIII. / *CAMBRIDGE EPHEMERIS.* / AN / ALMANACK / OF / Cœlestial Motions,Configurations &c. / For the year of the Christian Æra, / 1684. / Being from / The { *Creation of the World* 5633. / *Suffering of our Saviour* 1651 / *Restauration K. Charles* II.24 / *Leapyear (in our account)*

---

77. Hyphen missing in Yale copy.

1: / The Vulgar Notes / Cycle of ☉ & ☾ 13 / Dominical Letter E. / {} / Epact 23 / Numb.Direct. 09 / *Calculated for the Meridian of Cambridg in N.England,* / Lat. 42 degr. *about* 30 min. Long. 315 degr. / By N. Russel *Astrotyr.* / *Gen.* I. 14. And God said, let there be Lights / in the Firmament of Heaven. / *Psal.* 136. 8. The Sun to Rule by Day— / *Verse* 9. The Moon & Stars to Rule by night. / *Astra Regunt Homines ; sed Regit Astra Deus* / CAMBRIDGE. / Printed by *Samuel Green* 1684.

Samuel Green, Sr., Cambridge.

Sabin 62743. Evans 376. Nichols [51]. Drake 2870. Wing A2309.

Originals located at: Massachusetts Historical Society.

Readex microprint and photostat copies are available. The complete text of the Massachusetts Historical Society copy is available online at Evans Digital Edition. A full-page woodcut is reproduced by Marion Barber Stowell.[78]

Paging: Eight leaves.

Runs March to February.

Russell's years are 1659 to 1713. He graduated from Harvard in 1681.

This entry contains, on [8]r, the first full-page almanac illustration, according to Stowell, in the English colonies.[79] It is woodcut of King David with a harp. There is also a long passage on lightning, on [8]v, which is quoted by John Langdon Sibley.

Carlos de Sigüenza y Góngora.   A120
Repertorio.
México.

José Miguel Quintana doesn't list an almanac for Sigüenza for this year. However, Augustin de Vetancurt refers to a *Repertorio* of Sigüenza for 1684, and Quintana notes this on page 71.

Elías Trabulse used this year as an example of Sigüenza's using abbreviated histories at the head of his almanacs.[80]

1685

Juan Ramón Koenig.   A121
El Conocimiento de los Tiempos.
Lima.

---

78. Marion Barber Stowell, *Early American Almanacs*, New York, 1977: 51.
79. Ibid.:. 50–51.
80. Prologue to Carlos de Sigüenza y Góngora, *Libra Astronómica*, México, 2001: xii.

Nathanael Mather. A122

1685, / The BOSTON EPHEMERIS. / AN / ALMANACK / Of Cœlestial Motions of the Sun & / Planets,with some of the principal Aspects / For the Year of the Christian ÆRA / M DCLXXXV. / Being in our Account the second after / Leap-year, and from the Creation / 5634. / *The Vulgar Notes of which are* / Cycle of ☽ 14 / Roman Indiction 8 / Dominical Letter D / }{ / Cycle of ☉ 14 / Epact 4 / Fitted to the Meridian of *Boston* in *New-England,* / where the *Artick Pole* is elevated 42gr. 21 m. / BOSTON in NEW-ENGLAND / Printed by and for *Samuel Green.* 1685.

Samuel Green, Jr., Boston.

Sabin 46774, 62743. Evans 395. Nichols [52]. Drake 2871. Wing A1939.

Originals located at: Massachusetts Historical Society, New York Public Library.

Readex microprint and photostat copies are available. The complete text of the Massachusetts Historical Society copy is available online at Evans Digital Edition.

Paging: Eight leaves.

Handwritten "Nath Mather" on the New York copy (attributed to Samuel Sewall by John Langdon Sibley). The Massachusetts Historical Society copy has "By Nath. Mather Philom."

Runs March to February.

Mather's years are 1669 to 1688. He graduated from Harvard in 1685 and died at the age of 19. This almanac was published early in his senior year when he was 15.

A discussion, on [8]r, of recent discoveries made with the telescope mentions sunspots, the sun's rotation in twenty-six days, and the fact that Robert Hook and Giovanni Cassini have seen a spot on Jupiter with a rotation period of 9 hours and 56 minutes. Since the latter could be used as a universal clock, Mather notes that it is "better use for determining the Longitude of places than either the Eclipses of the Sun or Moon." Also mentioned are some moons of Saturn and the rotation of Mars in 24 hours, 40 minutes.

Mather's latitude for Boston is different from the 42°30′ that most other Massachusetts almanacs give.

Carlos de Sigüenza y Góngora. A123
[*Almanaque*].
México.

Several sources agree that Sigüenza produced an almanac every year from 1671 until his posthumous one in 1701. Following this, we will insert a Sigüenza title for years in which Quintana gives none.

Quintana notes in his entry for Sigüenza under 1681 that a Sigüenza lunario for 1685 was reported by Mariano Veytia. That may be a typographical error, since Leonard says that Veytia reported owning a copy of the 1681 issue.[81]

Sigüenza himself says in his almanac of 1690 that that almanac makes twenty in so many years. We take this to mean that no years were skipped even though no record is left for some years in the archives of the Inquisition.

William Williams. A124

MDC LXXXV. / *CAMBRIDGE EPHEMERIS* / AN / ALMANACK / OF / The Cœlestial Motions, For the Year / Of the Christian Æra, / 1685. / Being from / The { *Creation of the World 5634* / *Floud [sic] of Noah 3978* / *Suffering of* CHRIST *1652* / *Laying found.of Harv.Co.43* / *Leap year (in our account)* 2 / Whose Vulgar Notes are / Golden Number 14 / Cycle of the Sun 14 / Dominical Letter D / {} / Epact 4 / Numb. Direction 29 / Calculated for 315 *degr.* Longitude. And / Latitude 42 *degr.* 30 *min.* North. / By *W. Williams* Philopatr. / Isaiah 40. 26. *Lift up your eyes on high, / and behold who hath Created these things / that bringeth out their Host by number , And / calleth them all by names.* / CAMBRIDGE, / Printed by *Samuel Green* for the year / 1685.

Samuel Green, Sr., Cambridge.

Sabin 104395, 62743. Evans 399, 400. Nichols [53]. Drake 2872, 2873. Wing A2768, A2769.

Originals located at: American Antiquarian Society, Harvard University, Massachusetts Historical Society (Evans 399; Drake 2872; Wing A2768). A second edition with "Printed by *Samuel Green* for *Samuel Phillips,* / at the Exchange in Boston. 1685." is represented by a copy at Yale University (Evans 400; Drake 2873; Wing 2769).

Readex microprint and photostat copies are available. The complete texts of both variants, the American Antiquarian Society copy of the first and the Yale copy of the second, are available online at Evans Digital Edition.

---

81. Irving Leonard, *Ensayo Bibliográfico*, México, 1929: 13.

Paging: Eight leaves.
Runs March to February.
Williams's years are 1665 to 1741. He graduated from Harvard in 1683. His graduation question was "*An Terra Movetur?*" to which he gave an affirmative response.

Long comments on rainbows, on [8]r, and comets, on [8]v, are quoted in full by John Langdon Sibley. In the section on comets, Williams gives Aristotle's theory that comets are exhalations (see the entry in Part 1 under 1690), cites two objections, and refutes them. He claims that parallax measurements that put the comets above the moon are not a problem for the theory, and ends with, "That the nature of place above the Moon is falsly concieved [sic] of, if imagined to be really different from place below the Moon, Now if thus why may not Comets ascend above the Moon as well as up to the Moon."

1686

Samuel Atkins. **A125**

Kalendarium Pennsilvaniense, / OR, / America's Messinger. [sic] / BEING AN / ALMANACK / For the Year of Grace, 1686. / Wherein is contained both the English & Forreign / Account, the Motions of the Planets through the Signs, with / the Luminaries, Conjunctions, Aspects, Eclipses; the rising, / southing and setting of the Moon, with the time when she / passeth by,or is with the most eminent fixed Stars : Sun rising / and setting, and the time of High-Water at the City of *Phi-* / *ladelphia, &c.* / With Chronologies, and many other Notes, Rules, / and Tables, very fitting for every man to know & have ; all / which is accomodated to the Longitude of the Province of / *Pennsilvania,* and Latitude of 40 Degr. north, with a Table / of Houses for the same, which may indifferently serve *New-* / *England, New York, East & West Jersey, Maryland,* and most / parts of *Virginia.* / By SAMUEL ATKINS. / Student in the Mathamaticks [sic] and Astrology. / And the Stars in their Courses fought against Sesera, Judg.5. 29. / Printed and Sold by *William Bradford* at *Philadel-* / *phia* in *Pennsilvania,* 1685.

William Bradford, Philadelphia.

Evans 382. Drake 9460, 9461. Hildeburn 1. Wing A1303, A1304.

Original located at: Rosenbach Museum (Drake 9460; Wing A1303). A second impression, with the lines "Printed and Sold by *William Bradford,* sold also by / the Author and *H. Murrey* in *Philadelphia,* and / *Philip*

Richards in *New-York*; 1685." is represented by a copy at: Library Company (Historical Society of Pennsylvania copy) (Drake 9461; Wing A1304).

Readex microprint and photostat copies are available. The complete text of the Historical Society of Pennsylvania copy is available online at Evans Digital Edition. Early English Books Online shows a "Huntington" copy with leaves out of order and [2]r partially obscured.

Paging: Twenty leaves.

This is Pennsylvania's earliest printed work according to Charles Hildeburn. Hildeburn reprints Atkins's "To the Reader" and shows William Bradford's "The Printer to the Readers" in facsimile.

Isaiah Thomas wrote that the original location of Bradford's press is unknown.[82] But Thomas didn't know about this almanac, since he says that the Leeds almanac of 1687 is the earliest work printed by Bradford that he has knowledge of. In addition to the fact that Philadelphia is identified on the title page of the first impression as the location of the press, some features inside this almanac also give Bradford's location as Philadelphia. The printer's note to the reader is dated Dec. 28, Philadelphia. There is an advertisement for medicines "sold by William Bradford at Philadelphia" on [3]r. Atkins writes of ". . . the fairs here at Philadelphia . . ." on [3]v, although he may be referring to his own location only.

Joseph Smith and, later, Alexander Wall, Jr., both wrote that Bradford lived and worked in Oxford Township from the time he settled in Pennsylvania until he left for New York.[83]

The original printing offended the government in Pennsylvania by referring to "Lord Penn" in the chronology. This was blacked out on [3]r. The list of errata on [20]v says to make this "William Penn, Propriatary and Governour" in the first edition and this is changed to "William Penn, Propriator and Governour" in the second edition.

The verse from Judges in the title should be chapter 5, verse 20.

Note again that we list almanacs in Part 2 under the year of their calendar (1686 in this case) rather than the year of publication (1685).

---

82. Isaiah Thomas, *The History of Printing in America*, ed. Marcus McCorison, Weathervane, NY, 1970: 341.

83. Joseph Smith, *Short Biographical Notices of William Bradford, Reiner Jansen, Andrew Bradford, and Samuel Keimer, Early Printers in Pennsylvania*, London, 1891; Alexander J. Wall, Jr., *William Bradford, Colonial Printer*, reprinted from the *Proceedings of the American Antiquarian Society* (Oct. 1963), Worcester, 1963.

Runs from "The Eleventh Month (January)" to "The Tenth Month (December)" with no indication of New Year's Day.[84] This appears to be the first almanac in British North America to start the year with January and the first to list the feast days (John the Baptist on June 24, Christmas on December 25, etc.). These are both firsts that many authors credit to John Tulley's almanac in the following year.

There is a short biography of Bradford, the printer, in Charles Hildeburn's *Sketches of Printers and Printing in Colonial New York*. His years are 1663 to 1752. We will see later that after being the first printer in Pennsylvania, he moved his press in 1693 and became the first printer in New York.

At twenty leaves, this is the longest almanac printed in the English colonies so far. (John Foster had a few of sixteen leaves.) This makes lots of room for data. Each month gets twenty-one columns over two facing pages. These columns give the day of the month; the day of the week; feast days[85]; position of the sun, moon, and the other five planets; aspects of the moon with the planets and sun (e.g., on February 15 the moon has a conjunction with both Mars and Venus); aspects between planets; the moon's apogee and perigee; the ascending and descending nodes; high tide at Philadelphia; and the sun's rising and setting. There is a table of astrological houses that goes on for six pages. There is a chronology of notable events, a list of English kings, a list of symbols, and a list of errata. He even gives the dominical letter, and the dates of Easter and Whitsunday, in both the Julian and Gregorian systems.

Atkins credits the ephemerides of John Cadbury as his source for the positions of the planets.

Juan de Avilés Ramírez. **A126**
El Pronostico de Temporales con las Elecciones de Medizina y Navegaciones del año que biene de mill [sic] seiscientos y ochenta y seis.
México.
Quintana 52.

---

84. It is significant that on [3]r we see that 1686 is two years after bissextile. If the author truly saw the year as starting in March, as Samuel Danforth and Nathanael Mather do in their almanacs for this same year, then the last leap day would have been February 29, 1683 (see our discussion of this for the entry of 1650), and so 1686 should have been three years after leap year.

85. Curiously, Hildeburn says the feast days are omitted.

The record that pertains to this title at the Archivo General de la Nación is *Inquisición* 670, leaf 196.

Quintana says on p. 52 of *La Astrologiia* that Avilés Ramírez was a medical doctor at the University and submitted an anagram to a literary contest held by the University in 1682.

Samuel Danforth (1666–1727). A127

The New-England / ALMANACK / FOR / The year of our Lord. 1686. / And of the world. 5635. / Since the planting of Massachusets [sic] / Colony in New-England. 58. / Since the found. of Harv. Coll. 44. / *Whereof the Golden Number, Epact and Cycle of the Sun / are* 15, *And the Dominicall Letters* C B. *Being in / our*[86] *account the third from Leap Year.* / Kindly offering unto its Country a particular Prospect / of the places of the Planets, with the New-Moons / and other cælestial Configurations and Principall As- / pects, As also the Moons Latitude, Apogee, and Perigee, / with the Time of the rising and setting of both the lumi- / naryes and the hours and minutes of Twilight : All / calculated to 315 *gr.* longitude and 42 *gr.* 30. *m.* / North latitude. With the stated Court days of the / Three United Colonyes; As also Elections,[87] Commence- / ment, Artilleryes : Together with a Table shewing the / Time of High water in and near *Massachusets Bay.* and / an Account of Spring Tides With a brief Register of / severall memorable Events of Providence in *New- / England.* / *By S. D. Philomath.* / *Job.* 38. 33, *Knowest thou the ordinances of Hea- / ven ? canst thou set the dominion thereof in the Earth ?* / *Psal.* 90. 12. *So teach us to number our Days, that / we may apply our hearts unto*[88] *wisdom.* / CAMBRIDGE. / Printed by *Samuel Green. sen.* Printer to *Harvard / Colledge* in *New-England. A. D.* 1685.[89]

Samuel Green, Sr., Cambridge.

Sabin 52646, 62743. Evans 403, 404. Nichols [55]. Drake 2874, 2875. Wing A1578, A1579.

Originals located at: American Antiquarian Society, Massachusetts Historical Society, Yale University (Evans 403, Drake 2874, Wing 1578). Second impression (differing on the title page, in addition to the changes

---

86. o missing in Massachusetts Historical Society copy of Evans 404.
87. Comma is very faint in Massachusetts Historical Society copy of Evans 403.
88. n missing in Massachusetts Historical Society copy of Evans 404.
89. Someone has overwritten the 5 with a 6 in the Massachusetts Historical Society copy of Evans 403.

mentioned in the footnotes, with a "1686" at the end instead of "1685") at: Massachusetts Historical Society (Evans 404, Drake 2875, Wing A1579).

Readex microprint and photostat copies are available. The complete texts of both variants, the American Antiquarian Society copy for the first and the Massachusetts Historical Society copy for the second, are supposedly available online at Evans Digital Edition. In fact, they have posted the Massachusetts Historical Society copy twice, calling one of them the American Antiquarian Society copy. This mistake appears originally on the Readex microprint card and is copied from there.

Paging: Eight leaves.

Runs March to February.

In both impressions (generally called the first and second) there is the note, "Since the impression for February, wee hear of the deplorable decease of . . ." The first impression is dated 1685 and the second, 1686. For the sake of greater consistency, we are following the convention that almanacs will be listed under the year in the calendar that they contain.

Danforth has given two dominical letters in the title in spite of the fact that this is not a leap year.

There is a note on the title page of the Massachusetts Historical Society copy of Evans 403 to the effect that the owner (Samuel Sewall) received the book in $168\frac{5}{6}$. Sewall wrote on a later page, "The above $\mathrm{ac\overset{o}{c}}$ of y^e eclipse . . . was truer by much than Mr. Mather's." This refers to the Nathanael Mather almanac of the same year.

There is a table of tides for Massachusetts Bay indexed by the age of the moon.

The deaths of almanac writers Josiah Flint, Urian Oakes, John Foster, Thomas Shepard, John Sherman, and Nathaniel Chauncy are listed in various places.

Samuel Danforth's father, Samuel, and older brother, John, are also authors of almanacs on our list. The younger Samuel graduated from Harvard in 1683. He had some knowledge of the Massachuset (Algonquian) language, as he translated sermons by Increase Mather into that tongue and also compiled a dictionary that survives in manuscript form.

Juan Ramón Koenig. **A128**
El Conocimiento de los Tiempos.
Lima.

Nathanael Mather. **A129**

1686. / *The BOSTON EPHEMERIS.* / AN / ALMANACK / Of Cœlestial Motions of the Sun & / Planets,with some of the principal Aspects / For the Year of the Christian ÆRA/ M DCLXXXVI. / Being in our Account the third after / Leap-year, and from the Creation / 5635. / *The Vulgar Notes of which are* / Cycle of ☽ 15 / Dominical Letter C / }{ / Cycle of ☉ 15 / Epact 15 / Calculated for and fitted to the Meridian of *Boston* in / *New-England*, where the *North Pole* is elevated 42. / *gr.* 21 *m.* / By *Nathanael Mather.* / NEW-ENGLAND, / *Boston*, Printed and Sold by *Samuel Green*, 1686.[90]

Samuel Green, Jr., Boston.

Sabin 62743. Evans 418. Nichols [54]. Not in Drake. Wing A1940.

Originals located at: American Antiquarian Society, Boston Public Library, Harvard University, Massachusetts Historical Society, University of Virginia.

Readex microprint and photostat copies are available. The complete text of the American Antiquarian Society copy is available online at Evans Digital Edition.

Paging: Eight leaves.

Runs March to February.

There are some notes about discoveries made by telescope, including the rings of Saturn and sunspots on [8]r. Mather lists the periods, precise to the second, of the four Galilean moons of Jupiter on [8]v and notes that Robert Hook's discovery of the parallax of fixed stars favors the Copernican System on [1]v. He writes, "That Excellent Mathematician, Mr. Robert Hook hath lately discovered that there is a sensible parallax of the Earths Orb among the fixed Stars, which some think doth undeniably prove the Truth of the Copernican System. There was a time when it was judged heresie to maintain that there are Antipodes; and ignorant People still think that if there should be men on the other side of the World with their feet towards ours, and their heads downwards, they must needs either fall down into the Skies, or make their brains addle. There are very Learned Men who no more doubt that the Earths motion will be generally received within a small Compass of Years, then that now they believe the Antipodes."

Mather again uses the variant latitude for Boston from the previous year.

Handwritten "To the Rev. Mr. John Cotton." on the American Antiquarian Society copy. The Massachusetts Historical Society copy has "Recd Xr. 25. 1685.", probably written there by Samuel Sewall.

---

90. The Boston Public Library copy appears to end with ".1686."

Carlos de Sigüenza y Góngora. **A130**
El pronostico para el año que biene de ochenta y seis.
María de Benavides and José Rivera, Heirs to the widow of Bernardo Calderón, México.
Quintana 53.
The records that pertain to this title at the Archivo General de la Nación are *Inquisición* 670, leaves 190 to 191.
Quintana reports that the printers were the heirs to the widow of Bernardo Calderón. The book appeared under the name Juan de Torquemada.

1687

Antonio Sebastián de Aguilar Cantú. **A131**
El Pronostico para el año que biene de ochenta y siete.
México.
Quintana 54.
The records that pertain to this title at the Archivo General de la Nación are *Inquisición* 495, leaves 38 to 41.
Aguilar Cantú describes himself as doctor to the Bishop of Michoacán.

Juan de Avilés Ramírez. **A132**
Pronostico de Temporales para el año, que viene de 1687.
México.
Quintana 55.
The records that pertain to this title at the Archivo General de la Nación are *Inquisición* 495, leaves 42 to 43.

José de Campos. **A133**
El Pronostico para el Año que biene de ochenta y Siete.
Diego de Escobar, México.
Quintana 56.
The records that pertain to this title at the Archivo General de la Nación are *Inquisición* 670, leaves 369 to 372.
This appeared under the name Michael Henrico Romano. Quintana recounts in his very next entry that Juan de Avilés Ramírez protested this to the Inquisition, and this led to a ban on pseudonyms.
José de Campos was a medical student at the Royal University, although in the application the printer writes that the author was living in Michoacán.

Juan Ramón Koenig. A134
El Conocimiento de los Tiempos.
Lima.

Daniel Leeds. A135
AN ALMANACK / For the YEAR of Christian Account 1687. / Particularly respecting the Meridian and Latitude of *Burlington*, but may indifferently serve all places adjacent. / By Daniel Leeds, *Student in Agriculture.*

This is a broadside, or sheet almanac, with all the text on one side of a large page. Notes for the year are to the left of the title:
Vulgar Notes for this / Year 1687. / Golden Number 16 / Circle of the Sun 16 / Epact 26 / Dominical Letter B

There are three couplets to the right of the title:
No man is born unto himself alone, / Who lives unto himself he lives to none. / The blaze of Honour,Fortunes sweet Excess, / Do undeserve the Name of Happiness. / Place shews the man,& he whom honour mends / He to a worthy generous Spirit tends.

Printer's information is at the bottom of the text, and is followed by an advertisement:
Printed and Sold by *William Bradford*, near *Philadelphia* in *Pennsylvania, pro Anno* 1687. / There is now in the Press, *The Excellent Priviledge of Liberty and Property.* To which is added, *A Guide for the Grand and Petty Jury.*

William Bradford, Pennsylvania.

Sabin 39815. Evans 408, 39230. Drake 9462, 9463. Hildeburn 5. Wing A1864.

Originals located at: British Library (two fragments), Historical Society of Pennsylvania (bottom half), Library Company (three copies; copy 1 belongs to the Historical Society of Pennsylvania, copy 2 is a composite, copy 3 is the top half) (Evans 408, Drake 9462). A second impression, omitting the printer's information and the advertisement, is represented by a copy at: Boston Public Library (Evans 39230, Drake 9463).

The complete copy at the Library Company is pasted to a pasteboard cover from another book. Copy 2 is missing a strip along the right side which has been replaced with the corresponding part from the nineteenth century reprint with fragments from another original pasted over that. Copy 3 is the other half of the copy at the Historical Society of Pennsylvania (HSP). The Library Company and the HSP, right next door to each other, have an agreement in which the Library Company stores early printed matter belonging to both, and the HSP has the early manuscripts. The HSP copy of the bottom half of the Leeds almanac is pasted into John Watson's *Annals of Philadelphia, Volume I*, p. 277, and so is catalogued as a manuscript.

The British Library fragments and the Oxford copy of John Tulley's 1692 almanac are the only seventeenth-century American almanacs we know of that are held outside the United States.

A reprint of the first impression was done with a different typeface and some variations in spelling in 1854 or earlier, and copies of this can be found at the New York Public Library and at the British Library. The copy at the British Library has 1833 penciled on the back.

A facsimile of the British Library fragments is available. The back of the Boston Public Library copy is stamped, "Massachusetts Historical Society, March 10, 1922."

Readex microprint and photostat copies are available. The complete text of the Library Company copy of Evans 408 and a Massachusetts Historical Society facsimile (presumably of the Boston Public Library copy) of Evans 39230 are available online at Evans Digital Edition. Early English Books Online shows a copy of the nineteenth-century reprint, mistaking it for the original. They say, "Reproduction of the original in the British Library."

Runs "The xi. Month, January" to "The x. Month, December." Days of the week are indicated by 1 through 7 for Sunday through Saturday. Feast days are indicated.

Leeds's years are 1652 to 1720. (Alexander J. Wall says that Leeds was 68 and 11 months in September 1720, implying a birth year of 1651.)[91] Leeds was a surveyor, and he made the first map of Burlington, New Jersey, in 1696.[92] Later he bought a farm in Burlington County and became a farmer.[93]

Carlos de Sigüenza y Góngora.   **A136**
[*Almanaque*].
México.
See the note for Sigüenza under 1685.

---

91. Alexander Wall, *A List of New York Almanacs 1694–1850*, New York, 1921.
92. Henry H. Bisbee, *The Island of Burlington*, Burlington, 1977: 6.
93. E. M. Woodward, *History of Burlington County*, Philadelphia, 1883; reprinted, Burlington, NJ, 1984.

John Tulley. **A137**

Tulley 1687. / AN / ALMANACK / For the Year of Our LORD, / M DC LXXXVII. / Being the third after Leap-year, / and from the Creation / 5636. / *The Vulgar Notes of which are* / *Prime* 16 / *Epact* 26 / }{ / *Cicle* [sic] *of the* ☉ 16 / *Domin: Letter*. B / *Unto which is annexed a* Weather-Glass, *whereby the* / *Change of the Weather may be foreseen*. / Calculated for and fitted to the Meridian of *Boston* in / *New-England*, where the *North Pole* is elevated 42. / gr. 30 m.[94] / By *John Tulley*. / *Boston*, Printed by *S. Green* for *Benjamin* / *Harris*; and are to be Sold at his / Shop, by the *Town Pump* near the *Change* / 1687.

Samuel Green, Jr., Boston.

Sabin 62743, 97438. Evans 435. Nichols [57]. Drake 2877. Wing A2582.

Originals located at: American Antiquarian Society, Library Company (leaves [4]–[8], Historical Society of Pennsylvania copy), Library of Congress, Massachusetts Historical Society.

Readex microprint and photostat copies are available. The complete text of the American Antiquarian Society copy is available online at Evans Digital Edition.

In the Harvard library catalog there is a reference to an Essex Institute copy. "The Essex Institute copy has an imprimatur, printed on a preliminary leaf, not conjugate with the almanac: Novemb. 24th. 1686. I have perused the copy of an Almanack for the ensuing year, composed by John Tulley, and find nothing in it contrary to His Majesties laws, and therefore allow it to be printed, and published by Benjamin Harris book-seller in Boston. Edward Randolph secr." In fact, the Phillips Library at the Peabody Essex Museum now has only the 1698 Tulley, and there is no imprimatur leaf.

Paging: Eight leaves.

Tulley's years are 1638 to 1701. Samuel Abbott Green writes that Tulley came from England with his mother, settled in Saybrook, Connecticut, and died there.[95] He produced an almanac yearly until his posthumous one in 1702.

---

94. Period very faint in American Antiquarian Society and Massachusetts Historical Society copies.

95. Samuel Abbott Green, [Note] *Proceedings of the Massachusetts Historical Society* second series VII (1891–1892): 414–415.

The calendar section runs from January to December, the first almanac in New England to do so.[96] Note also "third after leap year" instead of "leap year." For the issues 1687 through 1689 Tulley labels January 1 as "New Year's Day." Lots of other special days are labeled, including Easter, Christmas-Day, and saints' days such as "Valentine" and "St. John Baptist."

In the Massachusetts Historical Society copy (which is also the one that was used for the photostat) a handwritten note on the title page says, "Recd Decr 6. 1686". Also in this copy someone (probably Samuel Sewall) has made notes of the actual weather conditions throughout the year, a few of which stand in contrast to the prognostications, e.g., "No Thunder no Tempest" for June.

William Williams. A138

MDCLXXXVII. / *CAMBRIDGE EPHEMERIS* / AN / ALMANACK / OF / Cœlestiall Motions and Configurations / for the Year of the Christian Epocha, / 1687. / Being (*in our account*) Leap year; / And from / The { Creation of the World — 5636. / Constitution of the Julian Year — 1731. / Suffering of CHRIST — 1654. / Correction of the Calend by P.Greg: 105. / Beginning of the *Reign* of our Soveraign / Lord JAMES II. — 03. / Whose Vulgar Notes are / Golden Number — 16. / Roman Indiction. — 10. / Dominicall Letters. B. A. / {} / Cycle of the Sun.16 / Epact. — 26. / N. Direction. — 05. / Calculated for Longitude 315. *gr.* and Latitude / 42. *gr.*

---

96. Many sources say that Tulley's almanac was the first in British North America to start the year with January, but the *Kalendarium Pennsilvaniense* by Samuel Atkins did so one year earlier in 1686. Stowell says, "Tulley in 1687 was. . . the first officially to use January as the first month of the year. . ." (*Early American Almanacs*, 1977: 59.) In the same year of 1687, Leeds put January first but called it "The xi. Month." We mentioned in an earlier footnote that Atkins's count of the years since leap year reinforces the fact that he starts his year with January in 1686, in spite of calling January the eleventh month.

Likewise it is often repeated that Tulley's almanac was the first to list the feast and fast days of the Church of England. (Puritans held that only Sunday was holy.) Stowell writes in continuation of the quote just cited, ". . . and the first to insert the feast and fast days of the English church (such as Valentine's Day, New Year's Day, Shrove Sunday, Easter), a notion considered anathema to the orthodox Puritan clergy." But, as already stated, the Atkins almanac does this one year earlier in 1686, and Leeds's almanac does this in the same year of 1687.

Both of these claims may have started with Charles Nichols ("Notes on the Almanacs of Massachusetts," *Proceedings of the American Antiquarian Society* new series 22 [1912]: 13).

30. min. North. / Nec frustra signorum obitus / speculamur , & ortus, / Temporibusque parem diversis / quatuor annum. / CAMBRIDGE. / Printed by S. G. Colledg Printer 1687.
Samuel Green, Sr., Cambridge.
Sabin 62743, 104395. Evans 436. Nichols [56]. Drake 2876. Wing A2770.
Original located at: Massachusetts Historical Society.
Readex microprint and photostat copies are available. The complete text of the Massachusetts Historical Society copy is available online at Evans Digital Edition.
Paging: Eight leaves.
Handwritten "Recd Feb. 1 $168\frac{6}{7}$: ex dono Authoris." This is similar to the way Samuel Sewall signed many almanacs in his collection.
Runs March to February, probably the last almanac in the Americas to do so.
This almanac was attributed to Williams by Charles Nichols.[97]
"The Explanation of the Ephemeris" appears on [8]r–v. The calendar uses the dominical letter B for Sundays for March through December, then A in January and February, e.g., A 2 3 4 5 6 7. The calendar's sixth column gives the moon's age, using 00 for 30. The seventh column is the time to add or subtract to the time the moon makes on a sundial to get the time of the night. On [8]v is an example of how to find the time of the night using the stars.

1688

Juan de Avilés Ramírez.   A139
Pronóstico de Temporales para el año que Viene de 88.
México.
Quintana 58.
The records that pertain to this title at the Archivo General de la Nación are *Inquisición* 670, leaves 99 to 101.

José de Campos.   A140
Pronostico de Temporales para 1688.
Heirs of Francisco Rodríguez Lupercio, México.
Quintana 57.

---

97. Charles Nichols, "Notes on the Almanacs of Massachusetts," *Proceedings of the American Antiquarian Society* new series 22 (1912): 33.

The records that pertain to this title at the Archivo General de la Nación are *Inquisición* 670, leaves 4 to 7. The author uses the assumed name Michael Henrico Romano. Leaf 5 is the petition by Juan de Avilés Ramírez protesting the use of pseudonyms. Leaf 6 is the reply by printer Gerónima Delgado y Cervantes, widow of Francisco Rodríguez Lupercio. Leaf 7 is the ban on pseudonyms that the Inquisition declared in October of 1687.

Edward Eaton.   A141
AN ALMANACK.
William Bradford, Philadelphia.
Evans 442. Drake 9464. Hildeburn 10.
Hildeburn: "Printed by order of the [Philadelphia] Quarterly Meeting, to take the place of the one by Leeds." (See the subsequent entry for Leeds.)
Eaton died in 1709.

Juan Ramón Koenig.   A142
El Conocimiento de los Tiempos.
Lima.

Daniel Leeds.   A143
AN ALMANACK.
William Bradford, Philadelphia.
Evans 430. Drake 9465. Hildeburn 6.
Charles Evans writes, "Condemned by the judgment of the Friends of the Philadelphia quarterly meeting for its light and airy view and the Printer ordered to surrender all that he hath, and call in all that were sent away."
Charles Hildeburn quotes a Mr. Kite: "In imitation of the Almanacks published in England, Daniel had added some light, foolish and unsavoury paragraphs, which gave great uneasiness and offence to Friends."
Bradford was paid £4 in compensation for his losses. A letter of apology from Leeds is reproduced by Hildeburn and also in *Pennsylvania Magazine*.[98]

Carlos de Sigüenza y Góngora.   A144
El Pronostico y Lunario para el año de 1688.
México.
Quintana 59.

---

98. Anon., "Notes and Queries," *The Pennsylvania Magazine of History and Biography* XXXVII/3 (1913): 379–384.

The records that pertain to this title at the Archivo General de la Nación are *Inquisición* 670, leaves 1 to 3.

Sigüenza got around the ban on pseudonyms for one last year, since his almanac was approved in September 1687, and the ban was declared the next month.

John Tulley.   A145
Other Author: Benjamin Harris?

Tulley 1688. / AN / ALMANACK / For the Year of Our LORD, / M DC LXXXVIII. / Being Bissextile or Leap-year, / And from the Creation / 5637. / *The Vulgar Notes of which are* / *Golden Number* 17 / *Epact* 07 / }{ / *Cicle* [sic] *of the* ☉ 17 / *Domin: Letters* AG / Calculated for and fitted to the Meridian of *Boston* in / *New-England*, where the *North Pole* is elevated 42 / *gr.* 30 m.[99] / By *John Tulley*. / *Imprimatur* Edw. Randolph. Secr: / BOSTON, / Printed by Samuel Green. / 1688.

Samuel Green, Jr., Boston.

Evans 454. Nichols [58]. Drake 2878, 2879. Wing A2583.

Originals located at: American Antiquarian Society, Library Company (leaves [1]–[6], Historical Society of Pennsylvania copy), Library of Congress (Drake 2878). A reissue with 12 leaves is held at the Massachusetts Historical Society (Drake 2879).

Readex microprint and photostat copies are available. The complete text of the Massachusetts Historical Society copy is available online at Evans Digital Edition.

Paging: Eight leaves (twelve in the reissue, the last seven *pages* of which are numbered 16–22).

A handwritten note in the Massachusetts Historical Society copy says, "Bought of Benj. Harris Jan.^{r} 4.th 168$\frac{7}{8}$." A handwritten note for January says, "A Clock going true, will gain this moneth 6.'34.'"" Similar notes for each month indicate what is generally known as the "equation of time" for that month. (Here, "equation" is used not in its mathematical sense, but in the sense of something added to create a balance.) Alfred B. Page says that this handwriting is that of Samuel Sewall.[100] On the last page is written, "No Cambridge Almanack this Year."

---

99. No period in Library of Congress or Massachusetts Historical Society copies. Very faint in Library Company copy.

100. Alfred B. Page, "John Tulley's Almanacs, 1687–1702," *Transactions of the Colonial Society of Massachusetts* (Dec. 1910): 207–223.

Runs January to December. January 1 is labeled "New Year's Day." Saints' days and holidays are still listed.

The extra four leaves are prognostications for the year. This is Tulley's most humorous issue, for all the prognostications seem to be made with tongue in cheek. Page thinks that this whole section was written by the printer, Benjamin Harris. (Note that Tulley's previous almanac was printed by Samuel Green for Harris.)

For February he says, "On the Twenty eight day of this month is like to be a very comfortable smell of Pancakes and Fritters. The Nights are still cold and long, which may cause great Conjunction betwixt the Male and Female Planets of our sublunary Orb, the effects whereof may be seen about nine months after, and portend great charges of Midwife, Nurse, and Naming the Bantling." The 28th of February in this year was Shrove Tuesday (Mardi Gras).

Each month's prognostications are introduced by a bit of verse. For January he offers:

The best defence against the Cold,
Which our Fore-Fathers good did hold,
Was early a full Pot of Ale,
Neither too mild nor yet too stale,
Well drenched for the more behoof
With Toast cut round about the Loaf.

Page refers to Harris as the "people's poet" of his time.[101] For more about Harris, see the entries for him under 1692.

1689

Antonio Sebastián de Aguilar Cantú. **A146**
El Pronostico para el Año que viene de Ochenta y Nuebe.
México.
Quintana 60.
The records that pertain to this title at the Archivo General de la Nación are *Inquisición* 670, leaves 91 to 95.

Juan de Avilés Ramírez. **A147**
Pronostico de Temporales, para el año que Viene de Mill, Seiscientos, y ochenta, y nueve.

101. Ibid.

México.
Quintana 61.
The records that pertain to this title at the Archivo General de la Nación are *Inquisición* 670, leaves 356 to 358.

José de Campos.   A148
El Pronostico para el Año que Viene de Ochenta y nueue.
México.
Quintana 62.
The records that pertain to this title at the Archivo General de la Nación are *Inquisición* 670, leaves 229 to 233.
Quintana reproduces several letters censuring de Campos for continuing to use the name Romano.

Juan Ramón Koenig.   A149
El Conocimiento de los Tiempos.
Lima.

Daniel Leeds.   A150
AN ALMANACK.
William Bradford, Philadelphia.
Evans 446. Drake 9466. Hildeburn 11.
Charles Hildeburn found a note by Jacob Taylor from 1706 saying that Leeds had been plagiarizing other people's work for nineteen years in his almanacs. Hildeburn takes this as supporting the assumption that Leeds put out an almanac every year in the gap between the extant copies in 1687 and 1693.

Carlos de Sigüenza y Góngora.   A151
El Lunario para el Año venidero de 1689.
México.
Quintana 63.
The records that pertain to this title at the Archivo General de la Nación are *Inquisición* 670, leaves 359 to 362.

John Tulley.   A152
Tulley 1689. / AN / ALMANACK / For the Year of Our LORD, / M DC LXXXIX. / Being First after Leap-year, / And from the Creation / 5638. /

The Vulgar Notes of which are / *Golden Number* 18 / *Epact* 18 / }{ / *Cicle* [sic] *of the* ☉ 18 / *Domin: Letters* F / Calculated for and fitted to the Meridian of *Boston* in / *New-England*, where the *North Pole* is elevated 42 / *gr.* 30 *m.* / By *John Tulley.* / *Imprimatur* Edw. Randolph. Secr: / BOSTON, / Printed by Samuel Green, / and are to be Sold at his house over- / against the *South-Meeting-House.* / 1689.

Samuel Green, Jr., Boston.

Evans 499. Nichols [59]. Drake 2880. Wing A2584.

Originals located at: American Antiquarian Society, Library of Congress, Massachusetts Historical Society.

Readex microprint and photostat copies are available. The complete text of the American Antiquarian Society copy is available online at Evans Digital Edition.

Paging: Eight leaves.

Runs January to December. January 1 is labeled "New Year's Day." Saints' days and holidays are still listed.

A tide table plotting the moon's age against the time of full sea is given as good for Boston and Saybrook.

1690

Antonio Sebastián de Aguilar Cantú.   A153

El Pronostico de los Temporales de el Año de mil seiscientos noventa.

México.

Quintana 64.

The records that pertain to this title at the Archivo General de la Nación are *Inquisición* 670, leaves 225 to 228.

Quintana quotes a postscript from the petition protesting the things Sigüenza had been saying against astrologers in his last two almanacs. This is on the bottom of leaf 226r.

Juan de Avilés Ramírez.   A154

El pronostico de Temporales para el año que viene de mil, seiscientos, i noventa.

México.

Quintana 65.

The records that pertain to this title at the Archivo General de la Nación are *Inquisición* 670, leaves 219 to 221.

Juan Ramón Koenig. **A155**
El Conocimiento de los Tiempos.
Lima.

Daniel Leeds. **A156**
AN ALMANACK.
William Bradford, Philadelphia.
Evans 473. Drake 9467. Hildeburn 18.

Henry Newman. **A157**
Non cessant anni, quamvis cessant homines. / HARVARD'S EPHEMERIS,[102] / OR / ALMANACK. / Containing an Account of the Cœlestial / Motions, Aspects &c. For the Year / of the Christian Empire. 1690[103] / Years[104] / And { Of the World — 5639. / Since the Floud [sic] — 3983. / — Building of *London* — 2797 / — Death of *Alexander* Mag. — 2013. / From the beginning of the *Julian* year — 1734 / / Suffering of CHRIST — 1657. / From the Correction of the Calendar — 108 / From the Planting of the Massachus.Colony.62 / Founding of Harvard Colledge — 48 / From Leap Year. — 2. / Whose Vulgar Notes be / The Prime or Cycle of the ☽ 19 / The Cycle of the ☉ — 19 / The Epact — 29 / }{ / Dominical Letter. E / Roman Indiction.13 / Number of Direct. 30 / Respecting the Meridian of Cambridge in N. E. / whose Latitude is 42. Degr. 27. *min.* Septen. / Longitude 315. Degr. / By *H. Newman* / CAMBRIDGE[105] / Printed by *Samuel Green.* / 1690.
Samuel Green, Sr., Cambridge.
Sabin 55009, 62743. Evans 544. Nichols [60]. Drake 2881. Wing A1987.
Originals located at: Massachusetts Historical Society, Yale University.
Readex microprint and photostat copies are available. The complete text of the Massachusetts Historical Society copy is available online at Evans Digital Edition.
Paging: Eight leaves.
Runs January to December.

---

102. Period in Yale copy.
103. Faint period in Yale copy.
104. This word appears as a heading for the column of years ending the lines below.
105. Period in Massachusetts Historical Society copy.

Newman's years are 1670 to 1743 in library catalogs. (John Langdon Sibley leaves his death date open.) He graduated from Harvard in 1687 and was librarian to the College from 1690 to 1693.

Newman's latitude for Boston varies from the usual 42°30', and it is also higher than the value Nathanael Mather used in 1685 and 1686.

In the postscript Newman says that the Earth moves around the Sun and that the tides had been explained by John Wallis.

Carlos de Sigüenza y Góngora. **A158**
Almanaque / Para el Año de 1690. / Compusolo. / D. Carlos de Siguenza y Gongora / Cosmographo y Cathedratico de / Mathematicas del Rey nro S̊ en su Real / Uniuersidad de Mexico. / Sacalo a luz Juan de Torquemada México.

Quintana 66.

The records that pertain to this title at the Archivo General de la Nación are *Inquisición* 670, leaves 200 to 209 and 211 to 214. The text is probably out of order, with the calendar on 202 to 207 and the introduction and *juycio* on 208 to 209.

Quintana reproduces the Introduction. It opens with, "Veynte son con este los Lunarios, Pronosticos, o Almanaques que con el nombre supuesto del Mexicano, ò el propio mio, o el de Ju° de Torquemada he impreso en otros tantos años aun mas por el util de la Republica que por el propio mío." This confirms the claim that Sigüenza has done an almanac every year since 1671, in spite of the gaps in the Inquisition record. Note that he says that in addition to the pseudonym, Juan de Torquemada, he has also called himself The Mexican. Could this be what happened in 1681, 1685, and/or 1687?

For this year on the Gregorian calendar, the epact, golden number, and solar cycle are all 19.

Sigüenza lists Easter for March 26. From Newman's almanac for this year, we can see that, since the number of direction is 30, those following the Julian calendar this year put Easter on April 20. That is April 30 on the Gregorian calendar, so in 1690 the readers of Newman's almanac celebrate Easter a full five weeks later than their counterparts in Latin America.

John Tulley. **A159**
Tulley, 1690. / AN / ALMANACK / For the Year of our LORD / MDCXC. / Being second after Leap-year; And / from the Creation, / 5639. / The Vulgar Notes of which are, / *Golden Numb.* 19 / *Epact* 29 / }{ / *Cicle* [sic] *of the Sun*

19 / *Domin. Letter* E.[106] / Calculated for, and Fitted to the Meridian of / *Boston* in *New-England*, where the *North* / *Pole* is Elevated 42 *gr.* 30[107] *min.* / By *John Tulley.* / *Boston* Printed and Sold by *Samuel* / *Green*, near the *South Church.* / 1690.

Samuel Green, Jr., Boston.

Evans 548. Nichols [61]. Drake 2882. Wing 2585.

Originals located at: American Antiquarian Society, Boston Public Library, Kislak, Library of Congress, Massachusetts Historical Society (two copies; copy 1 last leaf is $\frac{1}{2}$, copy 2 is [1] and [8]).

Readex microprint and photostat copies are available. The complete text of the American Antiquarian Society copy is available online at Evans Digital Edition. The title page is reprinted in the Christie's catalog for the Snider sale of 2005.

Paging: Eight leaves.

Runs January to December.

Saints' days, Christmas, and Easter, which appeared in earlier Tulley almanacs, have disappeared.

There are essays titled "Of the Rainbow" and "Of Thunder and Lightning." The tide table for Boston and Saybrook appears again.

Benjamin Harris advertises, "That Excellent *Antidote* against all manner of Gripings, called *Aqua antitorminalis*, which if timely taken, it not only cures the Griping of the Guts, and the Wind Chollick; but preventeth that woful Distemper of the *Dry Belly Ach*; With Printed Directions for the use of it. Sold by *Benjamin Harris* . . ." (Missing from Boston Public Library copy.)

Copy 1 at the Massachusetts Historical Society has handwritten weather notes (by Samuel Sewall?), e.g., "rain good deal." There are two notes about "small pocks."

1691

Antonio Sebastián de Aguilar Cantú.   **A160**

El Lunario y Pronostico de los Temporales de el Año de Mil Seiscientos y Nouēnta y vno.

---

106. Period very faint in Boston Public Library copy, and there is possibly a faint colon in the Kislak copy.

107. The 3 is inverted in the American Antiquarian Society copy and Massachusetts Historical Society copy 2.

México.
Quintana 67.
The records that pertain to this title at the Archivo General de la Nación are *Inquisición* 670, leaves 172 to 173 and 176 to 178.

Juan de Avilés Ramírez. **A161**
El Pronostico de Temporales para el año que Viene de Mil, Seiscientos, y Nouenta, y vn años.
México.
Quintana 68.
The records that pertain to this title at the Archivo General de la Nación are *Inquisición* 670, leaves 174 to 176.

Juan Ramón Koenig. **A162**
El Conocimiento de los Tiempos.
Lima.

Daniel Leeds. **A163**
AN ALMANACK.
William Bradford, Philadelphia.
Evans 518. Drake 9468. Hildeburn 24.

Henry Newman. **A164**
Ut Fluctus fluctum, sic annus annum trudit. / NEWS from the STARS. / AN / ALMANACK / Containing an Account of the *Cœlestial Mo-* / *tions, Aspects*, &c. For the Year of / the Christian Empire, 1691. / Years[108] / And { Of the *World*, 5640. / Since the *Floud*, [sic] 3984. / *Suffering of CHRIST*. 1658. / Planting *Massach. Colony*, 63. / *Founding of* Harvard *Colledge*, 49. / From *Leap-Year*, 3. / Whose *Vulgar Notes* be, / *Cycle* of the ☽ 1. / *Epact*. 11. / *Cycle* of the ☉ 20. / }{ / *Dominic. Let*. D. / *Roman Indict*. 14. / *Number of Direct* 22 / Respecting the Meridian of *Boston*, in *New-* / *England*, whose Latitude is, 42[109] *d*. 30. *min*. / Longitude, 315. Deg. / By *Henry Newman, Philomath*. / Printed by *R. Pierce* for *Benjamin Harris* at / the *London Coffee-House* in *Boston*, 1691.
Richard Pierce, Boston.

108. This word appears as a heading for the column of years ending the lines below.
109. Overwritten in the Boston Public Library and Massachusetts Historical Society copies.

Sabin 55010, 62743. Evans 574. Nichols [62]. Drake 2883. Wing A1988.
Originals located at: Boston Public Library (leaves 1–4, 11–12), Library of Congress, Massachusetts Historical Society.
Readex microprint and photostat copies are available. The complete text of the Massachusetts Historical Society copy is available online at Evans Digital Edition.
Paging: Thirteen leaves.
Runs January to December.
There is an essay titled "Of Telescopes." Throughout the book are pieces about discoveries made about the appearances of the planets. The rotation of Mars is just one example.
The epact has jumped by 12 over the previous year, something that happens every nineteen years. See the discussion of this for Jeremiah Shepard's almanac of 1672.

Carlos de Sigüenza y Góngora. **A165**
El Almanaque para el Año de 1691.
México.
Quintana 69, 70.
The records that pertain to this title at the Archivo General de la Nación are *Inquisición* 670, leaves 349 to 355. The permission form is signed "Juan de Torquemada."
It is not clear why José Miguel Quintana lists two different titles for Sigüenza in the same year. It would be more understandable to do that for 1695, when the records show two separate permission slips. Elías Trabulse says thirty-one almanacs were done by Sigüenza from 1671 to 1701, so that is one per year.

John Tulley. **A166**
Tulley. 1691. / AN / ALMANACK / For the Year of our *LORD,* / *MDCXCI.* / Being Third after Leap-Year ; and / From the CREATION / 5640. / *The Vulgar Notes of which are* / Golden Numb. 1 / The Epact 11 / } { / Cicle [sic] of the Sun 20 / Domin. Letter D / *Calculated for, and fitted to the Meridian of* / Boston in New England[110] *where the North* / *Pole is Elevated* 42 gr. 30. *min.* / By *John Tulley.* / *CAMBRIDGE.* / Printed by *Samuel Green,*

---

110. *New-England* in Harvard and Library of Congress copies and Massachusetts Historical Society copy 2.

and *B. Green.* / And are to be Sold, by *Nicholas Buttolph,* / at *Gutteridg's Coffee-House* in Boston. 1691.[111]

Samuel Green, Sr. and Bartholomew Green, Cambridge.

Evans 578. Nichols [63]. Drake 2884. Wing A2586.

Originals located at: American Antiquarian Society, Boston Public Library, Harvard University, Huntington Library, Library of Congress, Massachusetts Historical Society (two copies).

The Massachusetts Historical Society copy 1 is a different printing. The title page is as follows: *Tulley,* 1691. / AN / ALMANACK / For the YEAR of our *LORD,* / *MDCXCI.* / Being Third after Leap-Year ; and / From the CREATION / 5640. / *The Vulgar Notes of which are* / Golden Numb. 1 / The Epact 11 / }{ / Cicle [sic] of the Sun 20 / Domin. Letter D / *Calculated for, and fitted to the Meridian* / *of BOSTON in New-England where the* / *North Pole is Elevated* 42 gr. 30 min / By *John Tulley.* / *CAMBRIDGE:* / Printed by *Samuel Green,* & *Bartholomew Green.* / And are to be Sold by *Nicholas Buttolph,* / at *Gutteridg's Coffee-House* in Boston. 1691.

Readex microprint and photostat copies are available. The complete text of the American Antiquarian Society copy is available online at Evans Digital Edition. Early English Books Online shows the Huntington copy.

Paging: Eight leaves.

Runs January to December.

Samuel Green, Jr., died in July of 1690 of smallpox. The Tulley almanacs for 1691 and 1692 were printed by his father together with his younger brother Bartholomew. Bartholomew Green (1667–1732) had initially taken over his brother's press in Boston in 1690, but it was destroyed by fire, sending him back to his father's shop in Cambridge for the next two years.[112]

The years for the bookseller, Nicholas Buttolph, are 1668 to 1737.

"A Table of Expence" on [8v] takes daily amounts of 1 penny through 20 shillings and multiplies by 1 week, 1 month, and 1 year.

The Boston Public Library copy has handwritten notes about the weather, similar to those attributed to Samuel Sewall in other almanacs. The Huntington copy has many manuscript pages in some kind of code or shorthand. The Massachusetts Historical Society copy 1 has [8] bound in between [1] and [2] and has interleaved manuscript pages.

---

111. Final period missing in the Huntington and Library of Congress copies and Massachusetts Historical Society copy 2.

112. Isaiah Thomas, *The History of Printing in America,* New York, 1970.

1692

Charles Evans has an entry numbered 615 for this year, The Map of Man's Misery . . . being a perpetual almanack of spiritual meditations . . . by Patrick Ker, *previously published in London in 1690. However, as the title would suggest, it is not in the same genre as the astronomical almanacs listed here.*

Antonio Sebastián de Aguilar Cantú.   A167

PRONOSTICO / De los Temporales de el Año Bisiesto / De 1692. que / Ala Exma Señora Doña Elvira / De Toledo: Virreina de la Nu= / eva España. Le Consagra / El Bachiller Antonio Sebastian, de Agui / lar, Cantû: Medico / Aprovado. / Firmamentum dei, et stella nostra. / Maria diuina, en Guadalupe bella. / El Cielo biene,[113] de su hermoso dia: / i Yo, en el Firmamento de Maria, / todo el influxo de mi buena Estrella.

México.

Quintana 72.

The records that pertain to this title at the Archivo General de la Nación are *Inquisición* 670, leaves 313 to 333. The text is leaves 313 to 332, and there are corrections on leaf 319.

Both José Miguel Quintana and Carmen Corona give the complete text. (See our entry for Sigüenza for this year.)

Juan de Avilés Ramírez.   A168

Pronostico de Temporales / Con las elecciones de Medicina, Phlebotomia / Agricultura, Nauegacion, Segun loque indi= / can los mouimientos de los Astros, Este año / de 1692. / Bissiesto [sic] / Regulado al meridiano desta Ciu.^{d} de Me= / xico, por el Dr Juan de Auilez Ramires, Ca= / thedratico que fue de Prima de Medicina / en sostitucion, y lo es dela Cathedra de Me= / thodo, en esta Real Vniuersidad, y Medico / de la Real Carcel deste Corte. / Al Cap.^{n} Dr Domingo de Bettes, Larga= / che.

México.

Quintana 73.

The records that pertain to this title at the Archivo General de la Nación are *Inquisición* 670, leaves 294 to 295 and 299 to 311. The text is leaves 299 to 311.

Quintana gives the complete text.

---

113. Carmen Corona has "tiene".

Benjamin Harris? **A169**

BOSTON / ALMANACK / FOR THE / Year of our LORD GOD, 1692. / Being Bissextile or Leap-year. / Years[114] / And { Of the *World*, 5641 / Since the *Flood*,[115] 3985 / Suffering of *CHRIST* 1659 / Planting *Massach. Colony* 64 / Whose *Vulgar* Notes be, / Golden Numb. 2 / Epact 22 / }{ / Cicle [sic] of the Sun 21 / Dominic. Letter C.B / Calculated for the Meridian of *Boston* in / *New-England*, where the North Pole / is Elevated 42 *gr.* 30 *min.* / BY H B / Boston, *Printed by* Benjamin Harris, *and* / John Allen: *And are to be Sold at the* / London-Coffee-House. 1692.

Benjamin Harris and John Allen, Boston.

Sabin 6481, 62743. Evans 595. Nichols [64]. Drake 2885, 2886. Wing A1813, A1813A.

Originals located at: Library of Congress, Massachusetts Historical Society (Drake 2885; Wing A1813). A second impression is represented by copies at: New York Public Library, Owen (Drake 2886; Wing A1813A).

Most of the pages are printed in red and black in the first impression. (The Readex catalog card calls this the first color printing in America; they clearly mean Anglo-America since there were red and black title pages printed in Mexico in the sixteenth century.) The red and black letters do not line up well with each other. This was fixed in the second impression, which is printed all in black. The second impression has:

BOSTON / ALMANACK / FOR THE / Year of our LORD GOD. 1692. / Being Bissextile or Leap-year / Years[116] / And { Of the *World*, 5641 / Since the Flood. 3985 / Suffering of CHRIST. 1659 / Planting Massach. Colony 64 / Whose *Vulgar* Notes be, / Golden Numb. 2 / Epact 22 / }{ / Cicle [sic] of the Sun 21 / Dominic. Letter C.B / Calculated for the Meridian of *Boston* in / *New-England*, where the North Pole / is Elevated 42 *gr.* 30 *min.* / The Second Impression. / BY H B / *Boston*, Printed by *Benjamin Harris* , and / *John Allen* : And are to be Sold at the / *London-Coffee-House.* 1692.

Readex microprint and photostat copies of both impressions are available. The complete text of the Massachusetts Historical Society copy is available online at Evans Digital Edition. The title page of the second impression was reprinted in the Christie's catalog for the Snider sale of 2005.

114. This word appears as a heading for the column of years ending the lines that follow.
115. Possibly, *"Flovd,"*.
116. This word appears as a heading for the column of years ending the lines below.

Paging: Ten leaves.

Runs January to December.

This entry was attributed to Harris by Charles Nichols.[117] A poem on 1v signed "HB" in the first impression has these letters reversed to "BH" in the second. A poem on 8r that is unsigned in the first is likewise signed "BH" in the second.

According to Amory,[118] Harris had been known in England for his anti-Catholic writings and his antipathy to the heir to the throne. Robb Sagendorph adds to this that Harris was pilloried and fined £500 in 1681.[119] Harris probably came to Boston about 1685, because of the succession of James II. There he opened a coffee shop and bookstore.

The text contains some sample contracts: will, letter of attorney, indenture for an apprentice, etc. The advertisement for aqua antitorminalis was repeated from the John Tulley almanac of 1690.

Harris's years are 1673 to 1720.

Benjamin Harris? **A170**

Other Author: John Partridge.

Monthly Observations / AND / Predictions, / For this Present Year, 1692. / WITH / Astrological Judgments / On the whole Year. / All Taken from Mr. P*atridge's* [sic] / ALMANACK: / To which is Added, an Account of a / Plot / Which was lately Discovered in *England* : And / which was Foretold by the said *John* / *Patridge,* [sic] in his this Years Almanack. / Published for General Satisfaction. / Printed at *Boston* : And are to be Sold by / *Benj. Harris,* at the *London-Coffee-House.* / 1692.

Benjamin Harris, Boston.

Sabin 62743, 58966 (with year 1697). Evans 627. Drake 2887. Wing P621. Not in Nichols.

Original located at: Massachusetts Historical Society.

Readex microprint copies are available. The complete text of the Massachusetts Historical Society original is available online at Evans Digital Edition.

Paging: Sixteen pages, numbered except for [1].

---

117. Charles Nichols, "Notes on the Almanacs of Massachusetts," *Proceedings of the American Antiquarian Society* new series 22 (1912): 14.

118. Hugh Amory, "Printing and Bookselling in New England, 1638–1713," in *A History of the Book in America,* ed. Hugh Amory and David D. Hall, Cambridge, 2000.

119. Robb Sagendorph, *America and her Almanacs,* Dublin, NH, 1970: 42.

Runs January to December. (No calendar.)

This almanac has no tables, but it does discuss conjunctions and other astronomical events within the text. The "Account of a Plot" of the title tells of the rounding up of some persons alleged to be spying for the French. This entry is marginally a member of this list, but it seems important to draw attention to it to clear up the matter of its authorship.

It has been attributed to John Partridge by Joseph Sabin and Charles Evans, and this is repeated by Milton Drake and Donald Wing, but it reads as if Harris wrote it based on predictions in Partridge's almanac printed in London. The misspelling of Partridge as "Patridge" in the title appears later when other writers are referring to Partridge.[120]

There is some mention of the French king in the "Astrological Observations" and this seems to be an obsession of Harris's (see John Tulley's entries for 1694 and 1696) although it may also have been an obsession for Partridge. J. Churlton Collins calls Partridge a "Protestant alarmist" who fled to the Netherlands when James II came to the throne,[121] just as Harris fled to Boston for the same reason.[122]

Partridge's years are 1644 to 1715. He was born with the name Hewson. It appears that he had a run-in with Harris at some point, because his 1707 almanac says, "If there is anything added to this *Almanac* by B. Harris, either in the middle or end of it, besides these Three Sheets; it is a piece of knavery, and not mine."

The London printing of Partridge's almanac for this year had the following title page:

Merlinus Liberatus: / Being an / ALMANACK / For the Year of our Redemption, 1692. / And from the Creation of the World, according / to the best of History, 5641. / It being the *Bissextile,* or Leap year: / And the Fourth also of our Deliverance from / Popery and Arbitrary Government. / In which is contained things fit for such a Work: / As the Diurnal Motion of the Planets, Remarkable / Conjunctions, Lunations, Eclipses, Meteorological Obser- / vations. / A Table of *Sun* Rising and Setting to

---

120. Edward Arber writes, not of Harris but of later writers, that to lay grounds for a defense in case Partridge accused them of libel, they would undertake "the intentional misspelling of his name, as Partrige, or Patridge . . ." Edward Arber, ed., *An English Garner, Ingatherings from our History and Literature,* Vol. VI. Birmingham, 1883.

121. J. Churlton Collins, "Introduction," in *An English Garner, Critical Essays and Literary Fragments,* ed. Edward Arber, compiled by Thomas Seccombe, Westminster, 1903.

122. See the previous entry.

every sixth / day in the year. And a Table of Houses according to the / Doctrine of the Great PTOLOMY / to which is added, / *J.G.*'s Verses about the PRINCE OF WALES, / in his Almanack 1689. travesty'd. / Calculated for the Meridian of *London*. / Whose { Latitude / Longitude } is { 51 / 24 } deg. { 32 / 20 } Minutes. / By *JOHN PARTRIDGE*. / Student in Physick and Astrology. / *Diversos diversa juvant, non omnia Cunctos.* / *LONDON*: Printed by R.R. for the Company / of STATIONERS.

It was printed in red and black and had twenty-four leaves. Our entry does not appear to be a reprint of this so much as a sequel.

In 1708 Partridge was the victim of a practical joke. Jonathan Swift, using the name Isaac Bickerstaff, a name he saw on a locksmith's sign, wrote in his *Predictions for the Year 1708*: "My first Prediction is but a Trifle; yet I will mention it, to shew how ignorant those sottish Pretenders to Astrology are in their own Concerns: It relates to *Partrige* the Almanack-Maker; I have consulted the Star of his Nativity by my own Rules; and find he will infallibly die upon the 29th of *March* next, about eleven at Night, of a raging Fever: Therefore I advise him to consider of it, and settle his Affairs in Time." Then when the 29th of March came and passed, Swift wrote and had published a letter allegedly from a man who had visited Partridge just hours before his death. This letter stated that Partridge had died about 7 p.m., four hours earlier than Bickerstaff had predicted. Swift even wrote an elegy to Partridge. It ends with an epitaph in sonnet form, the first four lines of which are:

> HERE five Foot deep lyes on his Back
> A Cobbler, Starmonger, and Quack,
> Who to the Stars in pure Good-will,
> Does to his best look upward still.

The story, as usually told, now has Partridge beset by people who believed the report of his death. We find this at the Web site www.museumofhoaxes.com and its associated book,[123] for example.

In fact, another hoaxer got into the act. Someone, writing as John Partridge,[124] alleged in *Squire Bickerstaff Detected; or, The Astrological Imposter*

---

123. Alex Boese, *The Museum of Hoaxes*, New York, 2002.

124. Attributed to either Thomas Yalden, Nicholas Rowe, or William Congreve by the catalog of the University of Glasgow. Rowe is favored by the editor of the 1735 collected works of Swift. Rowe and Yalden are mentioned in the editor's notes to *An English Garner, Critical Essays and Literary Fragments*, and Collins mentions all three in his introduction.

Convicted that first the undertaker showed up to make measurements, then the sexton came to ask about arrangements for the funeral, and on and on. The account slid into absurdity as "Partridge" lamented that people stopped him in the street to ask why he had not paid his funeral expenses or to remark that the next time he died, he might have to toll his own bell. The spoof ended with an advertisement for a future pamphlet in which Partridge will prove "that France and Rome are at the bottom of this . . ."

In 1709, Swift returned with a pamphlet where Bickerstaff noted that he was being attacked by Partridge in his new almanac. In spite of this, claimed Bickerstaff, he could prove logically that in fact Partridge was dead. Both Partridge and Bickerstaff continued to issue their predictions for several more years, although far from everything with the Bickerstaff name was penned by Swift, and the Partridge almanacs disappeared for three years but returned for 1714 and 1715. In his 1714 almanac, Partridge warns, ". . . if there is anything in print in my name beside this *Almanack*, you may depend on it that it is a lie . . ."

Benjamin Franklin was so inspired by Swift's lampoon that he pulled the same trick in the first issue of his almanac. His *Poor Richard, 1733. An Almanack* (by Richard Saunders) predicted the death of Titan Leeds, son of Daniel Leeds, who had carried on the Leeds almanac tradition. On [2]v "Saunders" writes, "He dies, by my Calculation made at his Request, on Oct. 17. 1733 . . . By his own Calculation he will survive till the 26th of the same Month."[125]

Juan Ramón Koenig. **A171**
    El Conocimiento de los Tiempos.
    Lima.

Daniel Leeds. **A172**
    AN ALMANACK.
    William Bradford, Philadelphia.
    Evans 551. Drake 9469. Hildeburn 26.

Carlos de Sigüenza y Góngora. **A173**
    Almanaque / de D. C. d. S. y. G. / Para el año de 1692. Bisiesto. / Por diuersas suposiciones, y calculos, que los / errados, y defectuosissimos de

---

125. See also Robb Sagendorph, *America and Her Almanacs*, Dublin, NH, 1970: 75, and Marion Barber Stowell, *Early American Almanacs*, New York, 1977: 158–159.

Andres Argoli. / por quien todos hasta ahora se han / governado. / Sacalo á luz Juan de Torquemada.
México.
Quintana 71.
The records that pertain to this title at the Archivo General de la Nación are *Inquisición* 670, leaves 334 to 346. The text is 335 to 345.

José Miguel Quintana and Carmen Corona both reproduce the complete text.

The two almanacs for this year by Antonio Sebastián de Aguilar Cantú and Sigüenza were the subject of the book *Lunarios* by Corona. She reproduces the complete texts of both works in the appendices and gives a history of the early lunarios in Mexico, much of which overlaps with the content of Quintana's *La Astrología*. The emphasis in her book is on a war of words between Sigüenza and Aguilar.

John Tulley. **A174**

Tulley, 1692.[126] / AN / ALMANACK / For the YEAR of our LORD, / MDCXCII. / Being Bissextile or Leap-Year, / And from the CREATION, / 5641. / Amplified with Astronomical Observations / from the Suns Ingresse into Aries, and / the other Cardinal Points, and from / the Planets and their Aspects ; With / an Account of the Eclipses, Conjunctions, / and other Configurations of the / Cœlestial Bodies. / *Calculated for and fitted to the Meridian of* / BOSTON *in* New England,[127] *where the North* / *Pole is Elevated* 42. gr 30. min[128] *But may* / *indifferently serve any part of New-England.*[129] / By John Tulley. / CAMBRIDGE:[130] / Printed by *Samuel Green, & Bartholomew Green,* / for *Samuel Phillips,* and are to be Sold / at his Shop at the West end of the / Exchange in Boston. 1692.

Samuel Green, Sr. and Bartholomew Green, Cambridge.

Evans 630. Nichols [65]. Drake 2888, 2889. Wing A2587, A2587A.

Originals located at: American Antiquarian Society, Boston Public Library (leaves [1]–[11]; [9]–[11] are between [1] and [2] and have pieces

---

126. *1692.* in Massachusetts Historical Society copy.
127. *New-England* in Library Company and Massachusetts Historical Society copies. No comma in American Antiquarian Society, Harvard, or Oxford copies.
128. Period in Library of Congress, Library Company, Massachusetts Historical Society, and Oxford copies. Very faint period at Harvard and Morgan.
129. No period in Oxford copy. Hyphen very faint at Massachusetts Historical Society.
130. Period in Massachusetts Historical Society copy.

missing, [1] is ½), Harvard University, Huntington Library (missing leaf 11), Library of Congress, Massachusetts Historical Society, Morgan Library, Steinbock (leaves [2]–[8], [2] bound in backwards), University of Oxford (Drake 2888; Wing A2587). An alternative title page, with the last three lines replaced by "for *JOHN USHER* and are to be sold / at his Shop in Boston. 1692.," is represented by a copy at the Library Company (last leaf ¾, Historical Society of Pennsylvania copy) (Drake 2889; Wing A2587A).

Readex microprint and photostat copies are available. The complete text of the American Antiquarian Society copy is available online at Evans Digital Edition. Early English Books Online shows the Harvard copy. The Oxford copy and the British Library fragments of Leeds 1687 are the only seventeenth-century American almanacs we know of that are held outside the United States. The Morgan copy has diary pages of Rev. James Pierpont.

Paging: Twelve leaves. Pages [9]v–[12]r are numbered 2–7.

Runs January to December.

There is a separate table of rising and setting times of the sun for six days out of each month on [2]r, with an explanation of its use on [1]v. On [2]v is an explanation of the columns of the calendar pages. High tides are right in the calendar instead of a separate table.

1693

Milton Drake lists an anonymous almanac, for this year, as his number 2890. Because neither of the libraries that were supposed to have it (the Huntington and the Historical Society of Pennsylvania) can find it now, we have decided to remove it from the list.

Antonio Sebastián de Aguilar Cantú. **A175**
El Pronostico de los Temporales de el Año que viene de Nouenta y tres.
México.
Quintana 74.
The records that pertain to this title at the Archivo General de la Nación are *Inquisición* 670, leaves 296 to 298.

Juan de Avilés Ramírez. **A176**
El Pronostico de Temporales para el año que Viene de Nouenta i tres.
México.
Quintana 75.

The record that pertains to this title at the Archivo General de la Nación is *Inquisición* 670, leaf 161.

Juan Ramón Koenig.   A177
El Conocimiento de los Tiempos.
Lima.

Daniel Leeds.   A178
AN / ALMANACK / AND / *EPHEMERIDES* / For the Year of Christian Account 1693 / Whereunto are Numbered, / FROM THE[131] / Creation of the World, 5660 / Flood of *Noah*, 3986 / Building of *London*, 1800 / Building of *Rome*, 2084 / Death of *Alexander* the Great, 2016 / Constitution of the *Julian* Year, 1737 / Death and Passion of Christ, 1660 / *Hegyra*, or flight of *Mahomet*, 1102 / Correction of the Kalendar by P.*Gregory*, 111 / *Bessextile*, [sic] or Leap-year, 1 / Containing matters necessary & useful. / Being fitted to the Meridian of that part of *New-* / *Jersey* and *Pennsilvania*, where the Vertex is distant from / the Equator 40 Degrees ; but may, without sensible error / serve all parts adjacent, even from *Newfound-Land*, to the / Capes of *Virginia*. / By Daniel Leeds, *Philomath*. / Printed and Sold by *William Bradford*, 1693.

William Bradford, Philadelphia.

Evans 646. Drake 9470. Hildeburn 65. Wing A1865.

Originals located at: Huntington Library, Johns Hopkins University (leaves [1]–[20], first three partial), Library Company (Historical Society of Pennsylvania copy).

Microfilm and Readex microprint copies are available. The complete text of the Huntington copy is available at both Evans Digital Edition and Early English Books Online. Two pages for May are reprinted in *The Quaker*, May 14, 1920.

Paging: Twenty-six leaves.

Runs January to December.

This was the last Leeds almanac printed in Pennsylvania, as Bradford was to move his press to New York.

At twenty-six leaves, this is the second longest almanac on our list for which we have paging information (after John Clapp's 1697 almanac), and it is the longest with a complete copy.

---

131. These two words appear vertically and span the next ten lines.

Leeds is still using January, XI month, etc. He still lists Easter (April 16) which Tulley stopped doing after 1689.

There is a discussion of when to begin the year on [16]v to [17]r. Following the calendar is an essay entitled "Scripture-Names of the Twelve Months, with their Significations" on [14]v to [15]v. In the calendar section, in addition to the usual names, each month is marked with a "scripture" name in Hebrew characters. This evolved into the month names that Leeds used for his later New York almanacs.

On [14]v to [15]v there is a discussion of the months of the year with the Hebrew names in Roman characters. The months from March to February are identified, respectively, with Nisan, Ziv, Sivan, Tammuz, Ab, Ælul, Ethanim, Bul, Kisleu, Tebeth, [XI month not named], Adar. These names are slightly different from the ones we quote in the entry discussing Leeds's 1694 almanac.

Leeds does not mark the beginnings of Hebrew months in his calendar, although he does indicate the new moon. He really seems to be identifying Hebrew months with the Roman months that they overlap.

Leeds gives the earth-to-moon distance as 195,800 miles and the earth-to-sun distance as 5,021,890 miles and says it is "most rational to believe" that the earth moves around the sun on [17]v.

Carlos de Sigüenza y Góngora. **A179**

Almanaque / de D. C. de S. / para el año de / 1693. / Segun las nueuas ephemerides de / Flaminio de Mezzavachis. / Sacalo à luz Ju° de Torquemada

México.

Quintana 76.

The records that pertain to this title at the Archivo General de la Nación are *Inquisición* 670, leaves 281 to 293. The calendar is 282 to 287, and the rest of the text is 288 to 291. The application is 281, and other notes are on 292 to 293.

Quintana gives the complete text.

John Tulley. **A180**

Tulley 1693 / AN / ALMANACK / For the Year of our LORD, / MDCXCIII. / Being first after Leap-Year. / And from the Creation. / 5642. / Wherein is Contained, Astronomical Obser- / vations from the Suns Ingress into Aries, and / the other Cardinal Points, with an Account / of the Eclipses, Conjunctions, and other / Configurations of the Celestial Bodies. / With a brief Discourse of the natural causes of / Watry Meteors, as Snow, Hail, Rain, &c. / Calculated for and fitted to the Meridian of *BOSTON* in / *New-England*, where the North Pole is Elevated 42. / gr. 30 min. But may indifferently serve any part / of *New-England*. / By John Tulley / *Boston* Printed, by Benjamin Harris at the / London-Coffee-House. 1693.

Benjamin Harris, Boston.

Evans 682, 683. Nichols [66]. Drake 2891, 2892, 2893. Wing A2588, A2589.

Originals located at: American Antiquarian Society (two copies), Huntington Library, Library Company, Massachusetts Historical Society, Private Collection (Ann Arbor) (missing extra leaf and last leaf), Washington's Headquarters (Evans 682; Drake 2891; Wing A2588). An impression with the last two lines on the title page above replaced by "*Boston* Printed , by Benjamin Harris for / Samuel Phillips. 1693." is at: American Antiquarian Society, Boston Public Library, Harvard University, Library Company (Historical Society of Pennsylvania copy), Library of Congress, Massachusetts Historical Society (Drake 2892; Wing A2589). An impression with "for" in the latter replaced by "sold by" is reported (Evans 683; Drake 2893).

Readex microprint and photostat copies are available. The complete text of Copy 2 at the American Antiquarian Society (AAS) of Evans 682 is available online at Evans Digital Edition. The title page of the Massachusetts Historical Society (MHS) copy of Drake 2892 (misidentified as Evans 683) is also available. It is accompanied by three pages numbered 28 to 30, which are from some other text entirely. This mistake appeared originally on the Readex microprint card and has been copied from there. Early English Books Online appears to show two different Huntington copies of Evans 682, one of which is mistakenly identified as Evans 683. In fact, they show the same Huntington copy twice, but a different level of contrast makes the discolorations of the original appear different in the two images.

Paging: Twelve or thirteen leaves (see following).

Runs January to December.

The AAS catalog states, in reference to their two copies of Evans 682, "Multiple states noted. In some copies, a leaf is inserted between signatures B and C with 'A table of courts' and a prospectus. The state with the inserted leaf has a bookseller's advertisement on p. [24] by Benjamin

Harris. In the other, the same advertisement is by Samuel Phillips." Further down they say "American Antiquarian Society Copy 1 does not include the inserted leaf . . . . Copy 2 includes the inserted leaf. Pages [15–24] supplied from a second copy." For their copy of Drake 2892 (misidentified as Evans 683; let's call it "Copy 3") we see "American Antiquarian Society copy lacks p. [15–16] and [19–26]. Pages [9–14] and [17–18] supplied from a second copy."

It is plausible that there were originally only two variants. (Early English Books Online promotes this on their citation page, referring to "Variant A" and "Variant B.") One, printed by Harris for sale in his London Coffee House, has an advertisement for Cotton Mather's *The Wonders of the Invisible World*, with the note "Sold by Benjamin Harris," at the very end. He inserted into this version the extra leaf printed on one side only with the court dates and the advertisement for another book "Sold by Benjamin Harris." The second variant was to be sold by Samuel Phillips, and so it may or may not have had the extra leaf. It had the same advertisement for Mather's book included at the end, but with "Sold by Samuel Phillips." For this to be true, Copy 1 at the AAS must have been mixed so long ago (with the beginning from Variant 1 and the end from Variant 2) that the aging of the paper is consistent throughout.

Copy 1 at the AAS appears to be a mix of Variants 1 and 2. It has twelve leaves with [8]v blank. Copy 2 at the AAS actually appears to have eight leaves (not seven) from one copy and the last five from another, both of Variant 1. Copy 3 at the AAS is probably the first four leaves from some copy of Variant 2 together with leaves [5] through [7] and [9] (the extra leaf) from a copy of Variant 1. The Boston Public Library copy has eight leaves of Variant 2 followed by the last four leaves of Variant 1. The Harvard copy is Variant 2 throughout, without the extra leaf. The Huntington copy is Variant 1 throughout. The Library Company copy 1 is Variant 1 missing the last leaf with [9] (the extra leaf) bound in last. Their copy 2 is Variant 2 and has the extra leaf after [7]. The Library Company copy likewise is Variant 2 throughout but has the extra leaf after [7]. The MHS copy 1 is Variant 1 with the extra leaf last, interleaved with multiple blank leaves, and missing [10] and [11]. The MHS copy 2 is Variant 2, has the extra leaf bound in backwards, and has extra leaves with notes by Samuel Sewall. The Ann Arbor copy appears to be Variant 1 but is missing the extra leaf. The Washington's Headquarters copy is Variant 1 throughout with no extra leaf.

There is a diagram illustrating the moon's dominion over the body, assigning the zodiacal signs to certain body parts. This illustration is repeated in later Tulley almanacs. Robb Sagendorph writes that Tulley was the first to import this image into America from the almanacs of Europe,[132] but, in fact, it first appeared in John Foster's almanac of 1678. Charles Nichols and Marion Barber Stowell both note that Foster was first.

Sagendorph also says that this issue contains the first weather forecast in an American almanac,[133] but this is not true. Notations such as "cold and cloudy" or "rain or snow" appear for December of 1687 in Tulley's first issue.[134]

1694

Juan Ramón Koenig.   A181
El Conocimiento de los Tiempos.
Lima.

Daniel Leeds.   A182
AN / ALMANACK / For the Year of Christian Account / 1694. / And from the Creation of the World / 5661. / But by *Keplas* computation 5687. / Being the second after Leap-Year, / The Epact is 14. Golden Number 4. / and Dominical Letter G. / Containing Matters Necessary and / Useful, chiefly accomodated to the / Lat. of 40 Degrees, but may without / sensible Error serve the Places adjacent, / from *Newfound-Land* to the Capes of / Virginia. / By *Daniel Leeds*, Philomat. / A Motto, taught by the Sons of Urania. / *If to be born under* Mercury *disposeth us to be Witty,* / *and under* Jupiter *to be Wealthy, we de not owe* / *Thanks unto them, but unto that Merciful Hand* / *that ordered our indifferent & uncertain Nativities* / *unto such benevolent Aspects.* / Printed and Sold by *William Bradford* at / the Bible in *New-York*, 1694.
William Bradford, New York.
Evans 692. Drake 5526. Wing A1866.

---

132. Robb Sagendorph, *America and Her Almanacs*, Dublin, NH, 1970: 46.
133. Ibid.: 44.
134. Marion Barber Stowell also noted the earlier weather predictions in note 31 to Chapter 3 of *Early American Almanacs*, New York, 1977.

Originals located at: Huntington Library (missing last leaf).

Microfilm and Readex Microprint copies are available. A very clear title page is reproduced in Charles Hildeburn's *Sketches of Printing and Printers in Colonial New York*. The Huntington copy is available online at Evans Digital Edition.

Paging: Twelve leaves.

Runs January to December but the months are numbered XI, XII, then I through X. Months are named in the following way: *Sebat*, Cold *January*; *Adar*, Dirty *February*; *Abib*, Blustering *March*; *Zif*, Dripping *April*; *Sivan*, Pleasant *May*; *Tamuz*, Hot *June*; *Ab*, Parching *July*; *Elull*, Withering *August*; *Ethanim*, Temperate *September*; *Bull*, Mild *October*; *Chisleu*, Blustering *November*; *Tebeth*, Leaf-les *December*.

The modern Hebrew calendar uses the following months: Tishrei, Heshvan, Kislev, Tevet, Shevat, Adar, Nisan, Iyar, Sivan, Tammuz, Av, and Elul, so four of the names used by Leeds (Abib, Zif, Ethanim, and Bull) are no longer common. Abib comes from an early calendar whose source is unknown. It is the only known month name from this calendar. Abib means "ripened" and it is mentioned four times in the Bible. Ziv, Ethanim, and Bul are from a Phoenician calendar. The meanings are "month of flowers," "month of perennial streams," and "rain or showers," respectively.

The printer, Bradford, moved to New York after charges against him in Philadelphia were dismissed. The trial was, "perhaps, the first in America involving freedom of the press," according to Marion Barber Stowell.[135] A source friendlier to the colonial government of Pennsylvania emphasizes that this trial introduced the principle that a jury must decide whether or not a printed work is seditious.[136]

Ethel Metzger writes, "At his trial, the prosecution produced one of his printing forms set up with an inflammatory tract by George Keith. . . This was the piece of printing upon which the charge of sedition was based. The printing form was heavy and hard to move. One of the jurors, wishing to assist the court, shoved it with his cane. Whereupon, it fell to pieces and the evidence was gone. This forced Bradford's release."[137]

---

135. Marion Barber Stowell, *Early American Almanacs*, New York, 1977: 34.
136. Rufus Jones, *The Quakers in the American Colonies*, London, 1911: 54.
137. Ethel Metzger, "Supplement to Hildeburn's *Century of Printing*" (thesis, New York, 1930).

John F. Watson adds the detail that the jurors lifted the frame to view it because they were not used to reading backwards. A juror then pushed against the back of the type with his cane.[138]

Isaiah Thomas quotes at great length from Bradford's trial and then prefaces the story of the spilled printing frame with, "It is said, . . ."[139]

Joseph Blumenthal omits the story of the shove with the cane and says simply that, "Governor Fletcher of New York, temporarily with jurisdiction over Pennsylvania, stepped in and took Bradford to New York."[140]

The earliest source we have seen just says that Bradford was released because the jury couldn't agree, as it was composed of nine Quakers and three non-Quakers.[141]

Christian Lodowick?   A183

1694 / AN / ALMANACK / Of the Cœlestiall Motions, Aspects / and Eclipses, &c. For the Year of / our Lord GOD, *M DC XC IV*. / And of the World, / 5643. / Being the Second after Bissextile or Leap / Year, and of the Reign of Their Ma- / jesties William and Mary KING / and QUEEN of Great *Brittain,France,* / and *Ireland &c.* (which began / Feb. 13. 1688,9.) the Sixth Year. / *Calculated for the Meridian of Boston in* / *N.E.* 69.*deg.* 20. *min. to the Westward* / *of London, &* 42. *deg.* 30. *min. North* / *Latitude, but may indifferently serve the* / *most part of New= England.* / By *Philo—Mathemat.* / Boston, *Printed by* B. Green, *for* Samuel / Phillips, *near the South-East end of the Ex-* / *change, by the Rose & Crown Tavern.*1694.

Bartholomew Green, Boston.

Sabin 940, 62743. Nichols [67]. Evans 687. Drake 2894. Wing A1377.

Originals located at: American Antiquarian Society, Massachusetts Historical Society.

Readex microprint and photostat copies are available. The complete text of the American Antiquarian Society copy is available online at Evans Digital Edition.

Paging: Eight leaves.

Runs January to December.

---

138. John F. Watson, *Annals of Philadelphia*, vol. I, Philadelphia, 1900: 545.
139. Isaiah Thomas, *The History of Printing in America*, New York, 1970: 343–355.
140. Joseph Blumenthal, *The Printed Book in America*, Boston, 1977: 6.
141. Anon., *The Tryals of Peter Bos, George Keith, Thomas Budd, and William Bradford, Quakers, for Several Great Misdemeanors (as Was Pretended by Their Adversaries) before a Court of Quakers*, London, 1693.

This work has been attributed to William Brattle by Charles Evans and others. The Web sites of the American Antiquarian Society, the Massachusetts Historical Society, and Evans Digital Edition all contain notes that point out that there is a greater similarity to the almanac of Lodowick in the following year. It is suggested that Lodowick put out his first almanac anonymously to see if it would be sufficiently popular and only attached his name in later years.

The attribution to Brattle appears to begin with Samuel Abbott Green. In the *Proceedings of the Massachusetts Historical Society*,[142] Green reported that he had found a letter from almanac writer John Tulley to his printer, Benjamin Harris, thanking him for a Brattle almanac. The letter was dated Saybrook, May 7, 1694.

Lodowick's last name appears in catalogs variously as Ludwig and Ludovici. His years are 1660 to 1728. He also wrote a dictionary for English, German, and French.

In the almanac there is an essay on the difficulty of predicting the tides. He remarks that "Weather and other Astrologicall Prædictions" have been omitted.

Carlos de Sigüenza y Góngora.   A184
ALMANAQUE / Y / LUNARIO / de D. C. de S. y G. / Para el Año de 1694. / Segun el Meridiano de la Ciudad de Mex^{Co}. / Sacalo à Luz Juan de Torquemada.
México.
Quintana 77.

The records that pertain to this title at the Archivo General de la Nación are *Inquisición* 670, leaves 387 to 388. The sheets are large and probably there would have been two printed pages for each one in the manuscript. The calendar section is missing.

Quintana reproduces the text.

John Tulley.   A185
Other Author: Benjamin Harris.
Tulley, 1694 / AN / ALMANACK / For the Year of our LORD, / MDCXCIIII / Being Second after Leap Year,[143] / And from the Creation /

---

142. Samuel Abbott Green, [Note], *Proceedings of the Massachusetts Historical Society*, second series VII (1891–1892): 414–415.

143. Leap-Year, in Harvard copy.

ALMANACS, EPHEMERIDES, AND LUNARIOS 1695

5643. / Wherein is Contained, Astronomical Obser- / vations from the Suns Ingress into Aries, and / the other Cardinal Points, with an Account / of the Eclipses, Conjunctions, and other / Configurations of the Celestial Bodies. / To which is Added, The Nativity of the *French* / King . Together with other Things both Useful & / Profitable / Calculated for and fitted to the Meridian of *BOSTON* / in *New-England*, where the North Pole is Elevated / 42 gr. 30 min. But may indifferently serve any / part of *New-England.* / By John Tulley / Boston, Printed and Sold by *Benj. Harris*, / over-against the *Old Meeting.House.*[144] 1694

Benjamin Harris, Boston.

Evans 710. Nichols [68]. Drake 2895, 2896. Wing A2590.

Originals located at: Harvard University, Library of Congress (leaves [2]–[10], some partial) (Drake 2896). An impression printed by Bartholomew Green for Samuel Phillips is recorded but no copies are known (Drake 2895).

Readex microprint and photostat copies are available. The Library of Congress text is available online at Evans Digital Edition. (Readex says American Antiquarian Society photostat and gives a complete copy.) Early English Books Online shows the Harvard copy. There is a listing for leaves 3 to 12 only, priced at $100, in the sale of the Phelps collection.[145]

Paging: Twelve leaves.

Runs January to December.

The chart illustrating the moon's dominion over the body is repeated from 1693. Related to that, in the later pages, one sees advice as to the best times for various purges depending on the moon's position in the zodiac. For instance, we are told that it is best to purge the head by sneezing when the moon is in Cancer, Leo, or Virgo, and to "bath" when the moon is in Cancer, Libra, Aquarius, or Pisces.

There is an essay on the horoscope of the French king that contains anti-Catholic sentiments. Tulley disavows authorship of this in his 1696 almanac, saying that the whole thing was written by the printer, Benjamin Harris.

1695

Drake's entry for this year, numbered 2897, is an almanac by Increase Gatchell. Marion Barber Stowell argues persuasively that the correct date for Gatchell's

---

144. *Old-Meeting.House.* in Harvard copy.
145. David L. O'Neal, *Early American Almanacs, The Phelps Collection 1679–1900*, Petersborough, NH, 1978?

The Young American Ephemeris *is 1715. The error may have started with Samuel Briggs in 1888.*[146]

Antonio Sebastián de Aguilar Cantú.   A186
El Pronostico de los Temporales de el Año que Viene de Noventa y cinco.
México.
Quintana 79.
The records that pertain to this title at the Archivo General de la Nación are *Inquisición* 495, leaves 33 to 36. Two applications for permission were made.

Juan Ramón Koenig.   A187
El Conocimiento de los Tiempos.
Lima.

Daniel Leeds.   A188
AN / ALMANACK / For the Year of Christian Account / 1695. / And from the Creation of the World / 5662. / But by *Keplas* Computation 5688. / Being the Third after Leap-Year, / The Epact is 25. Golden Number 5. / and Dominical Letter F. / Containing Matters Necessary and / Useful, chiefly accomodated to the / Lat. of 40 Degrees, but may without / sensible Error serve the Places adjacent, / from *Newfound-Land* to the Capes of / *Virginia.* / By *Daniel Leeds,* Philomat. / Hermis Trismegistus, *pag.* 13. *speaking of the Na- / ture and Composition of Man, says, If he lay the cause / of evil upon fate or distiny,* [sic] *he will never abstain from / any evil Work. Wherefore we must look warily to such*[147] / *kind of People, that being in Ignorance, they may be less*[148] / *Evil for fear of that which is hidden and kept secret.* / Printed and Sold by *William Bradford* at / the Bible in *New-York,* 1694.
William Bradford, New York.
Evans 716. Drake 5527. Wing A1867.
Original located at: Huntington Library.
Microfilm and Readex microprint copies are available. The complete text of the Huntington copy is available online at Evans Digital Edition.

---

146. Samuel Briggs, "The Origin and Development of the Almanac," *Western Reserve Historical Society Tracts* II/69 (1888): 435–477.
147. See next footnote.
148. The h and the s at the ends of the lines have each slipped down a line.

Paging: Twelve leaves.
Months are named as in the 1694 issue.

Christian Lodowick.   **A189**
1695. / THE NEW-ENGLAND / ALMANACK / For the Year of our Lord CHRIST, / M DC XC V. / And of the WORLD, / 5644. / Being the third after Leap year, and of / the Reign of Their Majesties (which / began Feb. 13. 1688,9.) *the Seventh year.* / *Calculated for the Meridian of* Boston *in* / N.E. *69 deg. 20. min. to the Westward of* / London, *and 42. deg. 30. min. North* / *Latitude, and may serve for all* / New=England. / To which are added some seasonable / Cautions against certain Impieties and / Absurdities in *Tulley's* Almanacks, giv- / ing a truer Account of what may be / expected from Astrological *Prædictions*. / Together with some choice, experimented, / cheap, easy and parable Receipts, of a / General Benefit to *Country People*. / By C. *Lodowick*, Physician. / Boston, *Printed by* B. Green, *for* S. Phillips, *at* / *the Brick Shop near the Old Meeting-house.* 1695.
Bartholomew Green, Boston.
Sabin 62743. Evans 717. Nichols [69]. Drake 2898. Wing A1924.
Original located at: American Antiquarian Society.
Readex microprint and photostat copies are available. The complete text of the American Antiquarian Society copy is available online at Evans Digital Edition.
Paging: Eight leaves.
Runs January to December.
There is a brutal denunciation of astrology and John Tulley's use of same in his almanacs. Tulley responded in a relatively friendly way, considering the ferocity of the attack, in his almanac of 1696.
The "receipts" of the title are home remedies. Examples include ingesting quicksilver, i.e., mercury, for worms in children and dried and powdered earthworms for convulsions. To prevent toothaches, one is advised to frequently rub the teeth with tobacco ashes. For colic—horse dung steeped in wine or beer and strained.

Carlos de Sigüenza y Góngora.   **A190**
El Almanaque y Lunario que como acostumbre tiene dispuesto para el año que viene de 1695.
México.
Quintana 78.

The records that pertain to this title at the Archivo General de la Nación are *Inquisición* 670, leaves 389 to 392. Two applications for permission to print were made.

John Tulley. **A191**
Other Author: Benjamin Harris?
Tulley, 1695. / AN / ALMANACK / For the Year of our LORD, MDCXCV / Being Third after Leap-Year / And from the Creation 5644[149] / Wherein is Contained Astronomical Obser- / vations from the Suns Ingress into Aries, and / the other Cardinal Points, with an Account / of the Eclipses, Conjunctions, and other / Configurations of the Celestial Bodies. / To which is Added, An Account of the Cru- / elty of the Papists acted upon the Bodies / of some of the Godly Martyrs. / Calculated for and fitted to the Meridian of *BOSTON* / in *New-England*, where the North Pole is Elevated / 42 gr. 30 min. But may indifferently serve any / part of NEW-ENGLAND. / By John Tulley / *Boston*, Printed for *John Usher*, by *Ben- / jamin Harris*, who formerly lived over- / against the *Old-Meeting-House*, is now Re- / moved to the Sign of the *BIBLE*, over- / against the *Blew-Anchor*. 1695.

Benjamin Harris, Boston.

Evans 740, Nichols [70]. Drake 2899. Wing A2591.

Originals located at: American Antiquarian Society (two copies—copy 2 missing first leaf), Boston Public Library (leaves [1]–[8]), Massachusetts Historical Society.

Readex microprint and photostat copies are available. The complete text of the American Antiquarian Society copy 1 is available online at Evans Digital Edition.

Paging: Twelve leaves.

Runs January to December.

The American Antiquarian Society's online catalog notes differences between the advertisements in their two copies. Copy 1 has only an advertisement for Harris. Copy 2 also has an advertisement for the dentist Matthew Cary. "Mr. Mathew Cary near the Peavling Green in Boston, Cures all Pains in Teeth without Drawing." The Massachusetts Historical Society copy is also missing the ad for Cary.

There is more anti-Catholic propaganda, probably due once again to Harris.

---

149. There is a faint period in the Boston Public Library and Massachusetts Historical Society copies.

1696

Antonio Sebastián de Aguilar Cantú.   **A192**
El Pronostico de los Temporales de el Año que viene de Nouenta y seis.
México.
Quintana 80.
The record that pertains to this title at the Archivo General de la Nación is *Inquisición* 670, leaf 162.

Juan de Avilés Ramírez.   **A193**
El pronostico de Temporales para el año que viene de noventa, i seis.
México.
Quintana 81.
The record that pertains to this title at the Archivo General de la Nación is *Inquisición* 670, leaf 90.

Juan Ramón Koenig.   **A194**
El Conocimiento de los Tiempos.
Lima.

Daniel Leeds.   **A195**
AN / ALMANACK / For the Year of CHRISTIAN Account / 1696. / And from the Creation of the World / 5663. / But by *Keplas* Computation 5689. / Being the *Bessextile* [sic] or Leap-Year, / The Epact is 6. Golden Number 6. and / Dominical Letter ED. / Containing Matters Necessary and Use- / ful, chiefly accomodated to the Latitude / of 40 Degrees, and Longitude of about / 73 West from *London*, but may, without / sensible Error, serve all parts adjacent / from *New-found-Land* to the Capes of / *Virginia.* / By *Daniel Leeds*, Philomat. / *Whilst on the beauteous bountious Planets Rays,* / *My Muse her Contemplations did raise,* / *Asked* Urania, *When the time would be,* / *That we more peaceful settled Times should see?* / *She riddled,* When the fertile Grave doth yeild, [sic] / The harvest of an Ages Fallow field. / Printed and Sold by *William Bradford* at the / Bible in *New-York*, 1696.
William Bradford, New York.
Evans 744. Drake 5528. Wing A1868.
Originals located at: American Antiquarian Society (lacks first and last leaves), Huntington Library (last leaf missing).

Microfilm and Readex microprint copies are available. The Huntington copy is available online at Evans Digital Edition.
Paging: Twelve leaves.
Runs January to December.
Months are named as in the 1694 issue.

Carlos de Sigüenza y Góngora. **A196**
Almanaque y Lunario / de D. C. de S. y G. / Para el año Bisiesto de 1696. / Segun el meridiano / de Mex.^{co} / Sacalo a Luz Juan de Torquemada México.
Quintana 82.
The records that pertain to this title at the Archivo General de la Nación are *Inquisición* 670, leaves 234 to 242. The text is 235 to 242. The leaves are large and probably represent two printed pages per page of manuscript.
Quintana reproduces the complete text.

John Tulley. **A197**
Tulley, 1696. / AN / Almanack, / For the Year of our LORD, / *M DC XC VI*. / Being Bissextile or Leap-Year, / And from the *CREATION*, / 5645. / Wherein is Contained *Astronomical* Observa- / tions from the *SUNS* Ingress into *Aries*,and / the other *Cardinal Points*, with an account / of the Eclipses, *Conjunctions*, and other / Configurations of the Cælestial Bodies. / Calculated for and fitted to the Meridian of Boston, / in New=England, where the North *Pole* is / Elevated 42 *gr.* 30 *min.* But may indifferently / serve any part of New=England. / By John Tulley. / Licensed by Authority. / *BOSTON, N. E.* Printed by *Bartholomew Green*, / and *John Allen*, for John Usher, and / are to be Sold at his shop below the / Town-House, 1696.
Bartholomew Green and John Allen, Boston.
Evans 776. Nichols [71]. Drake 2900. Wing A2592.
Originals located at: American Antiquarian Society, Indiana University, Library of Congress, Massachusetts Historical Society, Private Collection (Ann Arbor) (missing last leaf), Rosenbach Museum.
Readex microprint and photostat copies are available. The complete text of the American Antiquarian Society copy is available online at Evans Digital Edition.
Paging: Eight leaves.
In response to the criticisms in Lodowick's almanac, Tulley says that astrology has been put into his almanacs in the past "by the desire of the

Printers." Specifically, he attributes the article about the French king in 1694 to Benjamin Harris and says that it was placed in his almanac without his knowledge and that he was quite disturbed to see such material of another printed under his own name.

1697

Antonio Sebastián de Aguilar Cantú.   A198
El Pronostico delos Temporales de el Año que viene de Nouenta y Siete.
México.
Quintana 83.
The record that pertains to this title at the Archivo General de la Nación is *Inquisición* 670, leaf 368.

John Clapp.   A199
NEW YORK ALMANACK.
William Bradford, New York.
Sabin 53519, Evans 779. Drake 5529. Wing A1413.
Originals located at: Library Company (leaves [2]–[27], [27] is $\frac{2}{3}$), New York Public Library (leaf [5], most of [6] in two fragments, $\frac{3}{4}$ of [7] in two fragments, $\frac{1}{2}$ of [8]).
Photostats of both copies and Readex microprint copies are available. The Library Company copy is available online at Evans Digital Edition.
Paging: Twenty-eight leaves. This is the longest almanac on our list for which the paging is known.
The title generally used now is based on Leeds's almanac for 1698: "As to what John Clapp wrote last year in his New York Almanack . . ." Joseph Sabin gave the title "An Almanack" and said that it came from Isaiah Thomas.
Runs January to December. Days of the week are indicated with the letters, a, b, C, d, e, f, g, with C standing for Sunday. This is because the year started on a Friday and so the dominical letter is C.
Clapp gives the earth-to-moon distance as 209,236 miles. He favors the heliocentric theory and calls Copernicus "that approved Philosopher & Prince of Astronomers."
For June on [8]v: "The 24 of this month is celebrated the Feast of St. *John Baptist*, in commemoration of which, (& to keep up a happy union & lasting friendship, by the sweet harmony of good society) a feast is held by the

Johns of this City, at *John Clapps* in the Bowry, where any Gentleman whose christian name is John, may find a hearty wellcome to joyn in confort with his Namesakes."

A quote from Dante on [17]r, within the discussion of eclipses, is said by Joseph Fucilla to be "the earliest bit of translation in America now known drawn from the *Divine Comedy*, . . . [and] possibly the earliest appearance in this country of words printed in the Italian Language."[150]

John Clapp (1651–1725? according to the NY Public Library catalog) was an innkeeper, and there is an advertisement for his inn on [26]v: "At the aforesaid Clapps, about two Mile without the City of New-York, at the place called the Bowry, any Gentlemen Travellers that are strangers to the City, may have very good Entertainment for themselves and Horses, where there is also a Hackney Coach and good Saddle Horses to be hired."

From the Fucilla article, we learn that Clapp was born in London and went to Charlestown, South Carolina, in 1680. By 1690 he was in Flushing and he moved to New York City in 1692. About 1705, he was in Westchester.

The Library Company copy came from the Michael Zinman collection, and from a member of the Clapp family before that, so it is probably the same copy that Fucilla refers to that was in the possession of the Clapp family.

Juan Ramón Koenig. **A200**
El Conocimiento de los Tiempos.
Lima.

Daniel Leeds. **A201**
AN / ALMANACK / For the Year of *CHRISTIAN* Account / 1697. / And from the Creation of the World / 5664. / But by *Dove's* Computation, 5701. / Being the first after *Bessextile* [sic] or Leap- / Year,the Epact being 17. / Golden Number / 7. and Dominical Letter *C.* / Containing Variety of Matter, Re- / specting the Latitude North 40 Degrees, / and Longitude West from *London,* 73 deg. / but may, without sensible Error, serve all / parts adjacent from *New-found-Land* to / the Capes of *Virginia.* / By *Daniel Leeds,*

---

150. Joseph Fucilla, "The First Fragment of a Translation of the *Divine Comedy* Printed in America," *Italica* XXV/1 (Mar. 1948): 1–4.

Philomat. / Of three liberal Sciences of a Christian. / 1. Arithmetick's *an Art that makes him raise,* / *And number out Gods Blessings and his Dayes.* / 2. Astronomy's *an Art takes out the Lead* / *From his dull Brows, and lifteth up the Head.* / 3. Geometry's *an Art that makes him have* / *The World in scorn, and measure out his Grave.* / Printed and Sold by *William Bradford* at the / Bible in *New-York*, 1697.

William Bradford, New York.

Evans 785. Drake 5530. Wing A1869.

Originals located at: Huntington Library (missing last leaf), New York Public Library (three leaves and fragments).

Microfilm and Readex microprint copies are available. The Huntington copy is available online at Evans Digital Edition. There is a listing for leaves 2 to 8 only, priced at $65, in the sale of the Phelps collection.[151]

Paging: 12 leaves.

Months are named as in the 1694 issue.

The mention of Dove's computation in the title is probably referring to Jonathan Dove, an English almanac writer. Actually Dove's almanac gives 5700 for this same year.

Carlos de Sigüenza y Góngora. A202
 [*Almanaque*].
 México.

John Tulley. A203

Tulley, 1697. / AN / ALMANACK / For the Year of our LORD, M DC XC VII. / Being First after Leap-Year, and from the / CREATION, 5646. / Wherein is Contained Astronomical Observations / from the Suns Ingress into *Aries*, & the other / Cardinal Points, with an Account of the / Eclipses, *Conjunctions*, and other Configura- / tions of the Cælestial Bodies : Unto which is / added a brief account of the late COMET / or Æthereal blaze. / Calculated for and fitted to the Meridian of Boston, / in New=England, where the North Pole is / Elevated 42. gr. 30. min. But may indifferently / serve any part of New=England. / By John Tulley. / Licensed by Authority / *BOSTON, N. E.* Printed by *Bartholomew Green,* / and *John Allen,* for John Usher, and / are to be Sold at his Shop below the / Town-House, 1697.

---

151. David L. O'Neal, *Early American Almanacs, The Phelps Collection 1679–1900*, Petersborough, NH, 1978?

Bartholomew Green and John Allen, Boston.
Evans 815. Nichols [72]. Drake 2901. Wing A2593.
Originals located at: American Antiquarian Society (two copies), Library of Congress, Massachusetts Historical Society, Old Sturbridge Village (leaves [6]–[8]).
Readex microprint and photostat copies are available. The complete text of the American Antiquarian Society copy 1 is available online at Evans Digital Edition with the note, "There is a second issue in which reference to astrologers on p. [15] has been deleted."
Paging: Eight leaves.
The last page discusses the comet of October 1695.
In regard to the deletion of material on leaf [8]r, negative aspects (i.e., "Wars, Tumults, Dissentions, Violence," etc.) of the lunar eclipse in Scorpio are expunged in the second issue, but negative aspects of the comet (i.e., "Slaughters, Chronick & long lasting Diseases, Snows, Earthquakes," etc.) are left in on the next page. In the American Antiquarian Society copies, [8]r ends with "of" in the first issue and with "Scorpio" in the second. The Massachusetts Historical Society copy is first issue but [8]r ends with "Scorpio." The Library of Congress copy is second issue but ends with "3 deg." on [8]r. The Sturbridge copy is second but ends with "the 29th of" on [8]r.

1698

Antonio Sebastián de Aguilar Cantú.   A204
El Pronostico de los Temporales de el Año que viene de nouenta y ocho.
México.
Quintana 84.
The record that pertains to this title at the Archivo General de la Nación is *Inquisición* 670, leaf 159.

Marco Antonio de Gamboa y Riaño.   A205
Lunario / y / Prognostico [sic] / de temporales; / Con las Elecciones de Medicina, Phle= / botomia, Nauegacion, y Agricultura / por lo que indican los Astros este / Año de / 1698 / Segundo despues de Bisiesto. / Calculalo [sic] para el Meridiano de / la Ciudad de Mexico / El B.^{r} Marcos Antonio de Gamboa / y Ryaño de quarto Curso de / Medicina

México.
Quintana 85.
The records that pertain to this title at the Archivo General de la Nación are *Inquisición* 670, leaves 103 to 113. The text is 104 to 113.
Quintana reproduces the text. He refers to Gamboa as a disciple of Carlos de Sigüenza y Góngora. Quintana says on page 53 of *La Astrología* that according to José Mariano Beristáin de Souza, Gamboa was from Havana, was a medical doctor, had a Chair of Mathematics, and was a reviewer of books for the Inquisition.

Juan Ramón Koenig.   A206
El Conocimiento de los Tiempos.
Lima.
Vargas Ugarte 937. (It is not clear why Vargas Ugarte gives a number to this issue of *El Conocimiento de los Tiempos* but not to any of the others. He repeats Gabriel Moreno's claim that this almanac appeared every year from 1680.)

Daniel Leeds.   A207
AN / ALMANACK / For the Year of *CHRISTIAN* Account / 1698. / But by Bishop *Usher*'s Account, 1702. / By the Account of others, 1700. / Being the second after *Bessextile* [sic] or / Leap-Year. / And from the Creation of the World 5665. / But by *Dove*'s Computation, 5702. / Containing a general Ephemerides of / the Planets Motions, with many other / Matters Useful & Necessary. / Chiefly accomodated to the Latitude / of 40 Degrees North, and Lougitude [sic] of / about 73 degr. west from *London*. But / may, without sensible Error serve all the / adjacent Places,even from *Newfound-Land* / to the Capes of *Virginia*.[152] / By *Daniel Leeds*, Philomat. / Printed and Sold by *William Bradford* at the / Bible in *New-York*, 1698.
William Bradford, New York.
Evans 821. Drake 5531. Wing A1870. (Proud, 1903, $400.)
Originals located at: American Antiquarian Society (first and last leaves supplied in facsimile), Huntington Library, Morgan Library, New York Public Library (one leaf and fragments).
Microfilm and Readex microprint copies are available. The complete text of the American Antiquarian Society copy is available online at Evans Digital Edition.

---

152. No period in Morgan copy.

Paging: Twelve leaves.

Runs January to December. The months are named and numbered as in the 1694 issue.

Bradford advertises, "*Lockyers Universal Pill, famous for the cure of Agues, Feavers, Scurvey, Gout, Dropsie, Jaundice, Bloody Flux, Griping in the Guts, Worms of all sorts. The Gravel, Stone, Collick, and many other Diseases, are to be sold by William Bradford. . .*"

There is a reference on [9]v to eclipses listed in Clapp's almanac of 1697. "As to what *John Clapp* wrote last year in his *New York* Almanack of the particular effects of Eclipses being hid from us mortals, I am of his Opinion therein. But where he infers that they have no Effects, because they are Natural, I widely differ from him there; for my experience has oft told me the contrary. 'Tis true, Eclipses are Natural, as he says. So likewise Meat and Drink is [sic] Natural, and that nourishes Mankind; so is Poyson Natural, but yet that destroys them. Let this suffice at present. I pass by the rest."

Carlos de Sigüenza y Góngora.   A208
 [*Almanaque*].
 México.

John Tulley.   A209

Tulley, 1698. / AN / ALMANACK / For the Year of our LORD, / M DC XC VIII. / Being Second after Leap-Year, / and from the CREATION, / 5647. / Wherein is Contained the Lunations, Courts, / Spring-tides,[153] Planets, Aspects and Weather, / the Rising and Setting of the SUN, to- / gether with the Sun and Moons place, and / time of Full Sea, or High-Water, with an / account of the Eclipses, Conjunctions, and / other Configurations of the Cælestial Bodies. / Calculated for and fitted to the Meridian of Boston / in New=England, where the North Pole is / Elevated 42. gr. 30. *min*. But may indifferently / serve any part of New=England. / By John Tulley. / Licensed by Authority. / *BOSTON, N. E*. Printed by *Bartholomew Green*, / and *John Allen*.[154] Sold at the Printing-House / at the South end of the Town. 1698.

Bartholomew Green and John Allen, Boston.

Evans 854. Nichols [73]. Drake 2902. Wing A2594.

153. Hyphen faint in Peabody Essex Museum copy.
154. Period faint in American Antiquarian Society, Library of Congress, and Peabody Essex Museum copies.

Originals located at: American Antiquarian Society, Boston Public Library, Harvard University (leaves [2]–[7]), Huntington Library, Library of Congress, Massachusetts Historical Society, Old Sturbridge Village, Peabody Essex Museum, Winterthur Museum.

Readex microprint and photostat copies are available. The complete text of the American Antiquarian Society copy is available online at Evans Digital Edition. Early English Books Online shows the Huntington copy.

Paging: Eight leaves.

The Winterthur copy is interleaved with pages of accounts attributed to Hezekiah Fuller. The Massachusetts Historical Society copy has notes by John Winthrop (probably the one whose years are 1681–1714) with holidays, deaths, and a drawing of a ship.

1699

Juan Ramón Koenig.   A210
El Conocimiento de los Tiempos.
Lima.

Daniel Leeds.   A211

AN / ALMANACK / For the Year of CHRISTIAN Account / 1699. / Being the Third after *Bessextile* [sic] or / Leap-Year. / And from the Creation of the World 5666. / But by *Dove*'s Computation 5703. / *Containing Matters Useful & Necessary*[155] / Chiefly accomodated to the Latitude / of 40 Degrees North, and Longitude of / about 73 degr. west from *London*. But / may, without sensible Error serve all the / adjacent Places,even from *Newfound-Land* / to the Capes of *Virginia*. / By *Daniel Leeds*, Philomat. / A Riddle from F. Q. / *The goods we spend, we keep, and what we save* / *We loose, and only what we loose we have* / Printed and Sold by *William Bradford* at the / Bible in *New-York*, 1699.

William Bradford, New York.

Drake 5532. Evans 39399. Wing A1871. (Proud, 1903, $10.)

Originals located at: Library Company (last leaf $\frac{3}{4}$ and supplied in facsimile, Historical Society of Pennsylvania copy), Morgan Library (two copies: copy 1 has first leaf missing; copy 2 is leaves [1]–[10], [9] is partial).

---

155. Period in Morgan copy.

Readex microprint copies exist in numerous libraries. The complete text of the Library Company copy is online at Evans Digital Edition.

Paging: Twelve leaves.

Times for high water in Philadelphia are given in the calendar. Adjustments for other places are given on [1]v.

If any readers can answer the riddle in the title, please send us your solution(s). F. Q. is Francis Quarles (1592–1644) and the quote is from his *Divine Fancies*, Book IV, article 70.

Carlos de Sigüenza y Góngora. **A212**
El Almanaque y Lunario para el año de 1699.
México.
Quintana 86.
The record that pertains to this title at the Archivo General de la Nación is *Inquisición* 670, leaf 102.

Jacob Taylor. **A213**
AN ALMANACK.
William Bradford, New York.
Wing A2538.

A printer's manuscript for a sheet almanac can be found at the New York Public Library. Days of the week are indicated by a through g, as in John Clapp's almanac of 1697. There is a record that A. S. W. Rosenbach once had a printed copy, but we have been unable to determine what became of it.

Taylor later began, in 1702, a series of almanacs printed in Philadelphia.

John Tulley. **A214**
Other Author: Cotton Mather.

Tulley, 1699. / AN / Almanack / For the Year of our Lord,*M DC XC IX.* / Being Third after Leap-Year, / and from the Creation, / 5648. / Wherein is Contained the Lunations, Courts, / Spring Tides,Planets,Aspects and Weather, / the Rising and Setting of the SUN, to- / gether with the Sun and Moons place, and / time of Full Sea, or High-Water, with an / account of the *Eclipses*, Conjunctions, and / otherConfigurations of the Cælestial Bodies. / Calculated for and fitted to the Meridian of Boston / in New=England, where the North Pole is / Elevated 42.gr.30. *min*. But may indifferently / serve any part of New=England. / By JOHN TULLEY. /

Licensed by Authority. / *BOSTON, N. E.* Printed by *Bartholomew Green,* / and *John Allen.* Sold at the Printing-House / at the South end of the Town. 1699.

Bartholomew Green and John Allen, Boston.

Evans 897. Nichols [74]. Drake 2903. Wing A2595.

Originals located at: American Antiquarian Society, Boston Public Library, Huntington Library (missing last leaf), John Carter Brown Library, Library Company (copy owned by the Historical Society of Pennsylvania), Library of Congress, Massachusetts Historical Society, Old Sturbridge Village.

Readex microprint and photostat copies are available. The complete text of the American Antiquarian Society copy is available online at Evans Digital Edition. Early English Books Online shows the Huntington copy. There is a listing for leaves 2 to 7 only, priced at $50, in the sale of the Phelps collection.[156]

Paging: Eight leaves.

Runs from January to December.

An essay on the last page is by Cotton Mather, "a few pungent Lines" according to an entry in Mather's diary, quoted by Marion Barber Stowell.[157]

1700

J. G. Riewald[158] *has pointed out that the single leaf at the Library Company that has been attributed (Drake 9471; Evans 39369; Wing A2538A) to a Jacob Taylor almanac for 1700 is really a leaf from the Leeds almanac.*

Antonio Sebastián de Aguilar Cantú.   A215

El Pronostico de los Temporales de el Año de mil y Setecientos, proxima venidero.

México.

Quintana 87.

The record that pertains to this title at the Archivo General de la Nación is *Inquisición* 670, leaf 116.

---

156. David L. O'Neal, *Early American Almanacs, The Phelps Collection 1679–1900,* Petersborough, NH, 1978?

157. Marion Barber Stowell, *Early American Almanacs,* New York, 1977: 26.

158. J. G. Riewald, *Reynier Jansen of Philadelphia,* Groningen, 1970: 173.

Samuel Clough. **A216**

Clough, 1700. / THE *New England* / ALMANACK / For the Year of our LORD, *M DCC*. / From the CREATION 5649. And / From the Discovery of America by *Chr.Columbus*,208. / Being Leap Year, & of the Reign of our Gra- / cious Sovereign, King WILLIAM the Third / (which began *Febr.* the 13.1688,9) the 12th. / Wherein is contained the several changes of the Moon, / and Aspects of the Planets, Courts, Spring Tides, Rise- / ing & Setting of the *Sun*, with the *Sun* & *Moons* place, / and time of *Full Sea* or High Water, with an ac- / count of the Eclipses : And a figure of the Visible E- / clipse as it will appear to us. / The Vulgar Notes of this year are, / Golden Number 10 / The Epact 20 / }{ / Cicle [sic] of the Sun 1 / Dominic lettersG.F. / Fitted to the Meridian of Boston in New=England / Lat. 42. *gr.* 30. *min.* But may serve any part of / New=England without any sensible Error. / By *Samuel Clough,* / A Lover of the Mathematicks. / Licensed by Authority. / *Boston,* Printed by *Bartholomew Green,* & *John Allen* / Sold at the Printing-House. 1700.

Bartholomew Green and John Allen, Boston.

Sabin 62743. Nichols [75]. Evans 861. Drake 2904. Wing A1417.

Originals located at: Huntington Library (some leaves torn, last leaf $\frac{2}{3}$), Library of Congress.

Microfilm and Readex microprint copies are available. The complete text of the Library of Congress copy is available online at Evans Digital Edition.

Paging: Eight leaves.

Runs January to December.

A table of weights says that an avoirdupois ounce is 438 grains (today's value is 437.5).

The title declares this to be leap year, and it was such under the Julian calendar. In the Gregorian calendar, the leap day is omitted this year and in any year divisible by 100 but not divisible by 400. The year 1700 was the first application of this rule, the first omission of a century-year leap day, since the Gregorian calendar was instituted. As of February 29, 1700 (Julian), the two calendars became eleven days apart instead of ten.

A corollary of this gap is that the epacts also differ by eleven days from this year on. The directions given by Diego García de Palacio in his work of 1587, which was discussed in Part 1, tell readers to start this change with 1700, giving a Gregorian epact of 9 for the year, even though the effect of the missing leap day on the age of the moon at the New Year doesn't show up until 1701.

Also, the Gregorian dominical letter for this year should be C (a single letter since it's not a leap year). In the next year, when the Julian dominical letter goes to E, the Gregorian will be B, and the Gregorian will continue to precede the Julian by three letters through the eighteenth century.

Clough's years are 1665 to 1707.

Juan Ramón Koenig.   A217
El Conocimiento de los Tiempos.
Lima.

Daniel Leeds.   A218

AN / ALMANACK / For the Year of *CHRISTIAN* Account / 1700. / Being *Bessextile* [sic] or Leap-Year. / And from the Creation of the World 5667. / But by *Dove's* Computation, 5704. / Containing Matters Useful & Necessary[159] / Chiefly accomodated to the Latitude / of 40 Degrees North, and Longitude of / about 73 degr. west from *London*. But / may, without sensible Error serve all the / adjacent Places,even from *Newfound-Land* / to the Capes of *Virginia*. / By *Daniel Leeds*, Philomat. / Long time hast thou, Great Babylon, stood high, / And thou hast lived very sumptuously.[160] / And hast Triumphed in thy lofty Pride, / As if thou shouldst as Queen forever ride : / But for all this, e'er long we shall have proof, / That thou'lt extinguish like a stinking Snuff. G. W. / Printed and Sold by *William Bradford* at the / Bible in *New-York*, 1700.

William Bradford, New York.

Evans 866. Drake 5533. Wing A1872.

Originals located at: Huntington Library, Library Company (eighth leaf), Morgan Library (leaves [1]–[9]).

Microfilm and Readex microprint copies are available. The complete text of the Huntington copy is available online at Evans Digital Edition.

Paging: Twelve leaves.

The single leaf at the Library Company has been misattributed in the past to Jacob Taylor. Because of this, this leaf may be viewed at Evans Digital Edition by searching under Taylor, or under the Evans number 39369. It has the calendar for December on one side and the eclipses for the year on the other. Jacob Taylor is mentioned as a source for the time of the eclipse of the moon. (This is interesting in view of Taylor's later

---

159. Period in Morgan copy.
160. Comma in Morgan copy.

feud with Leeds. In his 1706 almanac, Taylor accuses Leeds of plagiarism.)

Months are named as in Leeds's almanacs of 1694 and after.

Times for high water in Philadelphia are given in the calendar. Adjustments for other places are given on [1]v.

Leeds explains that he made an observation of a lunar eclipse on August 30, 1699, and compared it to John Gadbury's prediction for London. This gave him a time difference of 5 hours and 30 minutes, which corresponds to 82°30' of longitude. From 1696 he has been using 73° west of London for his position. The actual longitude of Burlington is fairly close to 75°.

Leeds adds that by the "Doctrine of Triangles," i.e., spherical trigonometry, his new estimate for the longitude puts the distance from Burlington to London at 4,150 miles. The distance D on a perfectly spherical Earth from a point with latitude y_1, longitude x_1, to a point with latitude y_2, longitude x_2, is given by

$$D = \frac{\pi R}{180} \cos^{-1}(\cos y_1 \cos y_2 (\cos(x_2 - x_1) - 1) + \cos(y_2 - y_1)),$$

where R is the radius of the Earth and the trigonometric functions are evaluated in degrees mode. If Leeds was using this formula, he must have been assuming a radius of 4,278 miles, a little higher than the 3,963 miles now given for Earth's equatorial radius.

Carlos de Sigüenza y Góngora. **A219**
El Almanaque para el año que viene de mill y setecientos.
México.
Quintana 88.

The record that pertains to this title at the Archivo General de la Nación is *Inquisición* 670, leaf 367.

Sigüenza died in 1700, having written the almanac for 1701. It appeared posthumously.

John Tulley. **A220**
Tulley, 1700. / AN / Almanack / For the Year of our Lord, 1700. / Being Bissextile or Leap-Year, and from / the Creation, 5649. And from the / Discovery of America by *Chr. Columbus*, 208. / Wherein is Contained the Lunations, Courts, / Spring Tides,Planets,Aspects and Weather, / the Rising and Setting of the SUN, to- / gether with the Sun and Moons place,

and / time of Full Sea, or High Water,[161] with an / account of the *Eclipses,* Conjunctions, and / otherConfigurations of the Cælestial Bodies. / Calculated for and fitted to the Meridian of *Boston* / in *New-England,* where the North Pole is Ele- / vated 42 *gr.* 30 *min.* But may indifferently / serve any part of *New-England.* / Unto which is added, Natural Prognosticks for / the judgment of the Weather. / *By* JOHN TULLEY. / Licensed by Authority. / BOSTON, Printed by *Bartholomew Green,* / & *John Allen.*[162] / Sold at the Printing-House / at the South end of the Town. 1700[163]

Bartholomew Green and John Allen, Boston.

Evans 955. Nichols [76]. Drake 2905. Wing A2596.

Originals located at: American Antiquarian Society, Boston Public Library, Harvard University (last leaf $\frac{3}{4}$), Huntington Library (some leaves torn), Library of Congress, Massachusetts Historical Society (2 copies; copy 2 is [8] only), Old Sturbridge Village ([1]–[7]), Owen.

Readex microprint copies are available. The complete text of the American Antiquarian Society copy is available online at Evans Digital Edition. Early English Books Online shows the Huntington copy, missing the last leaf, which is present in the original but torn in two. The title page was reprinted in the Christie's catalog for the Snider sale of 2005.

Paging: Eight leaves.

Tulley died in 1701. The posthumous almanac for 1702 was his last.

---

161. High-Water in Harvard, Massachusetts Historical Society, and Owen copies.

162. Period very faint in Boston Public Library and Sturbridge copies.

163. There is a period in the American Antiquarian Society and Library of Congress copies. Faint in Harvard.

APPENDIX A

A Guide to Astrological Symbols

Many astrological almanacs include a guide to symbols, but most do not. Anyone studying these almanacs may find the following table useful. The modern symbols are included here for the sake of completeness but, of course, seventeenth-century almanacs would not have used the symbols for earth, Uranus, Neptune, Pluto, or the asteroids.

The Zodiac:

Aries ♈	Leo ♌	Sagittarius ♐
Taurus ♉	Virgo ♍	Capricorn ♑
Gemini ♊	Libra ♎	Aquarius ♒
Cancer ♋	Scorpio ♏	Pisces ♓

The Planets and Asteroids:

Sun ☉	Jupiter ♃	Pallas ⚴
Moon ☽ ☾	Saturn ♄	Juno ⚵
Mercury ☿	Uranus ♅ ⛢	Vesta ⚶
Venus ♀	Neptune ♆	Chiron ⚷
Earth ⊕	Pluto ♇ ⯓	
Mars ♂	Ceres ⚳	

The Aspects and Other Symbols:

Ascending Node ☊	Semisextile ⚺	Square □
Descending Node ☋	Semi-square ∠	Trine △
Conjunction ☌	Sextile ⚹	Retrograde motion ℞
Opposition ☍	Quintile Q	

APPENDIX B

Seventeenth-Century Comet Books Printed in the New World

Since it may be useful to see the whole list of comet books in one place, and because James Howard Robinson's book, *The Great Comet of 1680*, has very little coverage of the Spanish-speaking world, here are the titles of comet books from the New World up through 1690 that we know about.

Diego Rodríguez, *Discurso Etheorológico del Nuevo Cometa, Visto en aqueste Hemisferio Mexicano; y generalmente en todo el mundo. Este Año de 1652* (México, 1652).

Juan Ruiz, *Discurso Hecho Sobre la Significación de Dos Impressiones Meteorológicas que se vieron el Año passado de 1652. La primera de un Arco que se terminava de Oriente a Occidente a 18. de Noviembre. Y la segunda del Cometa visto por todo el Orbe terrestre desde 17. de Diziembre del mesmo Año de 1652* (México, 1653).

Gabriel López de Bonilla, *Discurso, y Relación Cometographia del repentino aborto de los Astros, que sucedió del Cometa que apareció por Diziembre de 1653* (México, 1654).

Samuel Danforth, *An Astronomical Description of the Late Comet or Blazing Star as it appeared in New-England in the 9th, 10th, 11th, and in the beginning of the 12th Moneth, 1664. Together with a brief Theological Application thereof* (Cambridge, Massachusetts, 1665; reprinted in London, 1666).

Francisco Ruiz Lozano, *Tratado de Cometas, Observation, y iuicio del que se vio en esta ciudad de los reyes, y generalmente en todo el mundo, por los fines del año de 1664. Y principios deste de 1665* (Lima, 1665).

Carlos de Sigüenza y Góngora, *Manifiesto Filosófico contra los cometas despojados del imperio que tenían sobre los tímidos* (México, 1681).

Martín de la Torre, *Manifiesta Cristiano en favor de los cometas mantenidos en su natural significación* (México, 1681). (Elían Trabulse lately speculates that this was never printed. He says a manuscript copy exists.[1])

---

1. Prologue to *Libra Astronómica*, México, 2001.

APPENDIX B

Carlos de Sigüenza y Góngora, *Belerofonte Mathemático contra la quimera astrológica de Martín de la Torre* (México, 1681). (This may never have been printed, if we can read that into the introduction, by Sebastián de Guzmán y Córdova, to Sigüenza's *Libra Astronómica*.)

José de Escobar Salmerón y Castro, *Discurso Cometológico, y Relacion del Nuevo Cometa: visto en aqueste Hemispherio Mexicano, y generalmente en todo el Mundo: el Año de 1680; Y extinguido en este de 81: Observado, y Regulado en este mismo Horizonte de Mexico* (México, 1681).

Eusebio Francisco Kino, *Exposición Astronómica de el Cometa, Que el Año de 1680. por los meses de Noviembre, y Diziembre, y este Año de 1681. por los meses de Enero y Febrero, se ha visto en todo el mundo, y le ha observado en la Ciudad de Cadiz* (México, 1681).

Increase Mather, *Heaven's Alarm to the World or a Sermon, wherein is shewed, that Fearful Sights and Signs in Heaven, are the Presages of the great Calamities at hand. Preached at the Lecture of Boston in New-England; January, 20. 1680* (Boston, 1681).

Gaspar Juan Evelino, *Especulación Astrológica, y physica de la naturaleza de los Cometas, y juizio del que este Año de 1682. Se vè en todo el Mundo* (México, 1682).

Increase Mather, *The Latter Sign Discoursed of, in a Sermon Preached at the Lecture of Boston in New-England, August, 31. 1682. Wherein is shewed, that the Voice of God in Signal Providences, especially when repeated and Iterated, ought to be Hearkened unto* (Boston, 1682).

Gaspar Juan Evelino, *Disertación sobre los cometas y sus influencias sobre el aparecido nuevamente en México* (México, 1683—listed by José Toribio Medina and Trabulse [*Apéndices e Índices*]; not confirmed by other sources).

Increase Mather, ΚΟΜΗΤΟΓΡΑΦΙΑ. *or a Discourse Concerning Comets; Wherein the Nature of Blazing Stars is Enquired into: With an Historical Account of all the Comets which have appeared from the Beginning of the World unto this present Year, M.DC.LXXXIII. Expressing the Place in the Heavens, where they were seen, their Motions, Forms, Duration; and the Remarkable Events which have followed in the World, so far as they have been by Learned Men Observed. As also two Sermons Occasioned by the Blazing Stars* (Boston, 1683). (The two sermons mentioned in the title, *Heaven's Alarm to the World* and *The Latter Sign*, were printed earlier and separately but bound in with the *Kometographia*. The *Kometographia*, without the sermons, was reprinted in London in 1811.)

Carlos de Sigüenza y Góngora, *Libra Astronómica, y Philosóphica en que D. Carlos de Siguenza y Gongora Cosmographo, y Mathematico Regio en la Academia Mexicana, Examina no solo lo que a su Manifiesto Philosophico contra los Cometas opuso el R. P. Eusebio Francisco Kino de la Compañia de Jesus; sino lo que el mismo R. P. opinó, y pretendio haver demostrado en su Exposicion Astronomica del Cometa del año de 1681* (México, 1690).

APPENDIX B

Robinson also makes reference to Samuel Willard's *The Fiery Tryal no strange thing; Delivered in a Sermon Preached at Charlstown, February 15. 1681. Being a Day of Humiliation* (Boston, 1682), but this work only mentions comets at the very end as a cause of recent floods in the Low Countries.

Many of the almanac writers discussed in Part 2 also included general comments about comets in their almanacs. The first book printed in the New World devoted to a single astronomical phenomenon probably was the *Discurso sobre la magna conjuncion de los planetas Jupiter y Saturno, acaecida in 24 de Diciembre de 1603 en el 9 g. de Sagitario* by Enrico Martínez (México, 1604), which was included in his work of 1606 discussed in Part 1. See our entry in Part 2 under 1646 for more about Danforth, 1649 for more about López de Bonilla, 1651 for more about Ruiz Lozano, 1653 for more about Ruiz, 1655 (pseud. Martín de Córdoba) for more about Rodríguez, 1678 for more about Escobar Salmerón y Castro, and, of course, for more about Kino and Sigüenza see our entry in Part 1 under 1690.

APPENDIX C

Some Comments on the Place of Logic in Mathematics

We should address how logic books belong in a list of mathematical works. It would be appropriate to briefly review the history of logic from a mathematical point of view.

Aristotle is credited with developing the *syllogism*, a form of argument with two premises and a conclusion. An example would be the argument

> All men are animals.
> All animals are mortal.
> ∴ All men are mortal.

Aristotle cataloged all the various syllogism forms that constituted valid arguments.

Necessity was introduced by Aristotle to supplement his theory of syllogisms. It then became possible for him to distinguish between propositions that are true and those that are necessarily true. Necessity and possibility are called "modalities," and their logic is called "modal" logic. George Boolos says that, although Aristotle "developed the theory of the syllogism in almost perfect form," Aristotle's theory of modal syllogisms has been found "defective" even by "sympathetic commentators."[1]

Some time later, symbolic logic was developed. Implication was denoted by $\Rightarrow$ and negation by $\neg$. A statement of the form "all A are not B," which might form one line of a syllogism, could be represented by a string of symbols, $\forall x\, (Ax \Rightarrow \neg\, Bx)$.

The era from the 1870s to the 1930s saw revolutionary work on the foundations of mathematics.[2] The dual reductions that were achieved were (1) all mathematical objects came to be viewed as sets, and (2) all mathematical discourse was shown to be reducible (potentially at least) to formal logic. It became possible to answer the question, "What are we doing when we do mathematics?" by saying, "We use logic

---

1. George Boolos, *The Logic of Provability*, Cambridge, 1993: xv–xvi.
2. See the anthology by Jean van Heijenoort (*From Frege to Gödel, a Source Book in Mathematical Logic, 1879–1931*, Cambridge, 1967) of papers from this period.

APPENDIX C

to talk about sets."[3] Today, mathematics students at the graduate level have to become fluent in symbolic logic and basic set theory.

In symbolic logic, the modality of necessity was denoted by $\Box$ and $\Box p$ was read as "p is necessarily true." The statement "p is possibly true" may be rendered as $\neg\Box\neg p$ in this formalism.[4] Various systems of modal logic have been developed, which might include as axioms or theorems statements like $\Box p \Rightarrow p$, $\Box(p \Rightarrow q) \Rightarrow (\Box p \Rightarrow \Box q)$, or $\Box p \Rightarrow \Box\Box p$.

Many mathematicians, unlike philosophers, may not feel the need to distinguish between "p is true" and "p is necessarily true." (If p is a strictly mathematical statement, many philosophers may not feel the need either.) A recent reinterpretation of the symbol $\Box$, however, makes modal logic more relevant to the mathematical enterprise.

A 1931 paper by Kurt Gödel made the following argument. By reducing logical proof to a mechanistic process, Gödel showed that the provability of a statement in a formal system is equivalent to a statement in arithmetic. It follows that for a formal system that is sophisticated enough to contain arithmetic, the provability of a statement p in the system can be viewed as itself a statement in the system. Gödel ultimately used this to show that a formal system that encompassed at least arithmetic must allow propositions that are undecidable within the system, because they state, in so many words, their own unprovability. In today's mathematics, now that more advanced techniques are known, hundreds of undecidable statements have been found and published in the literature of such mathematical subfields as topology, analysis, combinatorics, and set theory.[5]

If we interpret $\Box p$ as "p is provable" in the sense above, then the system of modal logic that arises is similar but distinct from the systems that went before. Called GL by Boolos, this system has $\Box(p \Rightarrow q) \Rightarrow (\Box p \Rightarrow \Box q)$ as axiomatic and $\Box p \Rightarrow \Box\Box p$ as a theorem, but $\Box p \Rightarrow p$ is not always a theorem. GL has the rather odd inference rule that whenever $\Box A \Rightarrow A$ is a theorem, then one may derive A.

Also, it has come to our attention that modal logic is being employed by theoretical computer scientists to cope with issues arising from parallel processing.

Thus not only is formal logic at the very core of modern mathematical teaching and research, to the extent that it is required background for most graduate-level training in mathematics, but even modal logic, once just the tool of philosophers, is at least relevant to issues at the heart of current research in certain mathematical fields.

---

3. Not necessarily the best answer or most complete answer but at least a possible answer.

4. "Aristotle knew, and perhaps discovered, that these notions of *possibility* and *necessity* can each be expressed in terms of the other." Kenneth Konyndyk, *Introductory Modal Logic*, Notre Dame, IN, 1986: 1.

5. We have made a tiny contribution to this program in a paper of 2002.

APPENDIX C

Our point is that books such as Alonso de la Vera Cruz's first entry here, however philosophically minded they were at the time, are part of a thread that leads from Aristotle through scholastic philosophy and ultimately to modern mathematical logic.

When we do ethnomathematics, we do not ask the members of the culture in question whether they think they are doing mathematics. We judge that for ourselves, seeing whether the thing they are doing exhibits structures akin to what we think of as mathematics. Likewise, then, if we are deciding whether a Renaissance logic book is mathematics, we don't need to ask the author, the author's culture, or the librarians who preserve such books for us; we ask ourselves whether the book exhibits structures that remind us of mathematics. Therefore we feel that this book by Vera Cruz, and others like it, should be listed among mathematical texts.

Works Consulted

Our intention here is not only to list the references, but to include as many as possible of the works that might be useful to someone contemplating a similar project. For example, we have listed some books and articles that offer corrections to José Toribio Medina's bibliographies. While only one of them is cited in the body of this work, they all must be consulted, and so they are listed here. The rare works listed in the first section here are excluding main entries from parts 1 and 2.

Rare Works (Through 1850)

Anon., *Pragmática sobre los diez días del año*. Lima, 1584.

Anon., *The Tryals of Peter Bos, George Keith, Thomas Budd, and William Bradford, Quakers, for Several Great Misdemeanors (as Was Pretended by Their Adversaries) before a Court of Quakers*. London, 1693.

Chávez, Jerónimo de. *Chronographia o Reportorio de los tiempos, el mas copioso y preciso que hasta ahora ha salido á luz*. Sevilla: Alonso Escriviano, 1572.

Chávez, Jerónimo de. *Chronographia o Reportorio de los Tiempos, el mas copioso y precisso que hasta ahora ha salido a luz*. Sevilla: Alonso Escriviano, 1581.

Danforth, Samuel. *An Astronomical Description of the Late Comet or Blazing Star as it appeared in New-England in the 9th, 10th, 11th, and in the beginning of the 12th Moneth, 1664. Together with a brief Theological Application thereof*. Cambridge, MA: Samuel Green, 1665.

Danforth, Samuel. *An Astronomical Description of the late Comet or Blazing-Star as it appeared in New England in November, December, January, and in the beginning of February 1664. Being the first and greatest of the Three Comets which was seen at London and elsewhere in Europe. Together with a brief Theological Application thereof*. London: Peter Parker, 1666.

Escobar Salmerón y Castro, José de. *Discurso Cometologico y Relacion del Nuevo Cometa: visto en aqueste Hemispherio Mexicano, y generalmente en todo el Mundo: el Año de 1680; Y extinguido en este de 81: Observando, y Regulado en este mismo Horizonte de Mexico*. México: Viuda de Bernardo Calderón, 1681.

WORKS CONSULTED

Evelino, Gaspar Juan. *Especulacion Astrologica, y physica de la naturaleza de los Cometas, y juizio del que este Año de 1682. Se vè en todo el Mundo.* México: Viuda de Bernardo Calderón, 1682.

Holland, Richard. *Notes Shewing how to get the angle of Parallax of a Comet, or other Phænomenon at two observations, to be taken in any one station, or place of the earth, and thereby the distance from the earth.* Oxford: Lichfield, 1668.

López de Bonilla, Gabriel. *Discurso, y Relación Cometographia del repentino aborto de los Astros, que sucedió del Cometa que apareció por Diziembre de 1653.* México: Biuda [sic] de Bernardo Calderón, 1654.

Mather, Increase. ΚΟΜΗΤΟΓΡΑϕΙΑ. *or a Discourse Concerning Comets; Wherein the Nature of Blazing Stars is Enquired into: With an Historical Account of all the Comets which have appeared from the Beginning of the World unto this present Year, M.DC.LXXXIII. Expressing the Place in the Heavens, where they were seen, their Motions, Forms, Duration; and the Remarkable Events which have followed in the World, so far as they have been by Learned Men Observed. As also two Sermons Occasioned by the Blazing Stars.* Boston: Samuel Green, Jr., 1683.

Maurolico, Francesco. *Opusculo Mathematica.* Venice: Francisum Franciscium Senensem, 1575.

Moreno, Gabriel. *Almanaque Peruano y Guia de Forasterios para el Año de 1807.* Peru: La Real Casa de Niños Expósitos, 1807.

Partridge, John (pseud.). *Squire Bickerstaff Detected; or, the Astrological Imposter Convicted.* London, 1708.

Rodríguez, Diego. *Discurso Etheorológico del Nuevo Cometa, Visto en aqueste Hemisferio Mexicano; y generalmente en todo el mundo. Este Año de 1652.* México: Viuda de Bernardo Calderón, 1652.

Ruiz, Juan. *Discurso Hecho Sobre la Significación de Dos Impressiones Meteorológicas que se vieron el Año passado de 1652. La primera de un Arco que se terminava de Oriente a Occidente a 18. de Noviembre. Y la segunda del Cometa visto por todo el Orbe terrestre desde 17. de Diziembre del mesmo Año de 1652.* México: Juan Ruiz, 1653.

Serna, Juan Vázquez de. *Libro Intitulado Reduciones de Oro, y Señorage de Plata, con las reglas, y tablas generales de lo uno y de lo otro.* Cádiz: Juan de Borja, 1620.

Swift, Jonathan. *Bickerstaffe's [sic] Prediction Confirm'd in the Death of Partridge, The Almanack-Maker, the 29th Day of this Instant March, at 13 Minutes past 11 at Night.* London: J. Morphew, 1708.

Swift, Jonathan. *An Elegy on Mr. Patrige,* [sic] *the Almanack-Maker, who Died on the 29th of March last, 1708.* Reprint. Edinburgh, 1708.

Swift, Jonathan. *Bickerstaff Redivivus; or, Predictions for the Year 1709. With the Author's Defence of his last Year's calculation, particularly against Mr. Partridge's Assertion in an Almanack for this present Year; wherein 'tis averr'd, he is still alive: Which is plainly prov'd to be an Error.* London: B. Bragge, 1709.

Swift, Jonathan. *The Works of J. S., D.D., D.S.P.D. in Four Volumes. Vol I: The Author's Miscellanies in Prose.* Dublin: George Faulkner, 1735.

Vetancurt, Augustin de. *Teatro Mexicano descripción breve de los sucessos exemplares, historicos, politicos, Militares, y Religiosos del nuevo mundo Occidental de las Indias.* México: Maria de Benavides, Viuda de Juan de Ribera, 1698.

Reprints and Translations of Rare Works

Anon. "The Calendar for May—from a Quaker Almanac of Philadelphia, 1693." *The Quaker, a Fortnightly Journal Devoted to The Religious Society of Friends* I/2 (Fifth Month 14th, 1920): 12.

Arber, Edward, ed. *An English Garner, Ingatherings from our History and Literature,* Vol VI. Birmingham, 1883.

Arber, Edward, ed. *An English Garner, Critical Essays and Literary Fragments,* compiled by Thomas Seccombe, introduction by J. Churlton Collins. Westminster: Archibald Constable, 1903.

Campanus. *Campanus of Novara and Medieval Planetary Theory, Theorica planetarum,* ed. and trans. and with an introduction and commentary by Francis S. Benjamin, Jr. and G. J. Toomer. Madison: University of Wisconsin Press, 1971.

Díaz, Juan. *Itinerario de la armada del Rey Católico a la isla de Yucatán en la India, el año 1518 en la que fue por comandante y capitán general Juan de Grijalva. Escrito para su Alteza por el capellán mayor de la dicha armada.* Facsimile of the first printing in Italian with Spanish translation by Joaquín García Icazbalceta. México: Editorial Juan Pablos, 1972.

Díaz, Juan, et al. *La Conquista de Tenochtitlán.* Ed. Germán Vásquez. Madrid: Historia 16, 1988.

Díaz del Castillo, Bernal. *The Bernal Díaz Chronicles, the True Story of the Conquest of Mexico.* Trans. and ed. Albert Idell. New York: Doubleday, 1956.

Díaz del Castillo, Bernal. *The Conquest of New Spain,* trans. and with an introduction by J. M. Cohen. London: Penguin, 1963.

Diez Freyle, Juan. *Sumario compendioso de las quentas de plata y oro que in los reynos del Piru son necessarias a los mercaderes: y todo genero de tratantes. Con algunas reglas tocantes al arithmetica.* Facsimile edition. Madrid: Instituto de Cooperación Iberoamericana, 1985.

Eliot, John. *The Logic Primer, Reprinted from the Unique Original of 1672,* introduction by Wilberforce Eames. Cleveland: Burrows Brothers, 1904.

Escalante de Mendoza, Juan de. *Itinerario de Navegación de los Mares y Tierras Occidentales 1575,* commentary by Roberto Barreiro-Meiro. Madrid: Museo Naval, 1985.

Franklin, Benjamin. *Poor Richard, 1733. An Almanack for the Year of Christ 1733.* Bedford, MA: Applewood, 2002.

García de Palacio, Diego. *Carta Dirijida al Rey de España, por el Licenciado Dr. Don Diego Garcia de Palacio, Oydor de la Real Audencia de Guatmala; Año 1576. Being a Description of the Ancient Provinces of Guazacapan, Izalco, Cuscatlan, and Chiquimula, in the Audencia of Guatemala: with an account of the Languages,*

WORKS CONSULTED

Customs and Religion of their Aboriginal Inhabitants, and a Description of the Ruins of Copan, trans. and notes by E. G. Squier. Collection of Rare and Original Documents and Relations, Concerning the Discovery and Conquest of America, no. 1. New York: Charles Norton, 1860.

García de Palacio, Diego. *Diálogos Militares por el Doctor Diego García de Palacio*, facsimile edition with a prologue by Julio F. Guillén. Colección de Incunables Americanos, Siglo XVI, vol. VII. Madrid: Ediciones Cultura Hispánica, 1944.

García de Palacio, Diego. *Instrucción Náutica para Navegar por el Doctor Diego García de Palacio*, facsimile edition with a prologue by Julio F. Guillén. Colección de Incunables Americanos, Siglo XVI, vol. VIII. Madrid: Ediciones Cultura Hispánica, 1944.

García de Palacio, Diego. *Diálogos Militares—1.583, Facsímil del Libro Tercero: . . . de los Arcabuces y, [sic] Artillería . . .* Prologue by José Corderas Descárrega. Sevilla: Maestranza de Artillería, 1984.

García de Palacio, Diego. *Nautical Instruction*, trans. J. Bankston. Bisbee, AZ: The Press, 1986.

García de Palacio, Diego. *Nautical Instruction: A.D. 1587*, 2nd ed, trans. J. Bankston. Bisbee, AZ: Terranate Research, 1988.

García de Palacio, Diego. *Instrucción Nautica, Transcripción y estudio de Mariano Cuesta Domingo*. Madrid: Editorial Naval—Museo Naval, 1993.

García de Palacio, Diego. *Diálogos Militares. Estudio Preliminar de Laura Manzano Baena*. Madrid: Ministerio de Defensa, 2003.

González-Aller Hierro, José Ignacio, comp. *Obras Classicas de Navegación*. Serie II, Tematicas para la Historia de Iberoamérica, vol. 17. Compact disc. Madrid: La Fundación Histórica Tavera, 1998.

Kino, Eusebio Francisco. *Kino's Historical Memoir of Pimeria Alta, a Contemporary Account of the Beginnings of California, Sonora, and Arizona*, trans., ed., and annotated by Herbert Eugene Bolton. Berkeley: University of California Press, 1948.

Li, Andrés de. *Reportorio de los Tiempos*, ed. and with an introduction by Laura Delbrugge. London: Tamesis, 1999.

Martínez, Enrico. *Reportorio de los Tiempos e Historia Natural de Nueva España*. Introductory study by Francisco de la Maza. Bibliographic appendix by Francisco González de Cossío. México: Secretaría de Educación Pública, 1948.

Martínez, Enrico. *Reportorio de los tiempos y [sic] historia natural de Nueva España. Reproducción Facsimilar de la Primera Edición, México, 1606*. Prologue by Edmundo O'Gorman. Introduction by Francisco de la Maza, reprinted from the 1948 edition. Biographical sketch excerpted by Edmundo O'Gorman from the book *Enrico Martínez, cosmógrapho e impressor de Nueva España* by Francisco de la Maza. México: Grupo Condumex Chimalistac, 1980.

Martínez, Enrico. *Reportorio de los tiempos y [sic] historia natural de Nueva España. Reimpresión de la edición facsimilar de México Grupo Condumex, S. A. de C. V. 1980*. Prologue by Edmundo O'Gorman. Introduction by Francisco de la Maza, reprinted from the 1948 edition. Biographical sketch excerpted by Edmundo O'Gorman from the book *Enrico Martínez, cosmógrapho e impressor de Nueva España* by Francisco de la Maza. México: Centro de Estudios de Historia de México Condumex, 1981.

Martínez, Enrico. *Reportorio de los Tiempos e Historia Natural de Nueva España.* Introductory study by Francisco de la Maza. Bibliographic appendix by Francisco González de Cossío. México: Consejo Nacional para la Cultura y las Artes, 1991.

Morton, Charles, and William Brattle. *Aristotelian and Cartesian Logic at Harvard, Charles Morton's* A LOGICK SYSTEM *and William Brattle's* COMPENDIUM OF LOGICK, ed. Rick Kennedy. Publications of the Colonial Society of Massachusetts, vol. LXVII. Boston: Colonial Society of Massachusetts, 1995.

Peralta Barnuevo, Pedro de. *Obras Dramáticas con un Apéndice de Poemas Inéditos.* Introduction and notes by Irving A. Leonard. Santiago: Imprenta Universitaria Santiago de Chile, 1937.

Pisano, Leonardo [Fibonacci]. *Le Livre des Nombres Carrés*, trans. and with an introduction and notes by Paul ver Eecke. Bruges: Desclée de Brouwer et Cie., 1952.

Pisano, Leonardo [Fibonacci]. *The Book of Squares, An Annotated Translation into Modern English by L. E. Sigler.* Boston: Academic Press, 1987.

Plante, Rory. "The Libra Astronómica and Its Mathematics, with an English Translation of Selected Excerpts" (thesis, Bristol, RI: Roger Williams University), 2007.

Robledo, Antonio Gómez. *El Magisterio Filosófico y Jurídico de Alonso de la Veracruz, con una Antología de Textos.* Mexico: Editorial Porrúa, 1984.

Sigüenza y Góngora, Carlos de. *Obras, con una Biografía.* Biography by Francisco Pérez Salazar. México: Sociedad de Bibliofilos Mexicanos, 1928.

Sigüenza y Góngora, Carlos de. *Libra Astronómica y Filosófica.* Ed. Bernabé Navarro, with an introduction by José Gaos. México: Centro de Estudios Filosóficos, Universidad Nacional Autónoma de México, 1959.

Sigüenza y Góngora, Carlos de. *The Misadventures of Alonso Ramirez*, trans. Edwin H. Pleasants. México: Imprenta Mexicana, 1962.

Sigüenza y Góngora, Carlos de. *Libra Astronómica y Filosófica*, ed. Bernabé Navarro, with an introduction by José Gaos. Reprint of the 1959 edition. México: Universidad Nacional Autónoma de México, 1984.

Sigüenza y Góngora, Carlos de. *Seis Obras*, ed., notes, and chronology by William G. Bryant. Prologue by Irving A. Leonard. Caracas, Venezuela: Biblioteca Ayacucho, 1984.

Sigüenza y Góngora, Carlos de. *Libra Astronómica y Filosófica.* Facsimile edition. Prologue by Elías Trabulse. México: Sociedad Mexicana de Bibliófilos, 2001.

Smith, David Eugene. *The Sumario Compendioso of Brother Juan Diez, the Earliest Mathematical Work of the New World.* Boston: Ginn and Company, 1921.

Smith, David Eugene. *The Sumario Compendioso of Brother Juan Diez, the Earliest Mathematical Work of the New World.* Reprint of the 1921 edition. Ormond Beach, FL: Camelot, 1996.

Toledo, Francisco de. *Opera Omnia Philosophica (Köln 1615/16).* Introduction by Wilhelm Risse. Hildesheim: Georg Olms, 1985.

Vera Cruz, Alonso de la. *Investigación Filosófico-natural: Los Libros del Alma, Libros I y II,* ed. and trans. Owaldo Robles. México: Imprenta Universitaria, 1942.

Vera Cruz, Alonso de la. *Dialectica Resolutio cum Textu Aristotelis por el Reverendo Padre Alfonso Avera Cruce.* Facsimile edition. Colección de Incunables Americanos, Siglo XVI, vol. II. Madrid: Ediciones Cultura Hispánica, 1945.

Vera Cruz, Alonso de la. *The Writings of Alonso de la Veracruz: II, The Original Texts with English Translation Edited by Ernest J. Burrus. Defense of the Indians: Their Rights.* Sources and Studies for the History of the Americas: vol. IV. Rome: Jesuit Historical Institute, 1968.

Vera Cruz, Alonso de la. *Antología de Fray Alonso de la Veracruz.* Introduction and Selection of Texts by Mauricio Beuchot. Morelia: Universidad Michoacano de San Nicolás de Hidalgo, 1988.

Vera Cruz, Alonso de la. *Libro de los Elencos Sofísticos,* trans., introduction, and notes by Mauricio Beuchot. México: Universidad Nacional Autónoma de México, 1989.

Vera Cruz, Alonso de la. *Tratado de los Tópicos Dialécticos,* trans., introduction, and notes by Mauricio Beuchot. México: Universidad Nacional Autónoma de México, 1989.

Vera Cruz, Alonso de la. *Antología sobre el Hombre y la Libertad.* Introduction and Compilation by Mauricio Beuchot. Anejos de Novahispania 5. Universidad Nacional Autónoma de México, México, 2002.

Wyler, April. "A Translation and Mathematical Analysis of Breve Aritmetica by Benito Fernandez de Belo" (thesis, Western Connecticut State University, 1999).

Formally Cited Bibliographies

Drake, Milton. *Almanacs of the United States, Parts I–II.* New York: Scarecrow, 1962.

Evans, Charles. *American Bibliography, A Chronological Dictionary of All Books, Pamphlets and Periodical Publications Printed in the United States of America from the Genesis of Printing in 1639 Down to and Including the Year 1820, with Bibliographical Notes.* Vol. I 1639–1729. New York: Peter Smith, 1941.

García Icazbalceta, Joaquín. *Bibliografía Mexicana del Siglo XVI, Primera Parte, Catálogo Razonado de Libros Impressos en México de 1539 á 1600.* México: Andrade y Morales, 1886.

González de Cossío, Francisco. *La Imprenta en Mexico (1553–1820), 510 Adiciones a la Obra de Jose Toribio Medina en homenaje al primer centenario de su nacimiento.* México: Universidad Nacional de México, 1952.

Hildeburn, Charles Swift Riché. *A Century of Printing, The Issues of the Press in Pennsylvania, 1685–1784,* vol. I: 1685–1763. Philadelphia: Matlock and Harvey, 1885.

Karpinski, Louis C. *Bibliography of Mathematical Works Printed in America Through 1850.* Ann Arbor: The University of Michigan Press and Humphrey Milford; London: Oxford University Press, 1940.

Medina, José Toribio. *La Imprenta en México (1539–1821)*, tomos I–III. Santiago, Chile: Medina, 1912. Reprinted, Amsterdam: N. Israel, 1965.
Medina, José Toribio. *La Imprenta en Lima (1584–1824)*, tomos I–IV. Santiago, Chile: Medina, 1904. Reprinted, Amsterdam: N. Israel, 1965.
Nichols, Charles L. "Notes on the Almanacs of Massachusetts." *Proceedings of the American Antiquarian Society* New Series 22 (1912): 15–134.
Quintana, José Miguel. *La Astrología en la Nueva España en el Siglo XVII (De Enrico Martínez a Sigüenza y Góngora)*. México: Bibliófilos Mexicanos, 1969.
Sabin, Joseph. *A Dictionary of Books Relating to America from its Discovery to the Present Time*. New York, 1891. Reprinted, Amsterdam: N. Israel, 1962.
Vargas Ugarte, Ruben, S.J. *Impresos Peruanos (1584–1650)*. Biblioteca Peruana, tomos VII–VIII. Lima: Editorial San Marcos, 1954.
Wagner, Henry R. *Nueva Bibliografía Mexicana del Siglo XVI, Suplemento a las Bibliografías de don Joaquín García Icazbalceta, Don José Toribio Medina, y Don Nicolás León*, trans. Joaquín García Pimentel and Federico Gómez de Orozco. México: Editorial Polis, 1940.
Wing, Donald. *Short-Title Catalogue of Books Printed in England, Scotland, Ireland, Wales, and British America and of English Books Printed in other Countries, 1641–1700*, 2nd ed., vol. I–III, rev. and ed. John J. Morrison, et al. New York: Modern Language Association of America, 1994.

Other Bibliographic Reference Works and Articles

Anon. "Rare Volume of Arithmetic Donated to U. C. Museum." *San Francisco Chronicle* (Mar. 13, 1927).
Anon. "Rare Book Birthday Gift to St. Bona." *Olean Times Herald* (Oct. 31, 1951).
Anon. "¡Cosas de Gallegos!" *Cultura Gallega* Año II/29–32 (June–July 1937): 27. Facsimile ed., Santiago de Compostela: Xunta de Galicia, 1999.
Amory, Hugh. "Printing and Bookselling in New England, 1638–1713." In *A History of the Book in America, Volume One, The Colonial Book in the Atlantic New World*, ed. Hugh Amory and David D. Hall. Cambridge: Cambridge University Press, 2000.
Anders, Richard. *A Compilation of Material Relating to the Contents and Uses of Almanacs*, vols. 1–2. Manuscript at the American Antiquarian Society.
Antonio, Nicolás. *Biblioteca Hispana Nueva, o de los Escritores Españoles que Brillaron desde el Año MD hasta el de MDCLXXXIV*. Francisco Pérez Bayer, Madrid, 1788. Reprinted by Fundación Universitaria Española, Madrid, 1999.
Araujo Espinoza, Graciela. "Adiciones a 'La Imprenta en Lima (1584–1824).'" *Fenix* 8 (1952): 467–704.
Arbour, Keith. "The First North American Mathematical Book and Its Metalcut Illustrations: Jacob Taylor's TENEBRÆ, 1697." *The Pennsylvania Magazine of History and Biography* CXXIII/1/2 (Jan./Apr. 1999): 87–98.

Backer, Augustin de, and Aloys de Backer. *Bibliothèque de la Compagnie de Jésus*, new ed. Carlos Sommervogel. Bruxelles: Oscar Schepens, 1898.

Biblioteca Nacional de Chile. *Catálogo Breve de la Biblioteca Americana de J. T. Medina de la Nacional de Santiago*, tomos I–II, libros impresos I–II, suplemento. Santiago, Chile: Imprenta Universitaria, 1953–1954.

Blumenthal, Joseph. *The Printed Book in America*. Boston: Godine, 1977.

Bolaño e Isla, Amancio. *Contribución al Estudio Biobibliográfico de Fray Alonso de la Vera Cruz*. Prologue by Augustín Millares Carlo. México: Antigua Librería Robredo, 1947.

Briggs, Samuel. "The Origin and Development of the Almanac." *Western Reserve Historical Society Tracts* II/69 (1888): 435–477.

Brigham, Clarence Saunders. "An Account of American Almanacs and Their Value for Historical Study." *Proceedings of the American Antiquarian Society* (Oct. 1925).

Cajori, Florian. "The Earliest Arithmetic Published in America." *Isis* IX/3/31 (1927): 391–401.

Chapin, Howard M. "Checklist of Rhode Island Almanacs, 1643–1850." *Proceedings of the American Antiquarian Society* new series 25 (1915): 19–54.

Clement, Richard W. *The Book in America, with Images from the Library of Congress*. Golden, CO: Fulcrum, 1996.

Corona, Carmen. *Lunarios: Calendarios Novohispanos del Siglo XVII*. México: Publicaciones Mexicanas, 1991.

Cranz, F. Edward. *A Bibliography of Aristotle Editions, 1501–1600*. Bibliotheca Bibliographica Aureliana XXXVIII. Baden-Baden: Valentin Koerner, 1971.

Fernandez de Navarrete, Martín. *Biblioteca Marítima Española*, tomos I–IV. Barcelona: Palau & Dulcet, 1995.

Franklin, Benjamin, V, ed., *Boston Printers, Publishers, and Booksellers: 1640–1800*. Boston: G. K. Hall & Co., 1980.

Fucilla, Joseph G. "The First Fragment of a Translation of the *Divine Comedy* Printed in America: A New Find." *Italica* XXV/1 (Mar. 1948): 1–4.

Gallardo, Bartolomé José. *Ensayo de una Biblioteca Española de Libros Raros y Curiosos*, tomo segundo. Madrid: M. Rivadeneyra, 1866.

Gavito, Florencio. *Adiciones a la Imprenta en la Pueabla de los Ángeles de J. T. Medina*. Preface by Felipe Teixidor. México, 1961.

García Icazbalceta, Joaquín. *Bibliografía Mexicana del Siglo XVI, Catálogo Razonado de Libros Impressos en México de 1539 a 1600*, new ed. Augustín Millares Carlo. México: Fondo de Cultura Económica, 1954.

Green, Otis H., and Irving A. Leonard. "On the Mexican Book Trade in 1600: A Chapter in Cultural History." Reprinted, *Hispanic Review* 9 no. 1 (Jan. 1941).

Green, Samuel Abbott. [Note.] *Proceedings of the Massachusetts Historical Society* second series VII (1891–1892): 414–415.

Green, Samuel Abbott. "Early American Imprints." *Proceedings of the Massachusetts Historical Society* second series IX (1894–1895): 410–540.

Green, Samuel Abbott. *John Foster, the Earliest American Engraver and the First Boston Printer*. Boston: Massachusetts Historical Society, 1909.

Gray, Shirley B., and C. Edward Sandifer. "The Sumario Compendioso: The New World's First Mathematics Book." *Mathematics Teacher* 94/2 (Feb. 2001): 98–103.

Henkels, Stan V., compiler. *The Proud Papers* [catalogue]. Philadelphia: The Morris Press, 1903.

Hildeburn, Charles Swift Riché. *Sketches of Printers and Printing in Colonial New York*. New York: Dodd, Mead, & Co., 1895.

Karpinski, Louis C. "Books and Literature." *School and Society* XIV (Aug. 13, 1921): 83–84.

Karpinski, Louis C. "Colonial American Arithmetics." In *Bibliographical Essays, A Tribute to Wilberforce Eames*, ed. George Parker Winship. Cambridge: Harvard University Press, 1924: 242–248.

Karpinski, Louis C. "Early Arithmetics Published in America, a Concise History of the Rise of Figuring Things Out." *The Dearborn Independent* (Feb. 13, 1926): 20–21.

Karpinski, Louis C. "The Earliest Known American Arithmetic." *Science* 63 (Feb. 19, 1926): 193–195.

Karpinski, Louis C. "Review of Florian Cajori's *The Early Mathematical Sciences in North and South America*." *Isis* XII, I/37 (1929): 163–165.

Karpinski, Louis C. "The First Printed Arithmetic of Spain, Francesch Sanct Climent, *Suma de la Art de Arismetica Barcelona, 1482*." *Osiris* 1 (1936): 411–420.

Leonard, Irving A. *Ensayo Bibliográphico de Don Carlos de Sigüenza y Góngora*. Monografías Bibliográficas Mexicanos, Número 15. México: [Imprenta de la Secretaria de relaciones exteriors], 1929.

Leonard, Irving A. *Books of the Brave, Being an Account of Books and of Men in the Spanish Conquest and Settlement of the Sixteenth-Century New World*. Cambridge: Harvard, 1949. Reprinted, Berkeley, CA: University of California, 1992.

Littlefield, George Emery. *The Early Massachusetts Press, 1638–1711*, vol. I. Boston: The Club of Odd Volumes, 1907.

Lohmann Villena, Guillermo. "Un Libro Limeño Desconocido." *Fenix* 8 (1952): 462–466.

Martínez, Enrico. *La Obra de Enrico Martínez, Cosmógrafo del Rey, Intérprete del Santo Oficio de la Inquisición, Cortador y Fundidor de caracteres, Fallador de Grabados, Impresor de libros, Autor, Arquitecto y Maestro de la Obra del Desagüe del Valle de México*, Vol. 1, ed. Juan Pascoe. Santa Rosa: Martín Pescador, 1996.

Medina, José Toribio. *Bibliografía de la Imprenta en Santiago de Chile desde Sus Orígenes hasta Febrero de 1817*. Santiago, Chile: Medina, 1891.

Medina, José Toribio. *Historia y Bibliografía de la Imprenta en el Antiguo Virein-ato del Rio de la Plata*. La Plata: Museo de la Plata, 1892. Reprinted, Amsterdam: N. Israel, 1965.

WORKS CONSULTED

Medina, José Toribio. *La Imprenta en Arequipa, El Cuzco, Trujillo y Otros Pueblos del Perú durante las Campañas de la Independencia (1820–1825), Notas Bibliográficas*. Santiago, Chile: Elzeviriana, 1904. Reprinted, Amsterdam: N. Israel, 1964.

Medina, José Toribio. *La Imprenta en Bogotá (1739–1821), Notas Bibliográficas*. Santiago, Chile: Elzeviriana, 1904. Reprinted, Amsterdam: N. Israel, 1964.

Medina, José Toribio. *La Imprenta en Caracas (1808–1821), Notas Bibliográficas*. Santiago, Chile: Elzeviriana, 1904. Reprinted, Amsterdam: N. Israel, 1964.

Medina, José Toribio. *La Imprenta en Cartagena de las Indias (1809–1820), Notas Bibliográficas*. Santiago, Chile: Elzeviriana, 1904. Reprinted, Amsterdam: N. Israel, 1964.

Medina, José Toribio. *La Imprenta en Guadalajara de México (1793–1821), Notas Bibliográficas*. Santiago, Chile: Elzeviriana, 1904.

Medina, José Toribio. *La Imprenta en la Habana (1707–1810), Notas Bibliográficas*. Santiago, Chile: Elzeviriana, 1904. Reprinted, Amsterdam: N. Israel, 1964.

Medina, José Toribio. *La Imprenta en Mérida de Yucatán (1813–1821), Notas Bibliográficas*. Santiago, Chile: Elzeviriana, 1904. Reprinted, Amsterdam: N. Israel, 1964.

Medina, José Toribio. *La Imprenta en Oaxaca (1720–1820), Notas Bibliográficas*. Santiago, Chile: Elzeviriana, 1904. Reprinted, Amsterdam: N. Israel, 1964.

Medina, José Toribio. *La Imprenta en Quito (1760–1818), Notas Bibliográficas*. Santiago, Chile: Elzeviriana, 1904. Reprinted, Amsterdam: N. Israel, 1964.

Medina, José Toribio. *La Imprenta en Veracruz (1794–1821), Notas Bibliográficas*. Santiago, Chile: Elzeviriana, 1904. Reprinted, Amsterdam: N. Israel, 1964.

Medina, José Toribio. *Notas Bibliográficas Referentes á las Primeras Producciones de la Imprenta en Algunas Ciudades de la América Española (Ambato, Angostura, Curazao, Guayaquil, Maracaibo, Nueva Orleans, Nueva Valencia, Panamá, Popayán, Puerto España, Puerto Rico, Querétaro, Santa Marta, Santiago de Cuba, Santo Domingo, Tunja y Otras Lugares) (1754–1823)*. Santiago, Chile: Elzeviriana, 1904. Reprinted, Amsterdam: N. Israel, 1964.

Medina, José Toribio. *La Imprenta en la Puebla de Los Ángeles (1640–1821)*. Santiago, Chile: Cervantes, 1908. Reprinted, Amsterdam: N. Israel, 1964.

Medina, José Toribio. *La Imprenta en Guatemala (1666–1821)*. Santiago, Chile: Medina, 1910. Reprinted, Amsterdam: N. Israel, 1964.

Medina, José Toribio. *Catálogo Breve de la Biblioteca Americana que Obsequia a la Nacional de Santiago*, Tomos I–II, Libros Impresos I–II. Santiago, Chile: Imprenta Universitaria, 1926.

Medina, José Toribio. *Bibliografía de la Imprenta en Santiago de Chile desde Sus Orígenes hasta Febrero de 1817. Adiciones y Amplificaciones*. Santiago, Chile: Universidad de Chile, 1939.

Medina, José Toribio. "Adiciones Inéditas a 'La Imprenta en Lima.'" *Fenix* 8 (1952): 434–461.

WORKS CONSULTED

Metzger, Ethel. "Supplement to Hildeburn's *Century of Printing, 1685–1775*, with an Introductory Essay" (thesis, New York: Columbia University), 1930.

Millares Carlo, Augustín, and Julián Calvo. *Juan Pablos, Primer Impressor que a Esta Tierra Vino*. Documentos Mexicanos, vol. 1. México: Manuel Porrúa, 1953.

Morrison, Hugh Alexander. *Preliminary Checklist of American Almanacs 1639–1800*. Washington, DC: Library of Congress, 1907.

O'Neal, David L. *Early American Almanacs, The Phelps Collection 1679–1900*. Catalog 25. Petersborough, NH: O'Neal, 1978?

Page, Alfred B. "John Tulley's Almanacs, 1687–1702." *Transactions of the Colonial Society of Massachusetts* (Dec. 1910): 207–223.

Pérez Salazar, Francisco. "Dos Familias de Impresores Mexicanos de Siglo XVII." *Mémoires Société Scientifique "Antonio Alzate"* 43 (1925): 447–511.

Placer López, Gumersindo. *Bibliografia Mercedaria*. Publicaciones del Monasterio de Poyo 24. Madrid: Edita Reusta Estudios, 1968.

Riewald, J. G. *Reynier Jansen of Philadelphia, Early American Printer, A Chapter in Seventeenth-Century Nonconformity*. Groningen: Wolters-Noordhoff, 1970.

Roden, Robert F. *The Cambridge Press, 1638–1692, A History of the First Printing Press Established in English America, Together with a Bibliographical List of the Issues of the Press*. New York: Dodd, Mead, and Co., 1905.

Sagendorph, Robb. *America and her Almanacs, Wit, Wisdom & Weather, 1639–1970*. Dublin, NH: Yankee, 1970.

Sanchez Lamego, Miguel A. *El Primer Mapa General de México Elaborado por un Mexicano*. Pub. no. 175. México: Instituto Panamericano de Geografia e Historia, 1955.

Sandifer, C. Edward. "Spanish Colonial Mathematics: A Window on the Past." *The College Mathematics Journal* 33/4 (Sep. 2002): 266–278.

Santiago Vela, Gregorio de. *Ensayo de una Biblioteca Ibero-Americana de la Orden de San Agustin*. El Escorial, 1931.

Schons, Dorothy. *Bibliografía de Sor Juana Inés de la Cruz*. Monografía's Bibliográficas Mexicanas, Número 7. México, 1927.

Schwab, Federico. "Los Almanaques Peruanos ¿1680?–1874." *Boletín Bibliográfico* XIX/1–2 (1948).

Shipton, Clifford K., and James E. Mooney. *National Index of American Imprints through 1800, the Short-Title Evans*. Worcester, MA: American Antiquarian Society, 1969.

Smith, David Eugene. "The First Work on Mathematics Printed in the New World." *The American Mathematical Monthly* [28] (1921), 10–15. Reprinted, *Sherlock Holmes in Babylon, and Other Tales of Mathematical History*, ed. Marlow Anderson, Victor Katz, and Robin Wilson. Washington, D.C.: Mathematical Association of America, 2004.

Smith, David Eugene. "The First Printed Arithmetic (Treviso, 1478)." *Isis* 6 (1924): 311–331.

Smith, Joseph. *Supplement to a Descriptive Catalogue of Friends' Books, or Books Written by Members of the Society of Friends, Commonly Called Quakers, from Their Rise to the Present Time; Interspersed with Critical Remarks, and Occasional Biographical Notices, and Including All Writings by Authors before Joining and by Those after Having Left the Society, Whether Adverse or Not, as Far as Known.* London: Edward Hicks, 1893.

Soto, Pedro Blanco. "El Primer Libro de Filosofía Impreso in el Nuevo Mundo." *Beiträge zur Geschichte der Philosophie des Mittelalters* Supplementband 1 (1913): 365–391.

Stowell, Marion Barber. *Early American Almanacs: The Colonial Weekday Bible.* New York: Burt Franklin, 1977.

Tate, Vernon D. "The Instrvcion [sic] Nautica of 1587." *The American Neptune* (Apr. 1941): 191–195.

Thomas, Isaiah. *The History of Printing in America, with a Biography of Printers & an Account of Newspapers*, ed. Marcus A. McCorison. New York: Weathervane, 1970.

Thorndike, Lynn. *The* SPHERE *of Sacrobosco and Its Commentators.* Chicago: University of Chicago Press, 1949.

Trumbull, J. Hammond. *Origin and Early Progress of the Indian Missions in New England, with a List of Books in the Indian Language Printed at Cambridge and Boston, 1635–1721.* Worcester, MA: American Antiquarian Society, 1874.

Wagner, Henry R. "Sixteenth Century Mexican Imprints." In *Bibliographical Essays, A Tribute to Wilberforce Eames*, ed. George Parker Winship. Cambridge, MA: Harvard University Press, 1924: 249–268.

Wagner, Henry R. *Mexican Imprints, 1544–1600, in the Huntington Library.* San Marino: Huntington, 1939.

Wall, Alexander J. *A List of New York Almanacs 1694–1850.* New York: New York Public Library, 1921.

Winship, George Parker. *The Cambridge Press, 1638–1692, A Reëxamination of the Evidence Concerning* THE BAY PSALM BOOK and the ELIOT INDIAN BIBLE *as well as other Contemporary Books and People.* Philadelphia: University of Pennsylvania Press, 1945.

Wroth, Lawrence C. *The Way of a Ship, An Essay on the Literature of Navigation Science.* Portland, ME: Southworth-Anthoensen Press, 1937.

Wroth, Lawrence C. "Notes for Bibliophiles: Review of *Bibliography of Mathematical Works Printed in America Through 1850* by Louis C. Karpinski." *New York Herald Tribune Books* (Aug. 25, 1940).

History and Biography

Anon. "Notes and Queries." *The Pennsylvania Magazine of History and Biography* XXXVII/3 (1913): 379–384.

Arróniz, Othón. *El Despertar Científico en América, la vida de Diego García de Palacio, documentos inéditos del Archivo de Sevilla.* Facsimile edition. Prologue by Octavio Castro López. Xalapa-Enríquez: Universidad Veracruzana, 1994.

WORKS CONSULTED

Bancroft, Hubert Howe. *History of Mexico*, vols. I–VI. San Francisco: The History Company, 1890.

Beuchot, Mauricio, and Bernabé Navarro, eds. *Dos Homanajes: Alonso de la Veracruz y Francisco Xavier Clavigero*. México: Universidad Nacional Autónoma de México, 1992.

Bisbee, Henry H. *The Island of Burlington, Three Hundred Years of the City of Burlington and Burlington Township*. Burlington, NJ: Tom Cook, 1977.

Boese, Alex. *The Museum of Hoaxes, A Collection of Pranks, Stunts, Deceptions, and Other Wonderful Stories Contrived for the Public from the Middle Ages to the New Millennium*. New York: Dutton, 2002.

Bolton, Herbert Eugene. *Rim of Christendom, a Biography of Eusebio Francisco Kino, Pacific Coast Pioneer*. New York: Macmillan, 1936.

Bolton, Herbert Eugene. *The Padre on Horseback, a Sketch of Eusebio Francisco Kino, S.J., Apostle to the Pimas*. San Francisco: Sonora Press, 1932. Reissued, Chicago: Loyola University Press, 1963.

Bolton, Herbert Eugene. "The Epic of Greater America." In *Do the Americas Have a Common History? A Critique of the Bolton Theory*, ed. Lewis Hanke. New York: Alfred A. Knopf, 1964.

Braden, Charles S. *Religious Aspects of the Conquest of Mexico*. Durham, NC: Duke University Press, 1930.

Burrus, Ernest J. "Kino, Historian's Historian." *Arizona and the West* 4/2 (Summer 1962): 145–156.

Burrus, Ernest J. *Kino and the Cartography of Northwestern New Spain*. Tucson: Arizona Pioneers Historical Society, 1965.

Burrus, Ernest J. *Kino and Manje, Explorers of Sonora and Arizona, Their Vision of the Future, a Study of Their Expeditions and Plans*. Sources and Studies for the History of the Americas: Vol. X. Rome: Jesuit Historical Institute, 1971.

Bussey, W. H. "The Origin of Mathematical Induction." *The American Mathematical Monthly* XXIV/5 (May 1917): 199–207.

Cajori, Florian. *The Early Mathematical Sciences in North and South America*. Boston: The Gorham Press, 1928.

Cajori, Florian. *A History of Mathematical Notations, Volume I, Notations in Elementary Mathematics*. La Salle, IL: Open Court, 1928.

Cope, Gilbert. "Jacob Taylor, Almanac Maker." *Bulletins of the Chester County Historical Society* (1908): 10–28.

Cotter, Charles H. *A History of Nautical Astronomy*. London: Hollis and Carter, 1968.

Duncan, David Ewing. *Calendar, Humanity's Epic Struggle to Determine a True and Accurate Year*. New York: Bard, 1999.

Ennis, Arthur. "Fray Alonso de la Vera Cruz, O.S.A. (1507–1584), a Study of His Life and His Contribution to the Religious and Intellectual Affairs of Early Mexico," Introduction and Chapters I–III. *Augustiniana* V Fasc. 1–2 (Apr. 1955): 52–124.

Ennis, Arthur. "Fray Alonso de la Vera Cruz, O.S.A. (1507–1584), a Study of His Life and His Contribution to the Religious and Intellectual Affairs of Early Mexico," Chapter IV. *Augustiniana* V Fasc. 1–2 (Apr. 1955): 241–267.

Ennis, Arthur. "Fray Alonso de la Vera Cruz, O.S.A. (1507–1584), a Study of His Life and His Contribution to the Religious and Intellectual Affairs of Early Mexico," Chapter V. *Augustiniana* V Fasc. 1–2 (Apr. 1955): 362–399.

Ennis, Arthur. "Fray Alonso de la Vera Cruz, O.S.A. (1507–1584), a Study of His Life and His Contribution to the Religious and Intellectual Affairs of Early Mexico," Chapters VI–VII. *Augustiniana* VII Fasc. 1–2 (Apr. 1957): 149–195.

Ennis, Arthur. *Fray Alonso de la Vera Cruz, O.S.A. (1507–1584), a Study of His Life and His Contribution to the Religious and Intellectual Affairs of Early Mexico.* Offprint from *Augustiniana*. (Contains preliminary material and appendices not in aforementioned articles.) Louvain: E. Warny, 1957.

Farrés, Octavio Gil. *Historia de la Moneda Española.* Madrid: Adrados, 1976.

Hansen, Chadwick. *Witchcraft at Salem.* New York: George Braziller, 1969.

Haring, Clarence H. "American Gold and Silver Production in the First Half of the Sixteenth Century." *The Quarterly Journal of Economics* XXIX (May 1915): 433–479.

Heilbron, J. L. *The Sun in the Church, Cathedrals as Solar Observatories.* Cambridge, MA: Harvard University Press, 1999.

Iglesia, Ramón. "The Disillusionment of Don Carlos." In *Columbus, Cortés, and Other Essays*, trans. and ed. Lesly Byrd Simpson. Berkeley: University of California Press, 1969.

Instituto de Investigaciones Jurídicas. *Homenaje a Fray Alonso de la Veracruz en el Cuarto Centenario de su Muerte (1584–1984).* México: Universidad Nacional Autónoma de México, 1986.

Ives, S. A. "Alonso de la Vera Cruz: The Father of Scientific and Legal Study in America." *Rare Books, Notes on the History of Old Books and Manuscripts* 5/4 (June 1947): 1–5.

Jones, Rufus. *The Quakers in the American Colonies.* London: Macmillan, 1911.

Kandell, Jonathan. *La Capital, the Biography of Mexico City.* New York: Random House, 1988.

Karpinski, Louis C. *The History of Arithmetic.* Chicago: Rand McNally, 1925.

Katz, Victor J. *A History of Mathematics, An Introduction*, 2nd ed. Reading, MA: Addison-Wesley, 1998.

Lamb, Ursula. "Nautical Scientists and Their Clients in Iberia (1508–1624): Science from Imperial Perspective." *Revista da Universidade de Coimbra* XXXII (1985): 49–61.

Lamb, Ursula. "Cosmographers of Seville: Nautical Science and Social Experience." In *Cosmographers and Pilots of the Spanish Maritime Empire.* Aldershot: Variorum, 1995: 675–686.

Lamb, Ursula. "The Teaching of Pilots and the *Chronographía o Reportorio de Los Tiempos*." In *Cosmographers and Pilots of the Spanish Maritime Empire.* Aldershot: Variorum, 1995.

Leonard, Irving A. *Don Carlos de Sigüenza y Góngora, a Mexican Savant of the Seventeenth Century*. University of California Publications in History, vol. 18. Berkeley: University of California Press, 1929.

Leonard, Irving A. "Sigüenza y Góngora and the Chaplaincy of the Hospital del Amor de Dios." *The Hispanic American Historical Review* XXXIX/4 (Nov. 1956): 580–587.

Leonard, Irving A. *Baroque Times in Old Mexico. Seventeenth-Century Persons, Places, and Practices*. Ann Arbor: The University of Michigan Press, 1959.

Masotti, Arnaldo. "Maurolico, Francesco." In *Dictionary of Scientific Biography*, ed. Charles Coulston Gillispie. New York: Scribner, 1970–1980.

Mayer, Alicia, ed. *Carlos de Sigüenza y Góngora, Homenaje 1700–2000*. México: Universidad Nacional Autónoma de México, 2000.

Maza, Francisco de la. "Enrico Martínez." In *New Catholic Encyclopedia*. Washington, D.C.: The Catholic University of America, 1967.

Maza, Francisco de la. *Enrico Martínez, Cosmógrafo e Impresor de Nueva España*. Facsimile of the 1943 edition, together with a facsimile of Francisco González de Cossío's bibliographic appendix to the 1948 edition of the *Reportorio de los Tiempos*. México: Universidad Nacional Autónoma de México, 1991.

McCusker, John J. *Money and Exchange in Europe and America, 1600–1775, A Handbook*. Chapel Hill: The University of North Carolina Press, 1978.

O'Gorman, Edmundo. "Nuevos Datos sobre el Dr. Diego García de Palacio." *Boletín del Archivo General de la Nación* XVII/1 (Jan.–Mar. 1946): 1–31.

Paylore, Patricia P., et al. *Kino . . . A Commemoration*. Tucson: Arizona Pioneers Historical Society, 1961.

Polzer, Charles W. *Kino, A Legacy, His Life, His Works, His Missions, His Monuments*. Tucson: Jesuit Fathers of Southern Arizona, 1998.

Prescott, William Hickling. *History of the Conquest of Mexico and History of the Conquest of Peru*. New York: The Modern Library, 1936.

Ramírez López, Ignacio. *Tres Biografías, Fray Pedro de Gante, Fray Alonso de la Veracruz, Fray Juan Bautista Moya*. Biblioteca Enciclopédica Popular, tomo 201. México: Secretaria de Educación Pública, 1948.

Restall, Matthew. *Seven Myths of the Spanish Conquest*. Oxford: Oxford University Press, 2003.

Robinson, James Howard. *The Great Comet of 1680, a Study in the History of Rationalism*. Reprint of 1916 edition. Cleveland, OH: Zubal, 1986.

Ronan, Colin A. *Edmond Halley, Genius in Eclipse*. New York: Doubleday, 1969.

Schechner Genuth, Sara. *Comets, Popular Culture, and the Birth of Modern Cosmology*. Princeton, NJ: Princeton University Press, 1997.

Sibley, John Langdon. *Biographical Sketches of Graduates of Harvard University, in Cambridge, Massachusetts*, Volume I, 1642–1658; Volume II, 1659–1677; Volume III, 1678–1689. Cambridge: Charles William Sever, 1885.

Smith, David Eugene. *History of Mathematics, Volume I, General Survey of the History of Elementary Mathematics*. Boston: Ginn and Co., 1923.

Smith, David Eugene, and Jekuthiel Ginsberg. *A History of Mathematics in America Before 1900*. The Carus Mathematical Monographs, Number 5. Chicago: The Mathematical Association of America with the cooperation of the Open Court Publishing Company, 1934.

Smith, Fay Jackson, et al. *Father Kino in Arizona*. Phoenix: Arizona Historical Foundation, 1966.

Smith, Joseph. *Short Biographical Notices of William Bradford, Reiner Jansen, Andrew Bradford, and Samuel Keimer, Early Printers in Pennsylvania*. London: Edward Hicks, 1891.

Sobel, Dava. *Longitude, The True Story of a Lone Genius Who Solved the Greatest Scientific Problem of His Time*. New York: Penguin, 1996.

Swetz, Frank J. *Capitalism and Arithmetic, the New Math of the 15th Century, Including the Full Text of the* Treviso Arithmetic *of 1478, Translated by David Eugene Smith*. La Salle, IL: Open Court, 1987.

Trabulse, Elías. *Ciencia y Religión en el Siglo XVII*. Guanajuato: El Colegio de México, 1974.

Trabulse, Elías. *Historia de la Ciencia en México, Estudios y Textos, Siglo XVI*. México: Fondo de Cultura Económica, 1983.

Trabulse, Elías. *Historia de la Ciencia en México, Estudios y Textos, Siglo XVII*. México: Fondo de Cultura Económica, 1984.

Trabulse, Elías. *La Ciencia Perdida, Fray Diego Rodríguez, un Sabio del Siglo XVII*. México: Fondo de Cultura Económia, 1985.

Trabulse, Elías. *Historia de la Ciencia en México, Apéndices e Índices*. México: Fondo de Cultura Económica, 1989.

Trabulse, Elías. "Un Científico Mexicano del Siglo XVII: Fray Diego Rodríguez y su Obra." In *Historia de la Ciencia y la Tecnología*, ed. Elías Trabulse. México: El Colegio de México, 1991.

Trabulse, Elías. *Historia de la Ciencia en México*. Abridged version. México: Fondo de Cultura Económica, 1994.

Trueba Olivares, Alfonso. *El Padre Kino, Misionero Itinerante y Ecuestre*, 2nd ed. México: Editorial Jus, 1960.

Vacca, G. "Maurolycus, the First Discoverer of the Principle of Mathematical Induction." *Bulletin of the American Mathematical Society* XVI/2 (Nov. 1909): 70–73.

Vilanova Rodríguez, Alberto. "Díez, Juan." In *Gran Enciclopedia Gallega*. Santiago de Compostela: Silverio Cañada, 1974.

Wall, Alexander J., Jr. *William Bradford, Colonial Printer*. Reprinted from the *Proceedings of the American Antiquarian Society* (Oct. 1963). Worcester, MA: American Antiquarian Society, 1963.

Wallace, John William. "Early Printing in Philadelphia, The Friends Press—Interregnum of the Bradfords." *The Pennsylvania Magazine of History and Biography* IV/4 (1880): 432–444.

Watson, John F. *Annals of Philadelphia, and Pennsylvania, in the Olden Time; Being a Collection of Memoirs, Anecdotes, and Incidents of the City and Its Inhabitants, and of the Earliest Settlements of the Inland Part of Pennsylvania*, Vol. I. Enlarged, with many revisions and additions, by Willis P. Hazard. Philadelphia: Edwin S. Stuart, 1900.

Wilson, James Grant and John Fiske, eds. *Appleton's Cyclopedia of American Biography*. New York: D. Appleton and Co., 1887.

Winslow, Ola Elizabeth. *Samuel Sewall of Boston*. New York: Macmillan, 1964.

Woodward, E. M. *History of Burlington County, New Jersey, with Biographical Sketches of Many of Its Pioneers and Prominent Men*. Philadelphia: Everts and Peck, 1883. Reprinted, Burlington, NJ: Burlington County Historical Society, 1984.

Wyllys, Rufus Kay. *Pioneer Padre, the Life and Times of Eusebio Francisco Kino*. Dallas, TX: Southwest Press, 1935.

Miscellaneous

Beyer, Steven L. *The Star Guide, A Unique System for Identifying the Brightest Stars in the Night Sky*. Boston: Little, Brown and Co., 1986.

Boolos, George. *The Logic of Provability*. Cambridge: Cambridge University Press, 1993.

Chahal, Jasbir S. "Congruent Numbers and Elliptic Curves." *The American Mathematical Monthly* 113/4 (Apr. 2006): 308–317.

Conway, John H., and Richard K. Guy. *The Book of Numbers*. New York: Springer, 1996.

Dershowitz, Nachum, and Edward M. Reingold. *Calendrical Calculations*. Cambridge: Cambridge University Press, 1997.

Dudley, Underwood. *Mathematical Cranks*. Washington, D.C.: The Mathematical Association of America, 1992.

Flaste, Richard, et al. *The New York Times Guide to the Return of Halley's Comet*. New York: Times Books, 1985.

G., T., and W. S. B. W. "Calendar." In *The Encyclopedia Britannica, A Dictionary of Arts, Sciences, Literature and General Information*, 11th ed. New York: Encyclopedia Britannica, 1910.

Heath, Thomas. *Mathematics in Aristotle*. London: Oxford University Press, 1970.

Heijenoort, Jean van. *From Frege to Gödel, a Source Book in Mathematical Logic, 1879–1931*. Cambridge: Harvard University Press, 1967.

Koblitz, Neal. *Introduction to Elliptic Curves and Modular Forms*. 2nd ed. New York: Springer Verlag, 1993.

Konyndyk, Kenneth. *Introductory Modal Logic*. Notre Dame, IN: University of Notre Dame Press, 1986.

Real Academia Española. *Diccionario de la Lengua Española*, 21st ed. Madrid: Real Academia Española, 1992.

Wallenquist, Åke. *Dictionary of Astronomical Terms*, ed. and trans. Sune Engelbrektson. Garden City, NY: The Natural History Press, 1966.

General Index

This index is not intended to duplicate the listings in the indices of authors, libraries, and printers nor the list of short titles of the entries found in the Introduction. An entry is only referenced here under its author if it includes some biographical information for the same or if the author is listed under "Other Authors." Some cities are mentioned if they come up in the text of an entry but not if they are just a publication site.

Aguilar Cantú, Antonio Sebastián de, 291
algebra, 34, 64–65, 109, 134, 195; etymology of, 139–140
almanacs, 2, 3, 7–8, 17, 91n92, 117, 126, 157, 176, 183–186, 324; etymology of, 256; feast days mentioned in, 264, 270, 272n96, 307–308; finding the time by means of, 7, 224; hoax, 289–290; humor in, 221, 253, 276; illustrations in, 243–244, 259
almojarifazgo, 108
almucantar, 69–70
Alonso de la Vera Cruz. *See* Vera Cruz, Alonso de la
anti-Catholic propaganda, 221, 287, 301, 304
anti-English propaganda, 162n190
Aristotle, 6, 28, 31, 33, 73, 131, 155, 159, 161, 262, 325, 326n4, 327
arithmetic, 4, 5, 34, 54–64, 195, 309; use of large X in, 133, 135. *See also* modular arithmetic; squadrons
Arte de Navegar, 87
Asbaje, Juana. *See* Cruz, Juana Inés de la

astrology, 6, 117–119, 154n170, 155, 157, 184, 192, 226, 230, 243–244, 264, 289–290, 303, 306–307, 310
astronomy, 2, 6, 74, 77, 162, 164–165, 309. *See also* comets; moons of other planets; red spot on Jupiter; rings of Saturn; sunspots
Augustinians, 25, 43, 192
aurora borealis, 197
Avilés Ramírez, Juan de, 265, 268, 274
azimuth, 69–70

Belerofonte Mathemático, 158–159, 162–162, 323
Bickerstaff, Isaac. *See* Swift, Jonathan
bissextile, 189–190. *See also* leap year
Bolton, Herbert Eugene, 1–2, 153
Boyle, Robert, 143
Brackenbury, Samuel, 210, 217
Bradford, William, 263, 264, 274, 293, 298–299, 312
Bradstreet, Anne, 204, 219, 234
Bradstreet, Samuel, 204, 219
Brahe, Tycho, 155, 161, 162n188, 209
Brattle, Thomas, 3, 7, 139, 228, 239–241, 253

Brattle, William, 239, 253–254, 300
Brigden, Zecharia, 205
Browne, Joseph, 220, 245

calderón (U), 104, 110, 120, 133, 142, 173
calendar round, 8
Campanus of Novara, 66, 69, 75
Campos, José de, 268, 277
Carmelites, 43
castellano, 45, 47, 65, 66, 112, 179, 180
Castro, Felipe de, 192
Chauncey, Israel, 203, 209, 210
Chauncey, Nathaniel, 208, 266
Chávez, Alonso de, 92n94, 117n114
Chávez, Jerónimo de, 117, 126–127
Cheever, Samuel, 206
Chinese units of weight, 111
Clapp, John, 293, 307–308, 312, 314
Clough, Samuel, 244, 317
comets, 2, 3, 6, 139, 146, 155, 215, 220, 245, 251, 253, 255–256, 262, 310; Aristotle's views on, 155, 159, 161, 262; books on, 154–155, 156, 157–160, 161–163, 165–166, 189, 193, 195, 196, 198–199, 242, 255–256, 322–324; play on, 156. *See also* Great Comet of 1680–1681; Halley's Comet
congruent number problem, 59–64
Copernicus, Nicolaus, 75, 207, 158, 307
Cordoba, Martín de. *See* Rodríguez, Diego
corona, 52, 53
Cortés, Hernán, 37, 39, 41n38, 42, 43, 82
Cruz, Juana Inés de la, 167–168, 234
cubic equations, 199–201
cuento, 133, 173
cycle of the moon. *See* golden number
cycle of the sun, 100, 203, 280

Danforth, John, 188–189, 241, 266
Danforth, Samuel (1626–1674), 7, 154n170, 159n154, 187, 188–189, 241, 245, 266, 322
Danforth, Samuel (1666–1727), 7, 188–189, 241, 264n84, 266

D'Antonio, Lawrence, 59–60
Day, Matthew, 187, 190–191, 193
Daye, Stephen, 186–187, 190, 192
Díaz, Juan (secular priest), 37–38, 39–40, 43
Díaz, Juan (theologian), 37, 41
Díaz, Juan, Clérigo, 37, 38, 40–41
Diez Freyle, Juan, 3, 19, 37–45; Peruvian origin hypothesis for, 3, 42–43, 52, 65–66, 109
diezmo, 47, 52, 53, 54, 65, 66, 111, 129, 141, 177, 180
Dionysius Exiguus, 89, 255
Divine Comedy, 308
dominical letter, 100, 203, 205, 208, 209–210, 211, 214, 229, 235, 239, 245, 264, 266, 273, 307, 317
Dominicans, 28n18, 39, 43, 45, 184
Drake, Francis, 79
Drake, Milton, 7n13
Dudley, Joseph, 204, 210, 219
Dudley, Thomas, 204, 219
ducado, 45–46, 48n54, 52, 53, 110, 133
duplication of the cube and sphere, 171 172

Easter, 100, 245, 272, 281, 294; calculation of, 77, 88–89, 92, 97, 209–211, comparison of, for Julian and Gregorian calendars, 264, 280
Eliot, John, 143, 192
elliptic curves, 63n68
encomienda, 27
epact, 76, 88, 89, 90–98, 202, 210–211, 214, 224, 280, 283, 316
equation of time, 138, 275
escala altimentria, 81
Escobar Salmerón y Castro, José de, 157n174, 159–160, 226, 242, 245, 258, 323

Fagoaga, Francisco de, 179
false position, 109
Fermat's Last Theorem, 63n68
Fernández de Belo, Benito, 5, 82, 145, 174n207

GENERAL INDEX

Fibonacci (Leonardo of Pisa), 59–60
Flint, Josiah, 216, 266
Foster, John, 190, 229, 230–231, 251, 253, 256, 264, 266, 297
fractions: Babylonian, 98; currency denominations used for, 103, 106–107; decimal, 66, 150, 159, 163, 181; *infilzar* the, 134; large denominators in, 141, 173; large X with, 133, 135
Franciscans, 39n33, 43, 131
Franklin, Benjamin, 290
fray, frey, freile, 39
freedom of the press, trial involving, 298

Galileo, 8
galley method of division, 57, 87–88, 133, 150
Gamboa y Riaño, Marco Antonio de, 311
García de Palacio, Diego, 98–80, 174n207
geometry, 6, 7, 30, 32, 33, 74, 75, 149, 161, 171–172, 195, 199, 309
Gilbert and Sullivan, 45n44
golden number, 76, 87–88, 90–91, 94–95, 100, 202, 224, 225, 227, 280
grano, differing values of, 49, 54, 104, 150, 179
Great Comet of 1680–1681, 148, 153–154, 155–156, 157, 161–162, 168, 239, 251
Green, Bartholomew, 284
Green, Samuel, Jr., 239, 256–257, 276, 284
Green, Samuel, Sr., 143, 192, 207, 230, 256
Gregorian calendar reform, 75–77, 89, 90, 91–92, 117n114, 120, 211, 253, 316
Gutiérrez, Alonso. *See* Vera Cruz, Alonso de la
Guzmán y Córdova, Sebastian de, 158–159, 162–163, 323

Halley, Edmund, 157n176
Halley's Comet, 155, 157n176, 166, 253, 255
Harris, Benjamin, 77, 221, 276, 281, 287, 288, 296, 300, 301, 304, 307
Hebrew names for months, 294, 298
heliocentric theory, 205, 208, 216, 229, 251, 267, 280, 307. *See also* Copernicus, Nicolaus
Hindu-Arabic numerals, 55, 104, 110
Hobart, Nehemiah, 225
home remedies, 303

imaginary numbers, 199–201
induction, 75
interest, in word problems, 135
Iolanthe, 45n44
Itinerario (of Juan Díaz), 37, 38, 39, 41n38

Jefferson, Thomas, 28
Jesuits, 1, 43, 73, 156, 160–161, 166, 170
Johnson, Marmaduke, 143, 207, 230

Kepler, Johannes, 155, 157n176, 227, 228n42
King Philip's War, 143, 245, 246
Kino, Eusebio Francisco, 3, 6, 146, 149, 153, 154, 156, 160–162, 166–168, 239
Koenig, Juan Ramon, 170–171, 248n55, 249–250

latitude, 139; of Boston, 251, 260, 267, 280; of Mexico City, 123
leagues (distance), 8, 98
leap year, 89, 90, 91n93, 93, 94, 94n98, 95, 97, 100, 189, 192, 194, 203, 205, 210, 229, 235, 245, 246, 253, 264n84, 266, 272, 272n96, 316, 317. *See also* bissextile
Leeds, Daniel, 176, 190, 263, 270, 272n96, 274, 277, 290, 307, 315, 318
Leeds, Titan, 290
Leonard, Irving, 153
libraries, founding of, 25–26
Lodowick, Christian, 300, 306
logarithms, 159, 163–165, 199
logic, 5–6, 28, 73, 131, 143–144, 253–254, 325–327

347

longest day, 3, 239–241
longitude, 139; of Boston, 191; determination of, 87, 100–101, 126, 163, 260, 318
López de Bonilla, Gabriel, 154n170, 163, 193, 322

man of signs, 243–244. *See also* moon: dominion of, in man's body
maravedi, 45–46
marco, 45
"Mares Eat Oats," 7n13
Martínez, Enrico, 115–117, 127, 163, 164n193, 196, 197n15, 232, 241, 324
Massachusett (language), 142–144
Mather, Cotton, 256, 257, 296, 315
Mather, Increase, 154n170, 159n81, 166, 204, 227, 254, 256, 266, 323
Mather, Nathanael, 260, 264n84, 266, 280
Matta, Nicolás de, 221
Maurolico, Francesco, 74–75, 77
Metonic cycle, 90, 91, 93–94, 95, 211. *See also* golden number
Mercedarians, 39, 43, 198
México, D. F., i.e., Mexico City, flood of, 116
modular arithmetic, 63, 89–91, 93, 94, 202–203, 210–211, 214, 224
money of account, 47–48
moon: age of, i.e., phase, 88, 89, 92, 96, 97–98, 186, 191, 202, 222, 228, 266, 273, 278, 316; dominion of, in man's body, 243–244, 297, 301
moons of other planets, 260, 267

navigation, 86, 88, 98, 100–101, 184, 195
Newman, Henry, 7, 280
Newton, Isaac, 157n176, 239
Newton's Comet, 157n176. *See also* Great Comet of 1680–1681
Nowell, Alexander, 215, 225
number of direction, 209–211, 245, 280. *See also* Easter

Oakes, Urian, 7, 194, 256, 266
Olmedo, Bartolomé de, 39, 43

parallax, 149, 154n170, 155, 161, 199, 262, 267
Partridge, John, 77, 287–290
patacón, 47, 108, 128, 129
perihelion (Earth's), 138–139, 228
peso de minas, 42, 47–48, 50, 52, 53, 54, 65, 105, 110, 128, 129, 130, 142
peso de plata corriente, 47, 48, 105, 110, 128
peso de tipuzque, 47, 52, 53, 54, 65, 110, 128, 130, 141, 142, 180, 181
Philip II (King of Spain), 4, 47, 80n81
pi (π), 8, 162
piece(s) of eight, 48
Pierce, William, 187, 189, 191–192
plata del rescate, 52, 53, 54, 65, 111, 129, 130, 141
plata quintada, 52, 53, 54, 65, 66, 111, 129, 130, 141, 177
poems, 167–168, 183, 216, 243–244, 276, 289, 309
Porphyry, 28, 33, 130
printing press, first arrival in various cities, 2; Boston, 230; Cambridge, MA, 186–187; Lima, 4, 42, 73; New York, 264, 293, 298; Philadelphia, 263
pseudonyms, Inquisition ban on, 268, 274, 275
Puritan thought, 143, 190, 209, 214, 272n96

Quakers, 299
quinto (tax), 47, 53, 54, 57, 66, 108, 111, 177, 180

real (coin), 46
red spot on Jupiter, 260
Ricardo, Antonio, 73
Richardson, John, 221
riddles, 305, 313, 314
rings of Saturn, 267
Rodríguez, Diego, 154n170, 163, 195, 196, 198–199, 322
Roman indiction, 100, 202
Roman numerals, 54, 104, 216
Romano, Michael Henrico. *See* Campos, José de

Rousseau, Jean-Jacques, 28
Rowlandson, Mary, 234
Ruiz, Feliciana, 185, 232–234
Ruiz, Juan, 154n170, 157n174, 159n181, 185, 196–197, 199, 207, 225, 232, 322
Ruiz Lozano, Francisco, 171, 195, 196, 198, 199, 248, 249, 322
rule of three, 55, 108, 133, 135
Russell, Daniel, 222
Russell, Noadiah, 259

Santiago (religious order), 179
Salem witch trials, 239, 257
Saucedo, Juan de, 157n174, 226
seasons, lengths of, 138–139, 228, 239–240
Sewall, Samuel, 252, 253, 256–257, 258; ms notes by, in almanacs, 188, 229, 232, 239, 242, 245, 246, 251, 253, 257, 258, 260, 266, 267, 272, 273, 275, 281, 284, 296
Shepard, Jeremiah, 223, 227
Shepard, Thomas, 202, 245, 266
Sherman, John, 189, 227, 266
Sigüenza y Góngora, Carlos de, 3, 6, 146, 150n161, 153–154, 156–168, 177, 185, 193, 222–223, 226, 230, 235, 239, 252, 259, 261, 275, 278, 280, 291, 311, 318

Socrates, 33
Speculum Coniugiorum, 26
squadrons, 80, 81–82, 135, 139, 145
sunspots, 260, 267
Swift, Jonathan, 289–290
syllogism, 28, 143, 325

Taylor, Jacob, 7, 176, 277, 314, 315, 317
tides, 88, 92, 183, 280, 300
Toledo, Francisco de, 73
tomín, 48
Torre, Martín de la, 154n170, 158, 163, 322
trigonometry, 6, 81, 98–100, 120–126, 163–165, 174, 241, 318
Tulley, John, 190, 244, 264, 271, 272n96, 276, 297, 300, 301, 303, 306–307, 319

Valera, Jerónimo de, 131
Vera Cruz, Alonso de la, 25–28, 327
volvelle, 118

Washington, George, 256n73
Whitechurch, Edward, 33
Williams, William, 98, 262

Ytinerario de Navegación, 86

Index to Works in Part One

By Topics

Accounting (Gold and Silver, Currencies and Currency Reform, Percentages, etc.)

Juan Diez Freyle, *Sumario Compendioso*, 1556. **3.**
Joán de Belveder, *Libro General*, 1597. **10.**
Felipe Echagoyan, *Tablas de Reducciones*, 1603. **11.**
Francisco Juan Garreguilla, *Libro de Plata*, 1607. **13.**
Pedro de Aguilar Gordillo, *Alivio de Mercaderes*, 1610. **14.**
Álvaro Fuentes y de la Cerda, *Libro de Cuentas*, 1615. **16.**
Pedro Diez de Atienza, *Sobre la Reducción de Monedas*, 1650. **20.**
Francisco de Villegas, *Repuesta*, 1650. **21.**
Pedro Diez de Atienza, *Cerca de la Reformación de la Moneda*, 1652. **22.**
Luis Enríquez de Guzmán, *Para que Pudiese Correr la Moneda*, 1657. **23.**
Francisco de Villegas, *Reparos*, 1657. **24.**
Juan de Castañeda, *Reformación de las Tablas*, 1668. **27.**
Diego Pérez de Lazcano, *Proposición, y Manifiesto*, 1691. **33.**
José Martí, *Tabla General*, 1696. **36.**
Manuel de Zuaza y Aranguren, *Reducciones de Plata*, 1697. **38.**
Francisco de Fagoaga, *Reducción de Oro*, 1700. **39.**

Arithmetic and Algebra (Including Squadrons)

Juan Diez Freyle, *Sumario Compendioso*, 1556. **3.**
Diego García de Palacio, *Diálogos Militares*, 1583. **8.**
Joán de Belveder, *Libro General*, 1597. **10.**
Pedro de Paz, *Declaración de los Puntos*, 1621. **17.**
Pedro de Paz, *Arte para Aprender*, 1623. **18.**
Atanasius Reatón, Pasamonte, *Arte Menor*, 1649. **19.**
Antonio de Heredia y Estupiñán, *Teórica, y Práctica de Esquadrones*, 1660. **26.**

INDEX TO WORKS IN PART ONE

Benito Fernández de Belo, *Breve Aritmética*, 1675. **29**.
Martín de Echagaray, *Declaración del Quadrante*, 1682. **31**.

Astrology

Enrico Martínez, *Reportorio de los Tiempos*, 1606. **12**.
Juan de Figueroa, *Opusculo de Astrología*, 1660. **25**.

Astronomy and Aristotelian Physical Science

Alonso de la Vera Cruz, *Phisica, Speculatio*, 1557. **4**.
Francesco Maurolico, *De Sphæra*, 1578. **6**.

Astronomy—Comets

Eusebio Francisco Kino, *Exposición Astronómica*, 1681. **30**.
Carlos de Sigüenza y Góngora, *Libra Astronómica*, 1690. **32**.

Astronomy—Tables

Diego García de Palacio, *Instrucción Náutica*, 1587. **9**.
Enrico Martínez, *Reportorio de los Tiempos*, 1606. **12**.
Juan de Figueroa, *Opusculo de Astrología*, 1660. **25**.
Jacob Taylor, *Tenebræ*, 1697. **37**.

Calendrical Matters

Francesco Maurolico, *Computus Ecclesiasticus*, 1578. **7**.

Geometry

Alonso de la Vera Cruz, *Dialectica Resolutio*, 1554. **2**.
Francesco Maurolico, *De Sphæra*, 1578. **6**.
Diego García de Palacio, *Diálogos Militares*, 1583. **8**.
Eusebio Francisco Kino, *Exposición Astronómica*, 1681. **30**.
Juan Ramón Koenig, *Cubus*, 1696. **35**.
Jacob Taylor, *Tenebræ*, 1697. **37**.

Logic (Aristotelian Analysis)

Alonso de la Vera Cruz, *Recognitio, Summularum*, 1554. **1**.
Francisco de Toledo, *Introductio in Dialecticam Aristotelis*, 1578. **5**.
Jerónimo Valera, *Comentarii ac Quæstiones*, 1610. **15**.
John Eliot, *The Logic Primer*, 1672. **28**.
Nicolás de Olea, *Summa Tripartita—in Logicam*, 1693. **34**.

Military Matters

Diego García de Palacio, *Diálogos Militares*, 1583. **8**.
Diego García de Palacio, *Instrucción Náutica*, 1587. **9**.

Atanasius Reatón, *Arte Menor*, 1649. **19.**
Antonio de Heredia y Estupiñan, *Teórica, y Práctica de Esquadrones*, 1660. **26.**
Benito Fernández de Belo, *Breve Aritmética*, 1675. **29.**

Trigonometry

Diego García de Palacio, *Diálogos Militares*, 1583. **8.**
Diego García de Palacio, *Instrucción Náutica*, 1587. **9.**
Enrico Martínez, *Reportorio de los Tiempos*, 1606. **12.**
Carlos de Sigüenza y Góngora, *Libra Astronómica*, 1690. **32.**
Jacob Taylor, *Tenebræ*, 1697. **37.**

By Language

English

John Eliot, *The Logic Primer*, 1672. **28.**
Jacob Taylor, *Tenebræ*, 1697. **37.**

Latin

Alonso de la Vera Cruz, *Recognitio, Summularum*, 1554. **1.**
Alonso de la Vera Cruz, *Dialectica Resolutio*, 1554. **2.**
Alonso de la Vera Cruz, *Phisica, Speculatio*, 1557. **4.**
Francisco de Toledo, *Introductio in Dialecticam Aristotelis*, 1578. **5.**
Francesco Maurolico, *De Sphæra*, 1578. **6.**
Francesco Maurolico, *Computus Ecclesiasticus*, 1578. **7.**
Jerónimo Valera, *Comentarii ac Quæstiones*, 1610. **15.**
Nicolás de Olea, *Summa Tripartita—in Logicam*, 1693. **34.**
Juan Ramón Koenig, *Cubus*, 1696. **35.**

Massachusett (Algonquian)

John Eliot, *The Logic Primer*, 1672. **28.**

Spanish

Juan Diez Freyle, *Sumario Compendioso*, 1556. **3.**
Diego García de Palacio, *Diálogos Militares*, 1583. **8.**
Diego García de Palacio, *Instrucción Náutica*, 1587. **9.**
Joán de Belveder, *Libro General*, 1597. **10.**
Felipe Echagoyan, *Tablas de Reducciones*, 1603. **11.**
Enrico Martínez, *Reportorio de los Tiempos*, 1606. **12.**
Francisco Juan Garreguilla, *Libro de Plata*, 1607. **13.**
Pedro de Aguilar Gordillo, *Alivio de Mercaderes*, 1610. **14.**
Álvaro Fuentes y de la Cerda, *Libro de Cuentas*, 1615. **16.**

INDEX TO WORKS IN PART ONE

Pedro de Paz, *Declaración de los Puntos*, 1621. **17.**
Pedro de Paz, *Arte para Aprender*, 1623. **18.**
Atanasius Reatón, *Arte Menor*, 1649. **19.**
Pedro Diez de Atienza, *Sobre la Reducción de Monedas*, 1650. **20.**
Francisco de Villegas, *Repuesta*, 1650. **21.**
Pedro Diez de Atienza, *Cerca de la Reformación de la Moneda*, 1652. **22.**
Luis Enríquez de Guzmán, *Para que Pudiese Correr la Moneda*, 1657. **23.**
Francisco de Villegas, *Reparos*, 1657. **24.**
Juan de Figueroa, *Opusculo de Astrología*, 1660. **25.**
Antonio de Heredia y Estupiñan, *Teórica, y Práctica de Esquadrones*, 1660. **26.**
Juan de Castañeda, *Reformación de las Tablas*, 1668. **27.**
Benito Fernández de Belo, *Breve Aritmética*, 1675. **29.**
Eusebio Francisco Kino, *Exposición Astronómica*, 1681. **30.**
Martín de Echagaray, *Declaración del Quadrante*, 1682. **31.**
Carlos de Sigüenza y Góngora, *Libra Astronómica*, 1690. **32.**
Diego Pérez de Lazcano, *Proposición, y Manifiesto*, 1691. **33.**
José Martí, *Tabla General*, 1696. **36.**
Manuel de Zuaza y Aranguren, *Reducciones de Plata*, 1697. **38.**
Francisco de Fagoaga, *Reducción de Oro*, 1700. **39.**

Other Indices

Index of Authors

Antonio Sebastián de Aguilar Cantú. **A106, A131, A146, A153, A160, A167, A175, A186, A192, A198, A204, A215.**
Pedro de Aguillar Gordillo. **14.**
Samuel Atkins. **A125.**
Juan de Avilés Ramírez. **A126, A132, A139, A147, A154, A161, A168, A176, A193.**
Joán de Belveder. **10.**
Samuel Brackenbury. **A57.**
Samuel Bradstreet. **A30.**
Thomas Brattle. **A90.**
William Brattle. **A107.**
Zecharia Brigden. **A34.**
Joseph Browne. **A62.**
José de Campos. **A133, A140, A148.**
Juan de Castañeda. **27.**
Felipe de Castro. **A12.**
Israel Chauncy. **A44, A48.**
Nathaniel Chauncy. **A41.**
Samuel Cheever. **A37, A39.**
John Clapp. **A199.**
Samuel Clough. **A216.**
Martín de Córdoba. **A25, A42, A45, A49, A53.**
John Danforth. **A91, A95.**
Samuel Danforth (1626–1674). **A9, A10, A11, A13.**
Samuel Danforth (1666–1727). **A127.**
Pedro Diez de Atienza. **20, 22.**
Juan Diez Freyle. **3.**
Joseph Dudley. **A60.**
Edward Eaton. **A141.**
Martín de Echagaray. **31.**

Felipe de Echagoyan. 11.
John Eliot. 28.
Luis Enríquez de Guzmán. 23.
José de Escobar Salmerón y Castro. A92, A96, A99, A108, A112, A116.
Francisco de Fagoaga. 39.
Benito Fernández de Belo. 29.
Juan de Figueroa. 25.
Josiah Flint. A54.
John Foster. A79, A82, A86, A93, A97, A100, A103.
Álvaro Fuentes y de la Cerda. 16.
Marco Antonio de Gamboa y Riaño. A205.
Diego García de Palacio. 8, 9.
Francisco Juan Garreguilla. 13.
Benjamin Gillam. A117.
Benjamin Harris. A169, A170.
Antonio de Heredia y Estupiñán. 26.
Nehemiah Hobart. A71.
Eusebio Francisco Kino. 30.
Juan Ramón Koenig. 35, A101, A104, A109, A113, A118, A121, A128, A134, A142, A149, A155, A162, A171, A177, A181, A187, A194, A200, A206, A210, A217.
Daniel Leeds. A135, A143, A150, A156, A163, A172, A178, A182, A188, A195, A201, A207, A211, A218.
Christian Lodowick. A183, A189.
Gabriel López de Bonilla. A14, A27, A43, A46, A50, A55, A58, A61.
José Martí. 36.
Enrico Martínez. 12, A1.
Cotton Mather. A114.
Nathanael Mather. A122, A129.
Nicolás de Matta. A64.
Francesco Maurolico. 6, 7.
Henry Newman. A157, A164.
Alexander Nowell. A51.
Urian Oakes. A15.
Nicolás de Olea. A34.
Pedro de Paz. 17, 18.
Diego Pérez de Lazcano. 33.
William Pierce. A2.
Atanasio Reaton. 19.
John Richardson. A65.
Feliciana Ruiz. A83.
Juan Ruiz. A21, A35, A38, A40, A47, A52, A56, A59, A63, A66, A72, A75.
Francisco Ruiz Lozano. A17, A19, A23, A26, A28, A31, A33, A36.

OTHER INDICES

Daniel Russell. **A67**.
Noadiah Russell. **A119**.
Juan de Saucedo. **A73, A76, A87**.
Jeremiah Shepard. **A69**.
Thomas Shepard. **A29**.
John Sherman. **A77, A80, A84, A88**.
Carlos de Sigüenza y Góngora. **32, A68, A70, A74, A78, A81, A85, A89, A94, A98, A102, A105, A110, A115, A120, A123, A130, A136, A144, A151, A158, A165, A173, A179, A184, A190, A196, A202, A208, A212, A219**.
Jacob Taylor. **37, A213**.
Francisco de Toledo. **5**.
John Tulley. **A137, A145, A152, A159, A166, A174, A180, A185, A191, A197, A203, A209, A214, A220**.
Jerónimo Valera. **15**.
Alonso de la Vera Cruz. **1, 2, 4**.
Francisco de Villegas. **21, 24**.
William Williams. **A124, A138**.
Manuel de Zuaza Aranguren. **38**.

Index of Printers

Blas de Ayessa. Lima **36**.
Jeronimo Balli. México. **14**.
William Bradford. Philadelphia: **A125, A135, A141, A143, A150, A156, A163, A172, A178**. New York: **37, A182, A188, A195, A199, A201, A207, A211, A213, A218**.
Widow of Bernardo Calderón. México. **19, 29, A25, A42, A45, A49, A53, A68, A76, A87**.
Heirs of the widow of Bernardo Calderón. México. **32, A130**.
Francisco del Canto. Lima. **13, 15**.
Juan Joseph Guillena Carrascoso. México. **38, 39**.
Joseph de Contreras. Lima. **26, 34, 35**.
Matthew Day. Cambridge. **A10, A11**.
Stephen Daye. Cambridge. **A2, A3, A4, A5, A6, A7, A8, A9**.
Diego de Escobar. México. **A133**.
John Foster. Boston. **A82, A86, A93, A97, A100, A103**.
Bartholomew Green. Boston. **A183, A189**.
Bartholomew Green and John Allen. Boston. **A197, A203, A209, A214, A216, A220**.
Samuel Green, Jr. Boston. **A114, A117, A122, A129, A137, A145, A152, A159**.
Samuel Green, Sr. Cambridge. **A13, A15, A16, A18, A20, A22, A24, A29, A30, A32, A34, A37, A41, A51, A54, A57, A60, A69, A71, A77, A79, A80, A84, A88, A91, A95, A107, A111, A119, A124, A127, A138, A157**.
Samuel Green, Sr., and Bartholomew Green. Cambridge. **A166, A174**.
Samuel Green, Sr., and Samuel Green, Jr. Cambridge. **A90**.

Samuel Green, Sr., and Marmaduke Johnson. Cambridge. **A39, A44, A48, A62, A65, A67.**
Benjamin Harris. Boston. **A170, A180, A185, A191.**
Benjamin Harris and John Allen. Boston. **A169.**
Marmaduke Johnson. Cambridge. **28.**
Francisco Rodríguez Lupercio. México. **27, 30, A31.**
Heirs of Francisco Rodríguez Lupercio. México. **A140.**
Enrico Martínez. México. **11, 12, A1.**
Pedro Ocharte. México. **8, 9.**
Juan Pablos. México. **1, 2, 3, 4.**
Richard Pierce. Boston. **A164.**
Antonio Ricardo. México: **5, 6, 7.** Lima: **10.**
Feliciana Ruiz. México. **A83.**
Juan Ruiz. México. **16, 18, A17, A19, A21, A35, A38, A40, A47, A52, A56, A59, A63, A66, A72, A75.**

There are 103 items not attributed to a printer. The breakdown in Part 1 is one item from Mexico and seven items from Peru. In Part 2 there is no printer for any of the twenty-seven Peru almanacs or for sixty-eight of the Mexico almanacs.

Index of Libraries and Collections

American Antiquarian Society
185 Salisbury Street, Worcester, MA 01909-1634, USA; phone: 508-755-5221; fax: 508-753-3311; www.americanantiquarian.org.
Thomas Shepard, *An Almanac*, 1656.
Samuel Bradstreet, *An Almanac*, 1657.
Samuel Cheever, *An Almanac*, 1661.
Nathaniel Chauncy, *An Almanac*, 1662.
Israel Chauncy, *An Almanac*, 1663.
Israel Chauncy, *An Almanac*, 1664.
Alexander Nowell, *An Almanac*, 1665.
Josiah Flint, *An Almanac*, 1666.
Samuel Brackenbury, *An Almanac*, 1667.
Joseph Dudley, *An Almanac*, 1668.
Joseph Browne, *An Almanac*, 1669.
John Richardson, *An Almanac*, 1670.
Daniel Russell, *An Almanac*, 1671.
Jeremiah Shepard, *An Ephemeris*, 1672.
Nehemiah Hobart, *An Almanac*, 1673.
John Foster, *An Almanac*, 1675.
John Sherman, *An Almanac*, 1676. (Two copies.)
John Sherman, *An Almanac*, 1677.

John Foster, *An Almanac*, 1678.
John Danforth, *An Almanac*, 1679.
John Foster, *An Almanac*, 1680.
John Foster, *An Almanac*, 1681.
William Brattle, *An Ephemeris*, 1682.
Cotton Mather, *The Boston Ephemeris*, 1683.
Benjamin Gillam, *The Boston Ephemeris*, 1684.
William Williams, *Cambridge Ephemeris*, 1685. (Two copies.)
Samuel Danforth (1666–1727), *The New England Almanac*, 1686.
Nathanael Mather, *The Boston Ephemeris*, 1686.
John Tulley, *An Almanac*, 1687.
John Tulley, *An Almanac*, 1688.
John Tulley, *An Almanac*, 1689.
John Tulley, *An Almanac*, 1690.
John Tulley, *An Almanac*, 1691.
John Tulley, *An Almanac*, 1692.
John Tulley, *An Almanac*, 1693. (Three copies.)
Christian Lodowick, *An Almanac*, 1694.
Christian Lodowick, *The New England Almanac*, 1695.
John Tulley, *An Almanac*, 1695. (Two copies.)
Daniel Leeds, *An Almanac*, 1696.
John Tulley, *An Almanac*, 1696.
John Tulley, *An Almanac*, 1697. (Two copies.)
Daniel Leeds, *An Almanac*, 1698.
John Tulley, *An Almanac*, 1698.
John Tulley, *An Almanac*, 1699.
John Tulley, *An Almanac*, 1700.

Baylor University, Carroll Library, Texas Collection
One Bear Place 97142, Waco, TX 76798-7142, USA; phone: 254-710-1268;
  fax: 254-710-1368; www3.baylor.edu/Library/LibDepts/Texas/;
  e-mail: txcoll@baylor.edu.
Enrico Martínez, *Reportorio de los Tiempos*, 1606.

Biblioteca Cervantina ITESM
Eugenio Garza Sada 2501 Sur, Col. Tecnológico, CP 64849, Monterrey, N.L.,
  México; phone: (81)8358-2000 ext. 4065 to 4067; fax: (81)8358-2000 ext. 4067.
Alonso de la Vera Cruz, *Recognitio, Summularum*, 1554.
Alonso de la Vera Cruz, *Dialectica Resolutio*, 1554.
Alonso de la Vera Cruz, *Phisica, Speculatio*, 1557.
Francisco de Toledo, *Introductio in Dialecticam Aristotelis*, 1578.
Francesco Maurolico, *De Sphæra*, 1578.

Francesco Maurolico, *Computus Ecclesiasticus*, 1578.
Diego García de Palacio, *Instrucción Nautica*, 1687.
Enrico Martínez, *Reportorio de los Tiempos*, 1606.
Eusebio Francisco Kino, *Exposición Astronómica*, 1681.
Carlos de Sigüenza y Góngora, *Libra Astronómica*, 1690.

Biblioteca Nacional de Chile
Avenida Libertador Bernardo O'Higgins 651, Santiago, Chile; www.dibam.cl/biblioteca_nacional/home.asp; e-mail: biblioteca.nacional@bndechile.cl.
Alonso de la Vera Cruz, *Recognitio, Summularum*, 1554.
Alonso de la Vera Cruz, *Dialectica Resolutio*, 1554.
Alonso de la Vera Cruz, *Phisica, Speculatio*, 1557.
Francisco de Toledo, *Introductio in Dialecticam Aristotelis*, 1578.
Francesco Maurolico, *De Sphæra*, 1578.
Francesco Maurolico, *Computus Ecclesiasticus*, 1578.
Joán de Belveder, *Libro General*, 1597.
Jerónimo Valera, *Commentarii ac Quaestiones*, 1610.
Atanasius Reatón, *Arte Menor*, 1649.
Juan de Figueroa, *Opúsculo de Astrología*, 1660.
Antonio de Heredia y Estupiñán, *Teórica, y Práctica de Esquadrones*, 1660.
Martín de Echagaray, *Declaración del Quadrante*, 1682.
Diego Pérez de Lazcano, *Proposición, y Manifiesto*, 1691.
Nicolás de Olea, *Summa Tripartita—in Logicam*, 1693.
Juan Ramón Koenig, *Cubus*, 1696.
Manuel de Zuaza y Aranguren, *Reducciones de Plata*, 1697.

Biblioteca Nacional de España
www.bne.es/index.htm.
Diego Garcia de Palacio, *Diálogos Militares*, 1583.
Joán de Belveder, *Libro General*, 1597.
Enrico Martínez, *Reportorio de los Tiempos*, 1606.
Pedro de Aguilar Gordillo, *Alivio de Mercaderes*, 1610.
Juan de Figueroa, *Opúsculo de Astrología*, 1660.
Antonio de Heredia y Estupiñán, *Teórica, y Práctica de Esquadrones*, 1660.
Benito Fernández de Belo, *Breve Aritmética*, 1675.
Martin de Echagaray, *Declaración del Quadrante*, 1682.
Carlos de Sigüenza y Góngora, *Libra Astronómica*, 1690.
Francisco de Fagoaga, *Reducción de Oro*, 1700.

Biblioteca Nacional de México
biblional.bibliog.unam.mx/bib/biblioteca.html.
Alonso de la Vera Cruz, *Recognitio, Summularum*, 1554. (Two copies.)

Alonso de la Vera Cruz, *Dialectica Resolutio*, 1554.
Alonso de la Vera Cruz, *Phisica, Speculatio*, 1554.
Diego García de Palacio, *Diálogos Militares*, 1583.
Enrico Martínez, *Reportorio de los Tiempos*, 1606.
Eusebio Francisco Kino, *Exposición Astronómica*, 1681. (Two copies.)
Carlos de Sigüenza y Góngora, *Libra Astronómica*, 1690. (Two copies.)
Francisco de Fagoaga, *Reducción de Oro*, 1700.

Biblioteca Nacional del Perú
Av. Abancay 4ta cuadra, Lima, Perú; phone: 511-428-7690/428-7696;
fax: 511-427-7331; www.bnp.gob.pe/portalbnp/.
Joán de Belveder, *Libro General*, 1597.
Francisco Juan Garreguilla, *Libro de Plata*, 1607.
Jerónimo Valera, *Commentarii ac Quaestiones*, 1610.

Biblioteca Nazionale Centrale di Roma, Biblioteca Nazionale Centrale "Vittorio Emanuele II"
Viale Castro Pretorio, 105, 00185 Roma, Italia; phone: (+39) 06 49 891;
fax: (+39) 06 44 57 635; http://www.bncrm.librari.beniculturali.it;
e-mail: bncrm@bnc.roma.sbn.it.
Juan Ramón Koenig, *Cubus*, 1696.

Biblioteca Palafoxiana
5 Oriente no. 5, Centro Histórico, Puebla, México; phone: 246-3186, 246-6922.
Enrico Martínez, *Reportorio de los Tiempos*, 1606.
Eusebio Francisco Kino, *Exposición Astronómica*, 1681.
Carlos de Sigüenza y Góngora, *Libra Astronómica*, 1690.

Bibliothèque Nationale de France
Site François-Mitterrand, Quai François-Mauriac, 75706 Paris Cedex 13, France;
www.bnf.fr/.
Juan Ramón Koenig, *Cubus*, 1696.

Boston Public Library
700 Boylston Street, Boston MA 02116, USA; phone: 617-536-5400; www.bpl.org;
e-mail: info@bpl.org.
Samuel Brackenbury, *An Almanac*, 1667.
Joseph Dudley, *An Almanac*, 1668.
Joseph Browne, *An Almanac*, 1669.
John Richardson, *An Almanac*, 1670.
Daniel Russell, *An Almanac*, 1671.
Nehemiah Hobart, *An Almanac*, 1673.

John Sherman, *An Almanac*, 1674.
John Foster, *An Almanac*, 1675.
John Danforth, *An Almanac*, 1679.
John Foster, *An Almanac*, 1679.
Cotton Mather, *The Boston Ephemeris*, 1683.
Nathanael Mather, *The Boston Ephemeris*, 1686.
Daniel Leeds, *An Almanac*, 1687.
John Tulley, *An Almanac*, 1690.
Henry Newman, *News from the Stars*, 1691.
John Tulley, *An Almanac*, 1691.
John Tulley, *An Almanac*, 1692.
John Tulley, *An Almanac*, 1693.
John Tulley, *An Almanac*, 1695.
John Tulley, *An Almanac*, 1698.
John Tulley, *An Almanac*, 1699.
John Tulley, *An Almanac*, 1700.

British Library
96 Euston Road, London NW1 2DB, UK; www.bl.uk.
   Alonso de la Vera Cruz, *Recognitio, Summularum*, 1554.
   Alonso de la Vera Cruz, *Dialectica Resolutio*, 1554.
   Juan Diez Freyle, *Sumario Compendioso*, 1556.
   Alonso de la Vera Cruz, *Phisica, Speculatio*, 1557.
   Diego Garcia de Palacio, *Instrucción Náutica*, 1587.
   Enrico Martínez, *Reportorio de los Tiempos*, 1606.
   Juan de Figueroa, *Opúsculo de Astrología*, 1660.
   Juan de Castañeda, *Reformación de las Tablas*, 1668.
   John Eliot, *The Logic Primer*, 1672.
   Eusebio Francisco Kino, *Exposición Astronómica*, 1681.
   Daniel Leeds, *An Almanac*, 1687.

Brown University, John Hay Library
20 Prospect Street, Box A, Providence, RI 02912, USA; phone: 401-863-3723; fax: 401-863-2093; www.brown.edu/Facilities/University_Library/libs/hay/; e-mail: hay@brown.edu.
   Enrico Martínez, *Reportorio de los Tiempos*, 1606.

California Historical Society
678 Mission Street, San Francisco, CA 94105, USA; phone: 415-357-1848; fax: 415-357-1850; www.californiahistoricalsociety.org/; e-mail: reference@calhist.org.
   Eusebio Francisco Kino, *Exposición Astronómica*, 1681.

OTHER INDICES

California State Library, Sutro Library
480 Winston Drive, San Francisco, CA 94132, USA; phone: 415-731-4477; fax: 415-557-9325; www.library.ca.gov/collections/index.html#sutro; e-mail: sutro@library.ca.gov.
Francisco de Fagoaga, *Reducción de Oro*, 1700.

Centro de Estudios de Historia de México Condumex[1]
Plaza Federico Gamboa 1, Col. Chimalistac, San Angel, CP 01070, México D.F., México; phone: 53-26-51-71; fax: 56-61-77-97; www.cehm.com.mx; e-mail: cehmcond@prodigy.net.mx.
Alonso de la Vera Cruz, *Recognitio, Summularum*, 1554.
Alonso de la Vera Cruz, *Dialectica Resolutio*, 1554.
Alonso de la Vera Cruz, *Phisica, Speculatio*, 1557.
Francisco de Toledo, *Introductio in Dialecticam Aristotelis*, 1578.
Francesco Maurolico, *De Sphœra*, 1578.
Francesco Maurolico, *Computus Ecclesiasticus*, 1578.
Enrico Martínez, *Reportorio de los Tiempos*, 1606.
Eusebio Francisco Kino, *Exposición Astronómica*, 1681.

Duke University, Rare Book, Manuscript, and Special Collections Library
Durham, NC 27708-0185, USA; phone: 919-660-5822; fax: 919-660-5934; odyssey.lib.duke.edu/.
Juan Diez Freyle, *Sumario Compendioso*, 1556.
Joán de Belveder, *Libro General*, 1597.

El Escorial, el Real Monasterio de San Lorenzo de
Palacio Real, 28071 Madrid, Spain; phone: 91-890-59-02, -03; www.patrimonionacional.es/escorial/escorial.htm.
Diego García de Palacio, *Dialogos Militares*, 1583.
Diego García de Palacio, *Instrucción Náutica*, 1587.

Harvard University, Houghton Library
Cambridge, MA 02138, USA; phone: 617-495-2441; fax: 617-495-1376; hcl.harvard.edu/houghton/.
Jerónimo Valera, *Commentarii ac Quaestiones*, 1610.
Israel Chauncy, *An Almanac*, 1663.
Josiah Flint, *An Almanac*, 1666.
Nehemiah Hobart, *An Almanac*, 1673.
Cotton Mather, *The Boston Ephemeris*, 1683.
William Williams, *The Cambridge Ephemeris*, 1685.
Nathaniel Mather, *The Boston Ephemeris*, 1686.

1. We were not permitted to examine original copies at this library.

John Tulley, *An Almanac*, 1691.
John Tulley, *An Almanac*, 1692.
John Tulley, *An Almanac*, 1693.
John Tulley, *An Almanac*, 1694.
John Tulley, *An Almanac*, 1698.
John Tulley, *An Almanac*, 1700.

The Hispanic Society of America
613 West 155th Street, New York, NY 10032, USA; phone: 212-926-2234; fax: 212-690-0743; www.hispanicsociety.org/; e-mail: info@hispanicsociety.org.
Alonso de la Vera Cruz, *Recognitio Summularum*, 1554.
Alonso de la Vera Cruz, *Dialectia Resolutio*, 1554.
Alonso de la Vera Cruz, *Phisica, Speculatio*, 1557.
Francisco de Toledo, *Introductio in Dialecticam Aristotelis*, 1578.
Francesco Maurolico, *De Sphæra*, 1578.
Francesco Maurolico, *Computus Ecclesiasticus*, 1578.
Diego García de Palacio, *Instrucción Náutica*, 1587.
Enrico Martínez, *Reportorio de los Tiempos*, 1606.
Carlos de Sigüenza y Góngora, *Libra Astronómica*, 1690.

Historical Society of Pennsylvania
1300 Locust St., Philadelphia, PA 19107-5699, USA; www.hsp.org.
Daniel Leeds, *An Almanac*, 1687.

The Huntington Library, Art Collections, and Botanical Gardens
1151 Oxford Road, San Marino, CA 91108, USA; phone: 626-405-2100 (rare books: 626-405-2178); fax: 626-449-5720; www.huntington.org/LibraryDiv/LibraryHome.html.
Alonso de la Vera Cruz, *Recognitio, Summularum*, 1554.
Alonso de la Vera Cruz, *Dialectica Resolutio*, 1554.
Juan Diez Freyle, *Sumario Compendioso*, 1556.
Francisco de Toledo, *Introductio in Dialecticam Aristotelis*, 1578.
Francesco Maurolico, *De Sphæra*, 1578.
Francesco Maurolico, *Computus Ecclesiasticus*, 1578.
Diego García de Palacio, *Instrucción Náutica*, 1587.
Samuel Danforth (1626–1674), *An Almanac*, 1646.
Samuel Danforth (1626–1674), *An Almanac*, 1647.
Samuel Danforth (1626–1674), *An Almanac*, 1648.
Urian Oakes, *An Almanac*, 1650.
Israel Chauncy, *An Almanac*, 1663.
Eusebio Francisco Kino, *Exposición Astronómica*, 1681.
Carlos de Sigüenza y Góngora, *Libra Astronómica*, 1690.

John Tulley, *An Almanac*, 1691.
John Tulley, *An Almanac*, 1692.
Danial Leeds, *An Almanac*, 1693.
John Tulley, *An Almanac*, 1693.
Daniel Leeds, *An Almanac*, 1694.
Daniel Leeds, *An Almanac*, 1695.
Daniel Leeds, *An Almanac*, 1696.
Daniel Leeds, *An Almanac*, 1697.
Daniel Leeds, *An Almanac*, 1698.
John Tulley, *An Almanac*, 1698.
John Tulley, *An Almanac*, 1699.
Samuel Clough, *The New England Almanac*, 1700.
Daniel Leeds, *An Almanac*, 1700.
John Tulley, *An Almanac*, 1700.

Indiana University, The Lilly Library
1200 E. 7th Street, Bloomington, IN 47405-5500, USA; phone: 812-855-2452; fax: 812-855-3143; www.indiana.edu/~liblilly; e-mail: liblilly@indiana.edu.
Alonso de la Vera Cruz, *Recognitio, Summularum*, 1554.
Alonso de la Vera Cruz, *Dialectica Resolutio*, 1554.
Alonso de la Vera Cruz, *Phisica, Speculatio*, 1557.
Francisco de Toledo, *Introductio in Dialecticam Aristotelis*, 1578.
Francesco Maurolico, *De Sphæra*, 1578.
Francesco Maurolico, *Computus Ecclesiasticus*, 1578.
Joán de Belveder, *Libro General*, 1597.
Jerónimo Valera, *Commentarii ac Quaestiones*, 1610.
Martín de Echagaray, *Declaración del Quadrante*, 1682.
Carlos de Sigüenza y Góngora, *Libra Astronómica*, 1690.
John Tulley, *An Almanac*, 1696.

Institución Colombina, Biblioteca Capitular
Calle Alemanes, 41004 Sevilla, España; phone: 34-954-560769, -562721; fax: 34-954-211876; www.institucioncolombina.org; e-mail: direccionic@institucioncolombina.org.
Alonso de la Vera Cruz, *Dialectica Resolutio*, 1554.
Alonso de la Vera Cruz, *Phisica, Speculatio*, 1557.

The John Carter Brown Library
Box 1894, Providence, RI 02912, USA; phone: 401-863-2725; fax: 401-863-3477; www.brown.edu/Facilities/John_Carter_Brown_Library/index.html; e-mail: JCBL_Information@Brown.edu.
Alonso de la Vera Cruz, *Recognitio Summularum*, 1554.

Alonso de la Vera Cruz, *Dialectica Resolutio*, 1554.
Alonso de la Vera Cruz, *Phisica, Speculatio*, 1557.
Francisco de Toledo, *Introductio in Dialecticam Aristotelis*, 1578.
Diego García de Palacio, *Dialogos Militares*, 1583.
Diego García de Palacio, *Instrucción Náutica*, 1587.
Felipe de Echagoyan, *Tablas de Reducciones*, 1603.
Enrico Martínez, *Reportorio de los Tiempos*, 1606.
Francisco Juan Garreguilla, *Libro de Plata*, 1607.
Jerónimo Valera, *Commentarii ac Quaestiones*, 1610.
Benito Fernández de Belo, *Breve Aritmética*, 1675.
Eusebio Francisco Kino, *Exposición Astronómica*, 1681.
Carlos de Sigüenza y Góngora, *Libra Astronómica*, 1690.
John Tulley, *An Almanac*, 1699.
Francisco de Fagoaga, *Reducción de Oro*, 1700.

The Johns Hopkins University, John Work Garrett Library

Evergreen House, 4545 North Charles Street, Baltimore, MD 21210, USA; phone: 410-516-8662; www.library.jhu.edu/collections/index.html.
Daniel Leeds, *An Almanac*, 1693.

Jay and Jean Kislak

720 NE 69th Street, Apt. 21 W, Miami, FL 33138. Ann Stetser, Assistant to Mr. Jay Kislak; phone: 305-364-4208; e-mail: astetser@kislak.com.
John Tulley, *An Almanac*, 1690.

Library Company of Philadelphia

1314 Locust Street, Philadelphia, PA 19107-5698, USA; phone: 215-546-3181; fax: 215-546-5167; www.librarycompany.org; e-mail: refdept@librarycompany.org.
Samuel Atkins, *Kalendarium Pennsilvaniense*, 1686.
Daniel Leeds, *An Almanac*, 1687. (Three copies.)
John Tulley, *An Almanac*, 1687.
John Tulley, *An Almanac*, 1688.
John Tulley, *An Almanac*, 1692.
Daniel Leeds, *An Almanac*, 1693.
John Tulley, *An Almanac*, 1693. (Two copies.)
John Clapp, *New York Almanac*, 1697.
Jacob Taylor, *Tenebræ*, 1697.
Daniel Leeds, *An Almanac*, 1699.
John Tulley, *An Almanac*, 1699.
Daniel Leeds, *An Almanac*, 1700.

OTHER INDICES

Library of Congress
101 Independence Avenue, SE, Washington, DC 20540, USA; phone: 202-707-5000; lcweb.loc.gov.
Diego Garcia de Palacio, *Instrucción Náutica*, 1587.
Enrico Martínez, *Reportorio de los Tiempos*, 1606.
Zechariah Brigden, *An Almanac*, 1659.
Samuel Cheever, *An Almanac*, 1660.
Nathaniel Chauncy, *An Almanac*, 1662.
Israel Chauncy, *An Almanac*, 1663.
Alexander Nowell, *An Almanac*, 1665.
Daniel Russell, *An Almanac*, 1671.
William Brattle, *An Ephemeris*, 1682.
Cotton Mather, *The Cambridge Ephemeris*, 1683.
John Tulley, *An Almanac*, 1687.
John Tulley, *An Almanac*, 1688.
John Tulley, *An Almanac*, 1689.
John Tulley, *An Almanac*, 1690.
Henry Newman, *News from the Stars*, 1691.
John Tulley, *An Almanac*, 1691.
Benjamin Harris, *The Boston Almanac*, 1692.
John Tulley, *An Almanac*, 1692.
John Tulley, *An Almanac*, 1693.
John Tulley, *An Almanac*, 1694.
John Tulley, *An Almanac*, 1696.
John Tulley, *An Almanac*, 1697.
John Tulley, *An Almanac*, 1698.
John Tulley, *An Almanac*, 1699.
Samuel Clough, *The New England Almanac*, 1700.
Francisco de Fagoaga, *Reducción de Oro*, 1700.
John Tulley, *An Almanac*, 1700.

The Massachusetts Historical Society
1154 Boylston Street, Boston, MA 02215, USA; phone: 617-536-1608; fax: 617-859-0074; www.masshist.org/library/; e-mail: library@masshist.org.
Samuel Brackenbury, *An Almanac*, 1667.
John Richardson, *An Almanac*, 1670.
John Sherman, *An Almanac*, 1674. (Two copies.)
John Danforth, *An Almanac*, 1679.
John Foster, *An Almanac*, 1680.
John Foster, *An Almanac*, 1681. (Two copies.)
William Brattle, *An Ephemeris*, 1682.

Cotton Mather, *The Boston Ephemeris*, 1683.
Benjamin Gillam, *Boston Ephemeris*, 1684.
Noadiah Russell, *Cambridge Ephemeris*, 1684.
Nathanael Mather, *The Boston Ephemeris*, 1685.
William Williams, *Cambridge Ephemeris*, 1685.
Samuel Danforth (1666–1727), *The New England Almanac*, 1686. (Two copies.)
Nathanael Mather, *The Boston Ephemeris*, 1686.
John Tulley, *An Almanac*, 1687.
William Williams, *Cambridge Ephemeris*, 1687.
John Tulley, *An Almanac*, 1688.
John Tulley, *An Almanac*, 1689.
Henry Newman, *Harvard's Ephemeris*, 1690.
John Tulley, *An Almanac*, 1690. (Two copies.)
Henry Newman, *News from the Stars*, 1691.
John Tulley, *An Almanac*, 1691. (Two copies.)
Benjamin Harris, *Boston Almanac*, 1692.
Benjamin Harris, *Monthly Observations*, 1692.
John Tulley, *An Almanac*, 1692.
John Tulley, *An Almanac*, 1693. (Two copies.)
Christian Lodowick, *An Almanac*, 1694.
John Tulley, *An Almanac*, 1695.
John Tulley, *An Almanac*, 1696.
John Tulley, *An Almanac*, 1697.
John Tulley, *An Almanac*, 1698.
John Tulley, *An Almanac*, 1699.
John Tulley, *An Almanac*, 1700. (Two copies.)

The Morgan Library
29 East 36th Street, New York, NY 10016, USA; phone: 212-685-0610; www.morganlibrary.org; e-mail: readingroom@morganlibrary.org.
John Tulley, *An Almanac*, 1692.
Daniel Leeds, *An Almanac*, 1698.
Daniel Leeds, *An Almanac*, 1699. (Two copies.)
Daniel Leeds, *An Almanac*, 1700.

Museo Naval de Madrid, Biblioteca y Archivo
Calle Juan de Mena 2. 28071 Madrid, España; phone: 34 91 523 83 78; fax: 34 91 379 50 56; www.museonavalmadrid.com/biblioteca_archivo/index.asp; e-mail: direccion@museonavalmadrid.com.
Diego Garcia de Palacio, *Instrucción Náutica*, 1587.

OTHER INDICES

The New-York Historical Society
170 Central Park West, 2nd Floor, New York, NY 10024, USA; phone: 212-873-3400; fax: 212-875-1591; www.nyhistory.org/web/default.php?section=library; e-mail: reference@nyhistory.org.
  Enrico Martínez, *Reportorio de los Tiempos*, 1606.

The New York Public Library, Humanities and Social Sciences Library
5th Avenue and 42nd Street, New York, NY 10018-2788, USA; phone: 212-930-0830 x225, 226; www.nypl.org/research/chss/index.html.
  Alonso de la Vera Cruz, *Recognitio, Summularum*, 1554.
  Alonso de la Vera Cruz, *Dialectia Resolutio*, 1554.
  Alonso de la Vera Cruz, *Phisica, Speculatio*, 1557.
  Francisco de Toledo, *Introductio in Dialecticam Aristotelis*, 1578.
  Francesco Maurolico, *De Sphæra*, 1578.
  Francesco Maurolico, *Computus Ecclesiasticus*, 1578.
  Diego García de Palacio, *Instrucción Náutica*, 1587.
  Joán de Belveder, *Libro General*, 1597.
  Enrico Martínez, *Reportorio de los Tiempos*, 1606.
  Samuel Danforth (1626–1674), *An Almanac*, 1649.
  Nathanael Mather, *The Boston Ephemeris*, 1685.
  Carlos de Sigüenza y Góngora, *Libra Astronómica*, 1690.
  Benjamin Harris, *Boston Almanac*, 1692.
  John Clapp, *New York Almanac*, 1697.
  Daniel Leeds, *An Almanac*, 1697.
  Jacob Taylor, *Tenebræ*, 1697.
  Daniel Leeds, *An Almanac*, 1698.

Old Sturbridge Village Research Library
1 Old Sturbridge Village Road, Sturbridge, MA 01566, USA; phone: 508-347-0204; www.osv.org.
  John Tulley, *An Almanac*, 1697.
  John Tulley, *An Almanac*, 1698.
  John Tulley, *An Almanac*, 1699.
  John Tulley, *An Almanac*, 1700.

William Owen
E-mail: woowen@att.net.
  Benjamin Harris, *Boston Almanac*, 1692.
  John Tulley, *An Almanac*, 1700.

Peabody Essex Museum, Phillips Library
East India Square, Salem, MA 01970, USA; phone: 978-745-9500, 866-745-1876; pem.org/museum/library.php.
John Tulley, *An Almanac*, 1698.

Private Collection, Ann Arbor, Michigan, USA
John Tulley, *An Almanac*, 1693.
John Tulley, *An Almanac*, 1696.

Real Biblioteca
Bailén, 28071 Madrid, España; phone: 34-91-454-87-32, -33; fax: 34-91-454-88-67; realbiblioteca.es; realbiblioteca.patrimonionacional.es; e-mail: mluisa.lvidriero@patrimonionacional.es.
Joán de Belveder, *Libro General*, 1597.

Real Instituto y Observatorio de la Armada en San Fernando
Cecilio Pujazón, E-1110, San Fernando (Cádiz), España; phone: 34-956-599000 ext. 35599; fax: 34-956-599366; www.roa.es.
Juan de Figueroa, *Opúsculo de Astrología*, 1660.

The Rosenbach Museum and Library
2008–2010 DeLancey Place, Philadelphia, PA 19103, USA; phone: 215-732-1600; fax: 215-545-7529; www.rosenbach.org.
Samuel Atkins, *Kalendarium Pennsilvaniense*, 1686.
John Tulley, *An Almanac*, 1696.

Saint Bonaventure University, Friedsam Memorial Library.
St. Bonaventure, NY 14778, USA; phone: 716-375-2323; fax: 716-375-2389; web.sbu.edu/friedsam/.
Jerónimo Valera, *Commentarii ac Quaestiones*, 1610.

Southern Methodist University, DeGolyer Library
P.O. Box 750396, Dallas, TX 75275-0396, USA; phone: 214-768-2012; fax: 214-768-1565; www.smu.edu/cul/degolyer.
Alonso de la Vera Cruz, *Dialectica Resolutio*, 1554.
Carlos de Sigüenza y Góngora, *Libra Astronómica*, 1690.

Ted Steinbock
5803 Orion Road, Louisville, KY 40222, USA; e-mail: bibliotedd@aol.com.
John Tulley, *An Almanac*, 1692.

Trinity College, The Watkinson Library
300 Summit Street, Hartford, CT 06106-3100, USA; phone: 860-297-2268; library.trincoll.edu/research/watk/.
Daniel Russell, *An Almanac*, 1671.
Jeremiah Shepard, *An Ephemeris*, 1672.
Nehemiah Hobart, *An Almanac*, 1673.
John Sherman, *An Almanac*, 1674.
John Foster, *An Almanac*, 1675.
John Foster, *An Almanac*, 1676.
John Sherman, *An Almanac*, 1676.
John Sherman, *An Almanac*, 1677.
Thomas Brattle, *An Almanac*, 1678.
John Foster, *An Almanac*, 1678.
John Danforth, *An Almanac*, 1679.
John Foster, *An Almanac*, 1679.
John Foster, *An Almanac*, 1680.

Tulane University, Latin American Library, Howard-Tilton Memorial Library
Fourth floor, 7001 Freret Street, New Orleans, LA 70118, USA; phone: 504-865-5681; fax: 504-862-8970; lal.tulane.edu.
Enrico Martínez, *Reportorio de los Tiempos*, 1606.

Universidad Complutense de Madrid, Biblioteca Histórica Marqués de Valdecilla
Noviciado, 3, 28015 Madrid, España; phone: 913946612; fax: 913946599; www.ucm.es/bucm/foa/index.php; e-mail: buc_foa@buc.ucm.es.
Juan de Figueroa, *Opúsculo de Astrología*, 1660.

Universidad de Salamanca
c/ Libreros s/n, 37008 Salamanca, España; phone: 923-264400; fax: 923-294704; www.usal.es/web-usal/Menus/bibliotecas.shtml; e-mail: marga@gugu.usal.es.
Juan Diez Freyle, *Sumario Compendioso*, 1556.
Francisco de Toledo, *Introductio in Dialecticam Aristotelis*, 1578.
Francesco Maurolico, *De Sphæra*, 1578.
Francesco Maurolico, *Computus Ecclesiasticus*, 1578.
Diego Garcia de Palacio, *Instrucción Náutica*, 1587.
Enrico Martínez, *Reportorio de los Tiempos*, 1606.

Universidad de Santiago de Compostela, Biblioteca Universitaria, Colexio de Fonseca
Rúa do Franco, 15702 Santiago de Compostela, España; phone: 981-583800 or 563100 ext. 1080; busc.usc.es; e-mail: bxinfor@usc.es.
Eusebio Francisco Kino, *Exposición Astronómica*, 1681.

OTHER INDICES

University of Arizona, University Library, Special Collections
1510 E. University, Tucson, AZ 85721-0055, USA; phone: 520-621-6423; fax: 520-621-9733; www.library.arizona.edu/speccoll.
Carlos de Sigüenza y Góngora, *Libra Astronómica*, 1690.

University of California, Berkeley, The Bancroft Library
Berkeley, CA 94720-6000, USA; phone: 510-642-3781 (reference desk: 510-642-6481); fax: 510-642-7589; bancroft.berkeley.edu/;
e-mail: bancref@library.berkeley.edu.
Pedro de Paz, *Arte para Aprender*, 1623.
Carlos de Sigüenza y Góngora, *Libra Astronómica*, 1690.

University of Chicago, Special Collections Research Center, Regenstein Library
1100 E. 57th Street, Chicago, IL 60615, USA; phone: 773-702-8705; www.lib.uchicago.edu/e/spcl.
Carlos de Sigüenza y Góngora, *Libra Astronómica*, 1690.

University of Michigan, William L. Clements Library
909 S. University, Ann Arbor, MI 48109-1190, USA; phone: 734-764-2347; fax: 734-647-0716; www.clements.umich.edu; e-mail: clements.library@umich.edu.
Eusebio Francisco Kino, *Exposición Astronómica*, 1681.

University of Oxford, Department of Special Collections and Western Manuscripts, Bodleian Library
Broad Street, Oxford, OX1 3BG, UK; phone: +44 1865 277102; fax: +44 8165 287396; www.bodley.ox.ac.uk/; e-mail: rare.books@bodley.ox.ac.uk.
Pedro de Paz, *Arte para Aprender*, 1623.
John Tulley, *An Almanac*, 1692.

University of Texas at Austin, Nettie Lee Benson Latin American Collection
P.O. Box P, Sid Richardson Hall 1.109, The General Libraries, Austin, TX 78713-8916, USA; phone: 512-495-4520; fax: 512-495-4568; www.lib.utexas.edu/benson/; e-mail: blac@lib.utexas.edu.
Alonso de la Vera Cruz, *Recognitio, Summularum*, 1554.
Alonso de la Vera Cruz, *Dialectica Resolutio*, 1554.
Alonso de la Vera Cruz, *Phisica, Speculatio*, 1557.
Enrico Martínez, *Reportorio de los Tiempos*, 1606. (Two copies.)
Atanasius Reatón, *Arte Menor*, 1649.
Eusebio Francisco Kino, *Exposición Astronómica*, 1681. (Two copies.)
Carlos de Sigüenza y Góngora, *Libra Astronómica*, 1690.
Francisco de Fagoaga, *Reducción de Oro*, 1700. (Two copies.)

University of Virginia, Albert and Shirley Small Special Collections Library
P.O. Box 400113, Charlottesville, VA 22904-4113, USA; phone: 434-924-3021;
fax: 434-924-1431; www.lib.virginia.edu/small.
Cotton Mather, *The Boston Ephemeris*, 1683.
Nathanael Mather, *The Boston Ephemeris*, 1686.

Washington's Headquarters Historical Museum and Library
30 Washington Place, Morristown, NJ 07960-4299, USA; phone: 908-766-2821;
fax: 908-766-4589; nps.gov/morr.
John Tulley, *An Almanac*, 1693.

Wellcome Library for the History and Understanding of Medicine
183 Euston Road, London NW1 2BE, UK; phone: +44 (0)20 7611 8582;
fax: +44 (0)20 7611 8369; library.wellcome.ac.uk/; e-mail: library@wellcome.ac.uk.
Alonso de la Vera Cruz, *Phisica, Speculatio*, 1557.
Enrico Martínez, *Reportorio de los Tiempos*, 1606.

Williams College, Chapin Library
P.O. Box 426, Williamstown, MA 01267, USA; phone: 413-597-2462;
fax: 413-597-2930; www.williams.edu/resources/chapin/;
e-mail: chapin.library@williams.edu.
Alonso de la Vera Cruz, *Dialectica Resolutio*, 1554.
Juan de Figueroa, *Opúsculo de Astrología*, 1660.

Winterthur Museum, Garden, and Library, The Joseph Downs Collection of Manuscripts and Printed Ephemera
Winterthur, DE 19735, USA; phone: 800-448-3883; www.winterthur.org/about/library.asp; e-mail: webmaster@winterthur.org.
John Tulley, *An Almanac*, 1698.

Yale University, Beinecke Rare Book & Manuscript Library
P.O. Box 208240, New Haven, CT 06520-8240, USA; phone: 203-432-2972;
www.library.yale.edu/beinecke/brblinfo.htm; e-mail: beinecke.library@yale.edu.
Francisco de Toledo, *Introductio in Dialecticam Aristotelis*, 1578.
Diego García de Palacio, *Instrucción Náutica*, 1587.
Joán de Belveder, *Libro General*, 1597.
Enrico Martínez, *Reportorio de los Tiempos*, 1606. (Two copies.)
Juan de Figueroa, *Opúsculo de Astrología*, 1660.
Joseph Browne, *An Almanac*, 1669.
John Danforth, *An Almanac*, 1679.
John Foster, *An Almanac*, 1679.

OTHER INDICES

Eusebio Francisco Kino, *Exposición Astronómica*, 1681.
William Brattle, *An Ephemeris*, 1682.
Benjamin Gillam, *The Boston Ephemeris*, 1684.
William Williams, *The Cambridge Ephemeris*, 1685.
Samuel Danforth (1666–1727), *The New-England Almanac*, 1686.
Henry Newman, *Harvard's Ephemeris*, 1690.
Carlos de Sigüenza y Góngora, *Libra Astronómica*, 1690.
Juan Ramón Koenig, *Cubus*, 1696.
José Martí, *Tabla General*, 1696.

We have visited every library listed here except for Biblioteca Cervantina, Biblioteca Nacional del Perú, and Biblioteca Nazionale Centrale di Roma.

A number of other libraries held material that was useful for this project. They were not mentioned in this list because they did not have original copies of the main entries. These other libraries included the Academia de Artillería de Segovia, Archivo General de la Nación, Archivo General de las Indias, Biblioteca Catalunya, Burlington County Historical Society, Columbia University, Community College of Rhode Island, Friends House London, Haverford College, Mariners' Museum, Newberry Library, Northern Illinois University, Philadelphia Friends Center, Providence Atheneum, Providence College, Real Academia de las Ciencias Exactas, Rhode Island College, Rhode Island Historical Society, Roger Williams University, San Francisco Maritime National Historical Park, Smithsonian Institution, Stanford University, Swarthmore College, Universidad Autónoma "Benito Juárez" de Oaxaca, Universidad de Barcelona, Universidad de Granada, Universidad de Sevilla, Universidad Nacional Autónoma de México, University of Cambridge, University of Glasgow, University of Montreal, University of Pennsylvania, and University of San Francisco.

A great deal of thanks is due to all of the people at all of these libraries who went out of their way to help with this project.

About the Author

Bruce Stanley Burdick was born in Middletown, Connecticut, and grew up mostly in Ohio. After earning a Ph.D. in mathematics from the Ohio State University he eventually landed in Rhode Island, where he is Professor of Mathematics at Roger Williams University. His personal book collections include hundreds of examples of science fiction, comic books, books on chess, and eighteenth- and nineteenth-century almanacs. An avid traveler, he has visited fifty-two countries, all fifty of the United States, nine of the ten provinces of Canada, and twenty-four of the thirty-one states of Mexico. Frequent trips to Mexico and Spain help to ensure that his Spanish doesn't get much worse.

Corrections

Have you found errors in this book? Is there something we missed? E-mail bburdick@rwu.edu with your comments. We hope to eventually support a Web site with your comments sorted by page number. Thanks!

DATE DUE

SCI QA 33 .B87 2009

Burdick, Bruce Stanley.

Mathematical works printed
 in the Americas, 1554-1700